Mach 1 and Beyond

The Illustrated Guide to High-Speed Flight

Mach 1 and Beyond

The Illustrated Guide to High-Speed Flight

Larry Reithmaier

TAB Books
Division of McGraw-Hill, Inc.
New York San Francisco Washington, D.C. Auckland Bogotá
Caracas Lisbon London Madrid Mexico City Milan
Montreal New Delhi San Juan Singapore
Sydney Tokyo Toronto

Published by TAB Books, a division of McGraw-Hill, Inc.

pbk 2 3 4 5 6 7 8 9 10 11 DOC/DOC 9 9 8 7 6 5

Library of Congress Cataloging-in-Publication Data

Reithmaier, L. W. (Lawrence W.), 1921–
Mach 1 : the illustrated guide to high-speed flight / by Larry Reithmaier.
p. cm.
Includes bibliographical references (p. 267) and index.
ISBN 0-07-052021-6 (pbk.)
1. Aerodynamics, Supersonic. I. Title.
TL571.R45 1994
629.132'305—dc20 94-1421
CIP

Acquisitions editor: Jeff Worsinger
Editorial team: Robert Ostrander, Executive Editor
Norval Kennedy, Book Editor
Production team: Katherine G. Brown, Director
Lisa M. Mellott, Coding
Rose McFarland, Layout
Lorie L. White, Proofreading
Elizabeth J. Akers, Indexer
Design team: Jaclyn J. Boone, Designer
Brian Allison, Associate Designer
Cover photograph courtesy of NASA
Cover design: Holberg Design, York, Pa.

AV1
0520216

Acknowledgments

Extensive use was made of data, illustrations, and photographs supplied by the following:

Airbus Industries
Allied-Signal-Garrett
Beech Aircraft Corporation
Boeing
Cessna Aircraft Company
General Electric Aircraft Engine Groups
Learjet
Lockheed
McDonnell Douglas
National Aeronautics and Space Administration
Pratt & Whitney
Rockwell International
U.S. Air Force

Contents

Preface

THE YEARS SINCE the Wright brothers' first 12-second flight have witnessed phenomenal progress in the science and technology of aerodynamics. For those who possess an interest, the task of education encompassing all the various aspects of the subject is staggering. Although specialization is required, a background knowledge is essential.

Two extremes of scientific and technical literature are typically geared toward two audiences; one is highly specialized, and the other has broad interests. Between these two extremes there is a large group of people who either wish to or have to keep abreast of technological advances, and who often have to make decisions involving these advances.

The purpose of this book, therefore, is to explain the principles of flight near Mach 1 (the speed of sound) as well as supersonic and hypersonic flight, in language that a nontechnical layman can understand. This is the flight regime of turbine-engine-powered (propjet, turbojet, and fanjet) aircraft such as all airline and military aircraft.

High flight speeds and compressible flow dictate airplane configurations that are much different from the ordinary subsonic airplane. Wing sweepback, supercritical wings, and compressibility, for example, are explained along with their associated flight characteristics.

Mathematics and formulas are not used as a substitute for verbal presentation. Because aerodynamics is a complex science, most books on this subject are directed to technical students and professionals; however, students, pilots, mechanics, technicians, and others interested in aviation have a need or an interest in understanding the basics involving the design of high-speed airplanes that are a vital part of our modern society.

This volume concentrates on the aerodynamic and propulsion design of the high-speed airplane. Other aspects of design, such as structures, avionics, fuel systems, environmental control systems, hydraulic systems, and armament (for military airplanes), are not considered or only briefly discussed.

Obviously this book is not a design manual for engineers; however, it is of value to specialized technical people who wish to broaden their perspective.

METHOD OF PRESENTATION

The normal textbook method of presentation is to go into detail on apparently unrelated principles that gradually build up to the desired result. Because this is not a textbook, a different and possibly more interesting procedure is to present the "big picture" first and elaborate on it later. We don't want to miss the forest for the trees.

In order to avoid overuse of footnotes and supplementary explanations in the text, a glossary defines the more specialized terms in the book.

1

Introduction

THE AIRPLANE HAS COMPLETELY REVOLUTIONIZED long-distance travel to the extent that other modes are practically extinct. In many instances, such as transatlantic and transpacific travel, there is no alternative mode. In the United States, for example, the air carriers with their high-speed jet airliners have captured more than 95 percent of all passenger traffic from the railroads; therefore, the airline system is the primary mode of travel for distances more than a few hundred miles.

For the military forces, the airplane, in its various fighter, bomber, reconnaissance, and transport configurations, dominates all other weapons systems. Without air power, successful conclusion of any military conflict is impossible.

If there is one major characteristic that can define the airplane's importance to modern society, it is speed. Certainly there are other requirements besides flight speed for a successful airplane. But a fuel-efficient, comfortable, long-range commercial airliner with a speed of 150 mph is of no practical value. A 150-mph bomber carrying 100 tons of bombs for 10,000 miles is essentially useless. Speed, after all, is what an airplane is all about.

Prior to getting into the details of high-speed flight, a general discussion is in order to lay the groundwork for the detailed discussions.

Since the Wright brothers first 12-second flight in 1903, the quest for more speed has continued relentlessly. By the end of World War I in 1918, the airplane had evolved from a curiosity, a fairground attraction and daredevil's plaything, to a useful and dependable vehicle. Speeds of World War I fighters such as the French Spad XIII reached 138 mph and could climb to 21,000 feet altitude.

The period between the wars was the heyday of the air races. Record-setting flights were dominated by civil aviation. The National Air Races in the United States and the Schneider Trophy Races in Europe, along with various long-distance races and record attempts, slowly but ceaselessly led to the solution of technical problems. In 1932, the barrel-shaped Gee Bee racing airplane was flown to a record speed of 294 mph by Jimmy Doolittle.

The fastest military fighter airplanes (called *pursuit* in those early days) were exemplified by the Curtis Hawk P6E, which had a top speed of 197 mph, almost

100 mph slower than the civil Gee Bee. Military aviation received little emphasis for the first 15 years after World War I. Although engines had improved considerably, the airframe was similar to those of World War I. The strut-and-wire-braced biplane of wood, fabric, and steel-tube construction was still in vogue. Due to lack of governmental interest and economy, military aviation was in a rut.

Civil aviation continued to dominate technical progress. In 1933, the Italian Macchi-Costoldi M.C.72 seaplane was flown to a world's speed record at the unheard of speed of 440.7 mph. This record was finally broken in 1939 by a special racing Messerschmitt Bf109R of Germany, at a speed of 469.22 mph. This speed was the all-time high for propeller-driven aircraft and was finally broken in 1969, 30 years later, at 483 mph, by Darryl Greenamyer of the United States flying a highly modified surplus Grumman F8F-2 Bearcat. But these were all specially built racing machines, not operational airplanes.

By the mid-thirties, war rumblings in Europe finally awakened the military authorities. England, France, and Germany were developing modern, low-wing, all-metal, retractable landing gear monoplanes with speeds faster than 300 mph. It wasn't until 1939, however, at the outbreak of the war in Europe, that the United States began developing advanced airplanes in earnest.

Exactly as happened during World War I, World War II proved to be a potent force in the development of the airplane. Propeller-driven fighter aircraft speeds increased from about 300 mph for the prewar Spitfires and Messerschmitts to 450 mph for upgraded versions of these two workhorse fighters by the end of the war. World War II saw the ultimate development of the piston-engine-propeller airplane in the North American P-51. World War II produced the first jet engines that made possible high-speed flights faster than 500 mph.

It took 40 years for the airplane to painfully inch up to the 500–600 mph speeds of the first operational jet fighters: the Messerschmitt ME262 and the Gloster Meteor of World War II. The next 15 years, to the late 1950s, became the "era of the jet fighters." In that era, the "sound barrier" was hurdled and speeds surpassed twice the speed of sound (Mach 2.0), faster than 1,500 mph. The Mach-2.0 McDonnell F4 first flew in 1958 and became operational in the early 1960s.

Attainment of speeds much over Mach 2.0 are much more difficult than overcoming the former "sound barrier" at Mach 1. This time the problem is not completely a technical one, however. Although there are many technical problem areas, the biggest barriers are involved in cost and justification. The development of a Mach-3.0 airplane involves a fantastic engineering and test program with its attendant high costs. The Republic F-105 Mach-2.0 fighter of the 1950s required 26 engineering man hours for every single hour required by previous Mach 1 fighters built by them.

The tremendous engineering and development programs involved in attaining Mach 3.0 are due mostly to the extreme temperature environment at this speed. For example, at Mach 2.0 the air temperature surrounding the airplane reaches 250°F even at the –66°F outside temperature at 35,000 feet. This aerodynamic heating effect is due to the compression of air as the airplane moves through it. At sustained speeds of Mach 3.0, however, this air temperature shoots up to 640°F.

Because aluminum loses its strength at temperatures exceeding 250°F, entirely new structural design concepts must be developed using insulation, exotic composites, and cooling or other heavier metals. Also, all electronics, hydraulic systems, control mechanisms, and the crew must either operate in this high-temperature environment, or be cooled. Either design approach is tremendously costly when the entire airplane is involved.

What is high-speed flight? This is a relative term. For this discussion, *high-speed flight* is flight faster than about 350 miles per hour. Why 350 mph? Airplane design and performance is considerably simplified when flown slower than 350 mph. Issues related to "compressibility," "shock wave," "speed of sound," "Mach number," and "sound barrier" are not considered. Slower than 350 mph, air is considered incompressible, like water. Figure 1-1 shows that high-speed flight encompasses the speed range of 350 mph (Mach 0.45) to Mach 7 (seven times the speed of sound) and faster.

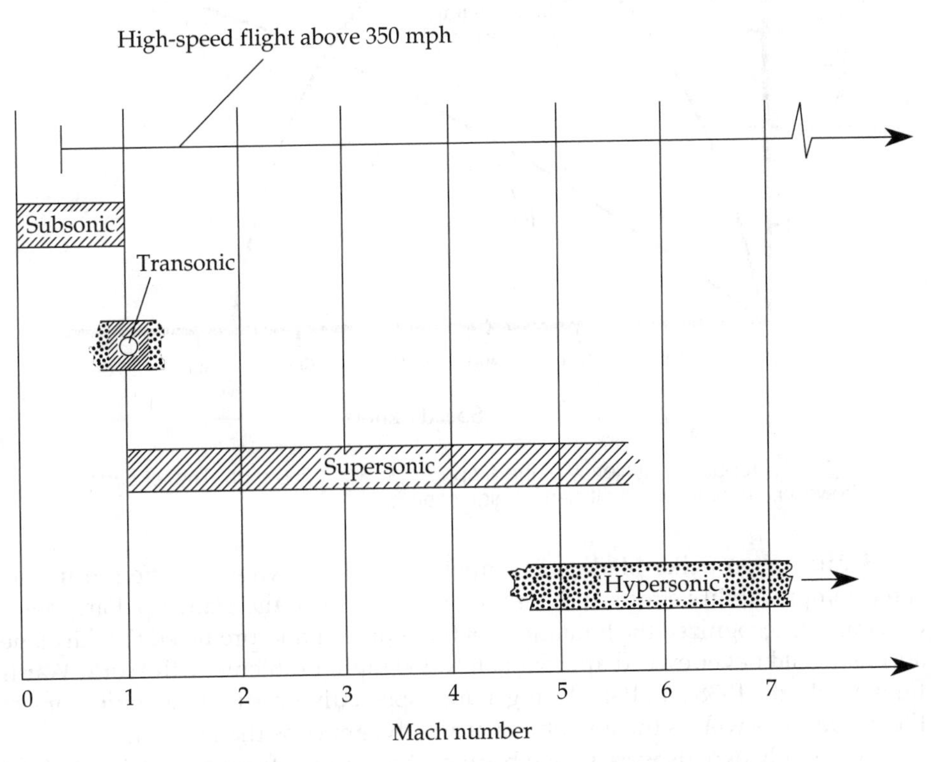

Fig. 1-1. Flight regime terminology verses Mach number. High-speed flight is considered to begin at 350 mph (Mach 0.45).

There was a time when the prospects of closely approaching or exceeding Mach one, the *sound barrier*, seemed impossible, owing to the lack of engines with the necessary power to overcome the rapid rise in drag that begins at the critical Mach number. Even without compressibility effects, the drag would rise with the square of the velocity, and the power, which is drag times velocity, with the cube of velocity. The effect of compressibility is to increase these values still further to astronomically high values.

The problem was real serious because it was assumed that aircraft would always be powered by propellers. The efficiency of a good propeller is about 80 percent at best, but it is best at speeds of 250–350 knots, after which its efficiency drops off rapidly (Fig. 1-2).

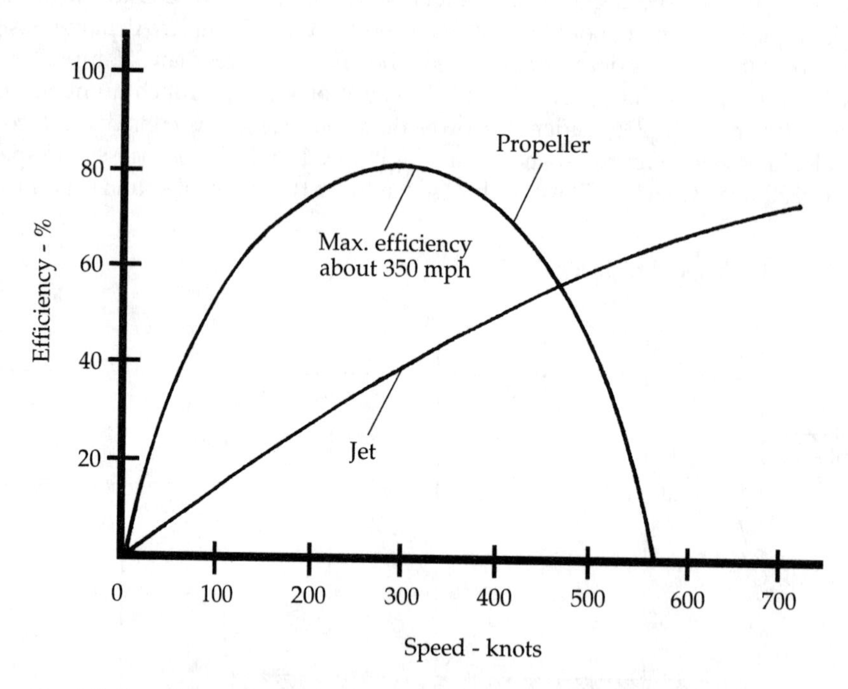

Fig. 1-2. Propulsion efficiency of the propeller peaks at 350 mph. The jet engine, however, becomes more efficient at high speeds.

Because a propeller blade is essentially a rotating wing, it suffers from the same compressibility effects and high drag, especially at the blade tip. Early aerodynamicists recognized the limitations of the propeller and predicted that airplane speeds would never exceed approximately 450 mph. Problems with World War II fighters like the P-38 and P-47 during a high-speed dive verified the limitations of the propeller as well as the limitations of aerodynamics as then known.

These early experts were wrong because they did not foresee the advent of the jet engine. This made all the difference, partly due to the elimination of the propeller and its compressibility problems, but also because the efficiency of the jet engine increases rapidly over 350 knots, when the efficiency of the propeller is rapidly falling off. The net result is that whereas the piston engine-propeller combination requires about 20 times as much power to fly at 500 knots compared to 250 knots, the jet engine only requires about five times the thrust, and it is thrust that counts in a jet engine (Fig. 1-3). Also, the weight of the jet engine, as well as its fuel consumption, are much less than that of the piston engine-propeller combination at this speed.

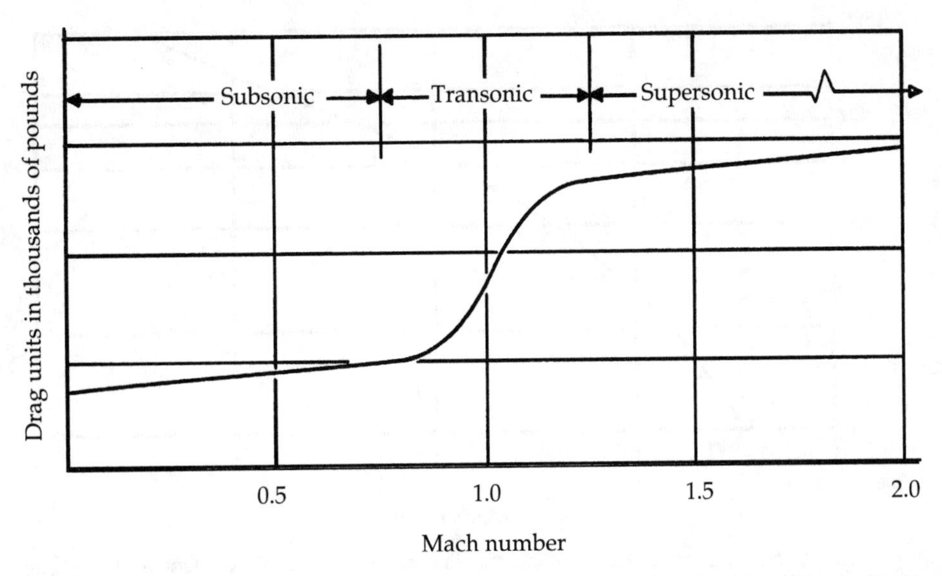

Fig. 1-3. Typical drag rise curve for a supersonic airplane. A lot of power (thrust) is required to overcome the "sound barrier."

The availability of practical jet engines in the late 1940s, as well as captured German wartime research data, ushered in high-speed flight. The engine and propeller combination, however, remains the most efficient propulsion system below about 350 mph. The jet engine is the propulsion system for high-speed flight as shown in Fig. 1-2.

Most of today's airplanes, however, fly slower than Mach one. Primarily military airplanes fly supersonically. Why don't more airplanes fly faster than Mach 1? Recall that there is a tremendous increase in drag that requires more power as Mach 1 is approached and exceeded. Also, at numbers faster than Mach 2.0, air temperatures due to compression and skin friction exceed 200°F, requiring complex design solutions such as exotic materials and cooling for crew, passengers, and equipment (Fig. 1-4).

In other words, Mach 1, the speed of sound, is now an "economic barrier." Consequently, many efforts at achieving supersonic flight are directed at delaying it, or in effect, decreasing drag so as to approach Mach 1 as close as possible. Commercial airplanes of the 1990s, such as the Boeing 777 and McDonnell Douglas MD11, fly only a little faster but more efficiently than airplanes of the 1960s, such as the Boeing 727 and McDonnell Douglas DC-8; however, all fly below Mach 1.

Figures 1-5 through 1-11 show examples of high-speed airplanes from the 350-mph Beech King Air to the Lockheed F-22 Mach 3.0+ fighter for the 1990s and early twenty-first century.

This book is not concerned with space flight, but a quick comparison of space flight and atmospheric flight is in order.

We all know that air has weight, and therefore density, and that air gets thinner as altitude is increased. As we go up in altitude, drag decreases due to the thinner air; however, lift and thrust also decrease. An airplane depends on the air's

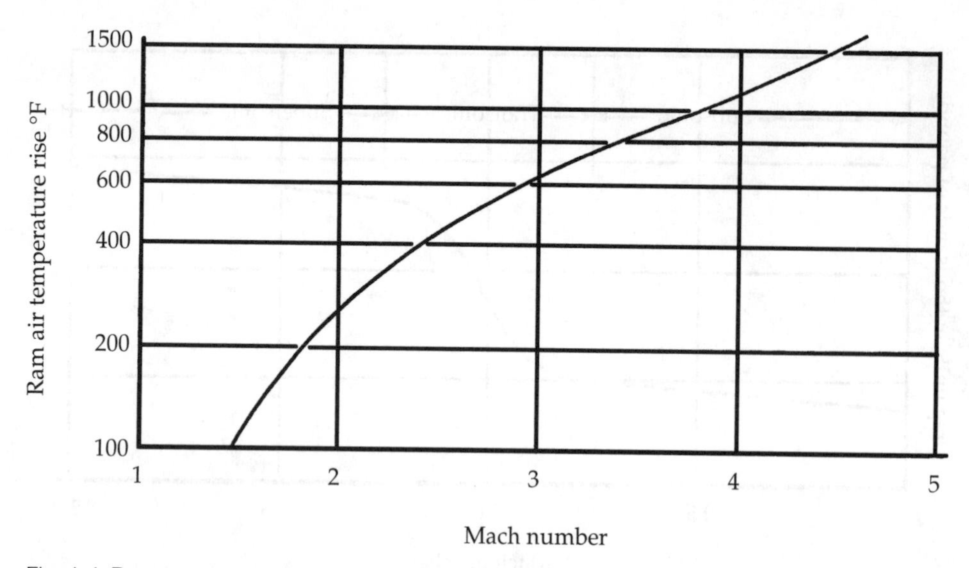

Fig. 1-4. Due to compression and skin friction, the air temperature surrounding the airplane rises dramatically over Mach 2.0. Conventional aluminum alloys begin to lose strength over 200°F or Mach 2.0.

density to stay up and the jet is an air-breathing engine. Higher than about 100,000 feet, sustained flight is not practicable. We've simply run out of air; however, there is still plenty of air to create skin friction so that meteors, projectiles, or spacecraft can burn up when entering the atmosphere at high speed. Above approximately 100 miles altitude, air drag becomes negligible. This is where space flight begins.

Sustained atmospheric flight is not practical in the region between approximately 100,000 feet and 100 miles because the air is too thin for lift and air-breathing engines. Conversely, because the air density in the same region creates too much drag and skin friction, space flight is not practical. This is the transitional region from atmospheric flight to space flight.

Fig. 1-5. The Beechcraft Super King Air 300 is a 350-mph turboprop airplane that represents the high end of the low-speed range (or the low end of the high-speed range). High-speed aerodynamics is considered to begin at 350 mph.

Fig. 1-6. The Boeing 747 turbofan-powered airliner is capable of high subsonic speed (M = 0.82 or 550 mph).

Fig. 1-7. The North American F-100 turbojet-powered fighter was the first operational transonic airplane (1954) with a short-time top speed (in afterburner) of M = 1.3 (860 mph at 35,000 feet) and a cruising speed of M = 0.8.

Fig. 1-8. The Boeing 707 was the first practical operational jet airliner.

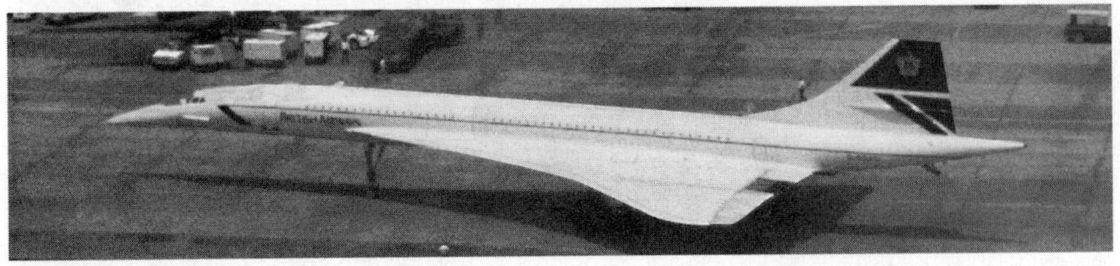

Fig. 1-9. The British-French Concorde was the first and remains the only operational supersonic cruise airliner (M = 2.2).

Fig. 1-10. Lockheed SR-71 Blackbird, a supersonic Mach 3.0 reconnaissance airplane.

Fig. 1-11. The F-22 advanced tactical fighter of the 1990s with a top speed faster than Mach 3.0 and supersonic cruise without afterburner. The F-22 is a Lockheed/Boeing design.

2

The atmosphere

ALL FLIGHT IS CONFINED to the lower levels of the atmosphere below about 100,000 feet (19 miles), as shown in Fig. 2-1. In this region, the atmosphere consists of a homogeneous mixture of nitrogen (78 percent) and oxygen (21 percent) plus about 1 percent of trace gases. This homogeneous mixture is called *air*, the fluid that concerns the aerodynamicists and aircraft designer.

The *troposphere* (Fig. 2-1) is the most important atmospheric layer to aeronautics because most aircraft fly in this region. Most weather occurs here and man lives here also. Without the beneficial ozone layer in the *stratosphere* absorbing harmful solar ultraviolet radiation, life as we know it would not have developed. The *ionosphere* begins in the *mesosphere* and extends indefinitely outward. The ionosphere represents the region in which ionization of one or more of the atmospheric constituents is significant.

THE STANDARD ATMOSPHERE

Knowledge about the vertical distribution of pressure, temperature, density, and speed of sound is required for pressure altimeter calibrations and the performance and design of aircraft. Because the real atmosphere never remains constant at any particular time or place, a hypothetical model must be employed as an approximation to what can be expected. This model is known as the *standard atmosphere*. The air in the model is assumed to be devoid of dust, moisture, and water vapor and to be at rest with respect to the Earth: no winds or turbulence.

Figure 2-2 is a multiple plotting of pressure, density, temperature, and speed of sound from sea level to 62 miles. It is intended merely to indicate the general variation of these parameters. The temperature-defined atmospheric shells are also included. Table 2-1 shows these parameters in more detail.

In the troposphere (from sea level to 6–12 miles in the standard atmosphere), the temperature decreases linearly with altitude. In the stratosphere, it first remains constant at –56.5°C (–69.7°F) before increasing again. The speed of sound shows a similar type of variation. Both the density and pressure are seen to de-

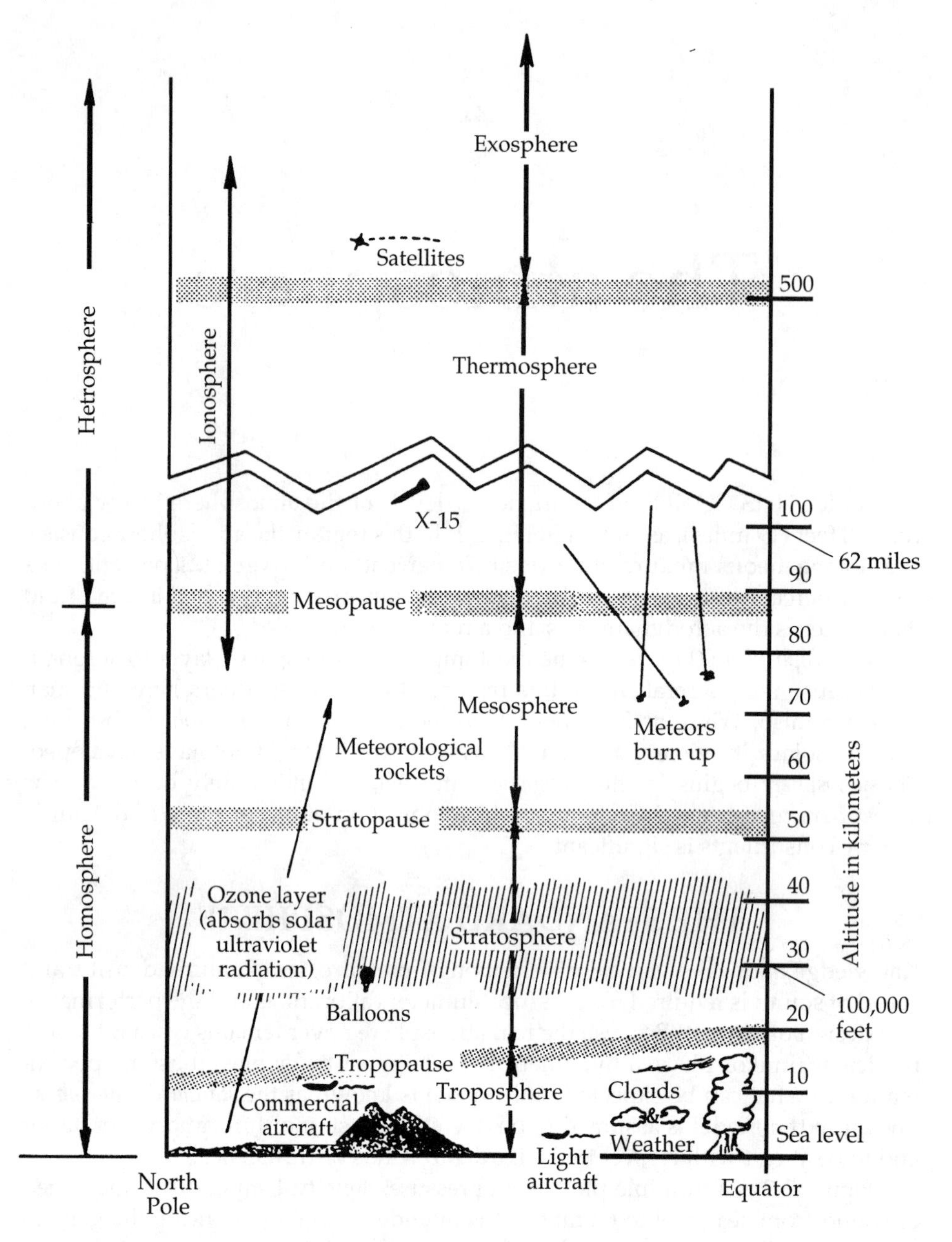

Fig. 2-1. The atmospheric structure. Sustained flight of all airplanes is confined to the lower atmosphere below 100,000 feet. Commercial airliners normally fly at about 35,000 feet.

crease rapidly with altitude. The density curve is of particular importance because the lift of an airfoil is directly dependent on the density.

The rate at which atmospheric pressure decreases with altitude is much greater near the surface of the Earth (Fig. 2-2 and Table 2-1). Between sea level and

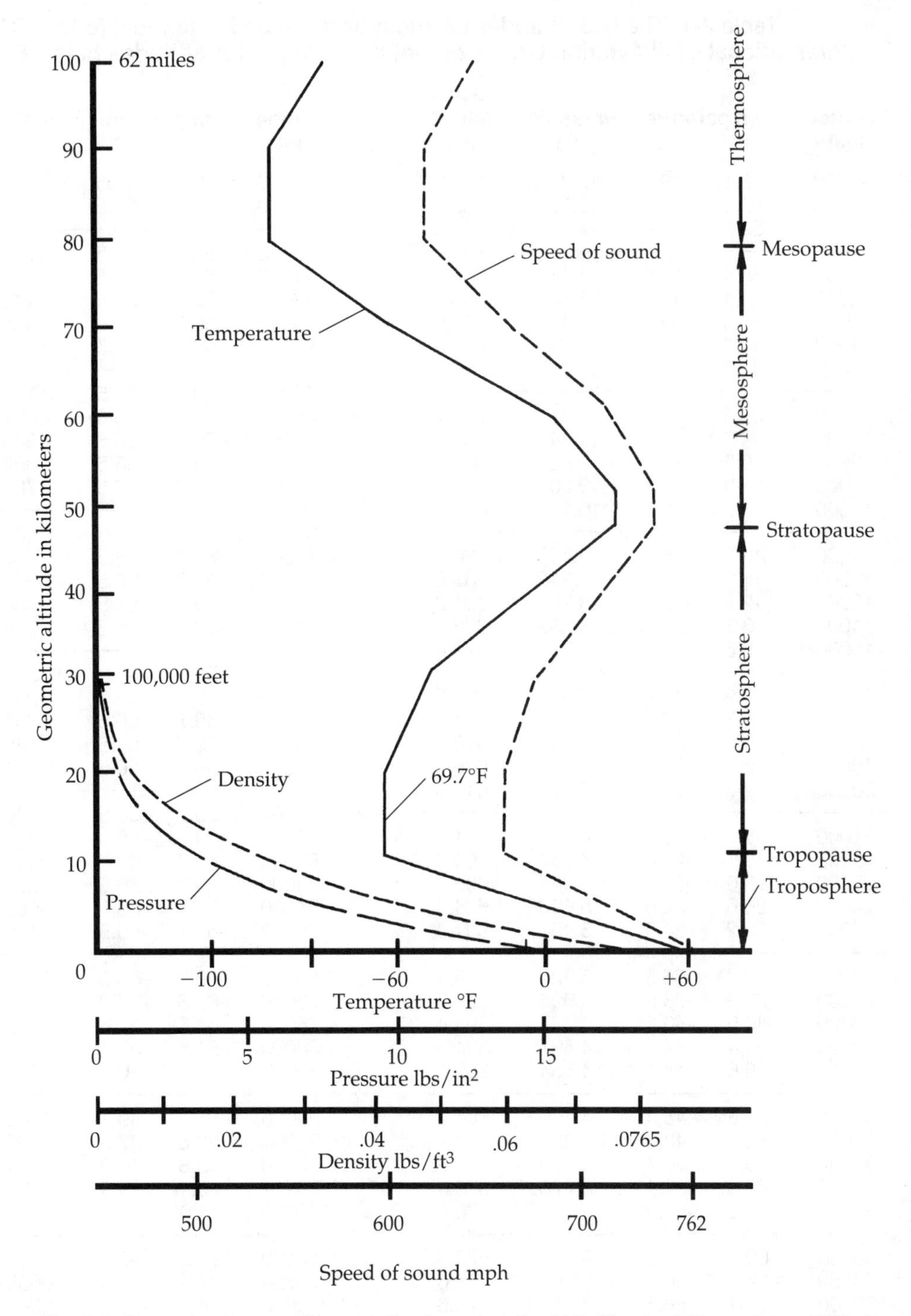

Fig. 2-2. Atmospheric properties variation based on the U.S. Standard Atmosphere, 1962. At sea level, the standard atmospheric temperature is 59°F (15°C) and the pressure is 14.7 psi or 29.92" Hg.

Table 2-1. The U.S. Standard Atmosphere which is identical to the ICAO (International Civil Aviation Organization) atmosphere for altitudes below 65,617 feet.

Altitude feet	Temperature °F	°C	Pressure psia	Sonic Velocity kts
–2000	66.1	19.0	15.79	666.0
–1000	62.5	17.0	15.23	663.7
0	59.0	15.0	14.70	661.5
1000	55.4	13.0	14.17	659.2
2000	51.4	11.0	13.66	656.9
3000	48.3	9.1	13.17	654.6
4000	44.7	7.1	12.69	652.3
5000	41.2	5.1	12.23	650.0
6000	37.6	3.1	11.78	647.7
7000	34.6	1.1	11.34	654.4
8000	30.5	–0.8	10.92	643.0
9000	26.9	–2.8	10.50	640.7
10000	23.3	–4.8	10.11	638.3
11000	19.8	–6.8	9.720	636.0
12000	16.2	–8.8	9.346	633.6
13000	12.6	–10.7	8.984	631.2
14000	9.1	–12.7	8.633	628.8
15000	5.5	–14.7	8.294	626.4
16000	1.9	–16.7	7.965	624.0
17000	–1.6	–18.7	7.647	621.6
18000	–5.2	–20.7	7.339	619.2
19000	–8.8	–22.7	7.041	616.7
20000	–12.3	–24.6	6.754	614.3
21000	–15.9	–26.6	6.475	611.9
22000	–19.5	–28.6	6.207	609.4
23000	–23.0	–30.6	5.947	606.9
24000	–26.6	–32.5	5.696	604.4
25000	–30.2	–34.5	5.454	601.9
26000	–33.7	–36.5	5.220	599.4
27000	–37.3	–38.5	4.994	596.9
28000	–40.9	–40.5	4.777	594.4
29000	–44.4	–42.4	4.567	591.9
30000	–48.0	–44.4	4.364	589.3
31000	–51.6	–46.4	4.169	586.8
32000	–55.1	–48.4	3.981	584.2
33000	–58.7	–50.4	3.800	581.6
34000	–62.3	–52.4	3.626	579.0
35000	–65.8	–54.3	3.458	576.4
36000	–69.4	–56.3	3.297	573.8
*36089	–69.7	–56.5	3.282	573.6
37000	–69.7	–56.5	3.142	573.6

Altitude feet	Temperature °F	°C	Pressure psia	Sonic velocity kts
38000	–69.7	–56.5	2.994	573.6
39000	–69.7	–56.5	2.854	573.6
40000	–69.7	–56.5	2.720	573.6
41000	–69.7	–56.5	2.592	573.6
42000	–69.7	–56.5	2.471	573.6
43000	–69.7	–56.5	2.355	573.6
44000	–69.7	–56.5	2.244	573.6
45000	–69.7	–56.5	2.139	573.6
46000	–69.7	–56.5	2.039	573.6
47000	–69.7	–56.5	1.943	573.6
48000	–69.7	–56.5	1.852	573.6
49000	–69.7	–56.5	1.765	573.6
50000	–69.7	–56.5	1.682	573.6
51000	–69.7	–56.5	1.603	573.6
52000	–69.7	–56.5	1.528	573.6
53000	–69.7	–56.5	1.456	573.6
54000	–69.7	–56.5	1.388	573.6
55000	–69.7	–56.5	1.323	573.6
56000	–69.7	–56.5	1.261	573.6
57000	–69.7	–56.5	1.201	573.6
58000	–69.7	–56.5	1.145	573.6
59000	–69.7	–56.5	1.091	573.6
60000	–69.7	–56.5	1.040	573.6
61000	–69.7	–56.5	.9913	573.6
62000	–69.7	–56.5	.9448	573.6
63000	–69.7	–56.5	.9005	573.6
64000	–6.97	–56.5	.8582	573.6
65000	–69.7	–56.5	.8179	573.6
* 65617	–69.7	–56.5	.7941	573.6
70000	–67.3	–55.2	.6437	575.3
75000	–64.6	–53.6	.5073	577.3
80000	–61.8	–52.1	.4005	579.3
85000	–59.1	–50.6	.3167	581.3
90000	–56.3	–49.1	.2509	583.3
95000	–53.6	–47.5	.1990	585.3
100000	–50.8	–46.0	.1581	587.3
*104987	–48.1	–44.5	.1259	589.2
150000	21.0	–6.1	.01893	636.8
*154199	27.5	–2.5	.01609	641.1
*170604	27.5	–2.5	.00557	641.1
200000	–4.8	–20.4	.02655	619.5
*200131	–4.9	–20.5	.02641	619.4

*Boundary between atmosphere layers of constant thermal gradient

10,000 feet, the pressure has decreased from 14.7 pounds per square inch (psi) to 10.1 psi with a corresponding decrease in density from 0.0765 to 0.0565 pounds per cubic foot. From 10,000 feet, the pressure gradually decreases down to slightly more than 1 psi at 60,000 feet; thus, the atmospheric pressure is greater at sea level and decreases to almost insignificant amounts at 100,000 feet (19 miles). This pressure gradient is due to the compressibility of air. The air near the Earth's surface is compressed by the air above it.

Temperature of the air also decreases with an increase in altitude. The radiant heat from the Sun passes through the atmosphere without appreciably raising its temperature; however, the Earth absorbs the Sun's radiant heat, raising its temperature as well as the air in contact with it. This lower layer of heated air expands with subsequent decrease in density and rises, setting up convection currents in the atmosphere. This rising column of air expands as it reaches lower pressure and, according to laws of thermodynamics, will in itself cause a reduction in temperature; however, even without considering convection currents, the more dense air near Earth's surface will absorb more heat than the thinner air at high altitudes.

The rate of temperature change does not change until about 36,000 feet, where the temperature remains practically constant at –69.7°F (56.5°C). This is the tropopause, which is the beginning of the stratosphere. Most flying takes place in the area before the tropopause, or the troposphere (Fig. 2-2); however, the stratosphere, above 36,000 feet, is the realm of the supersonic airplane.

Sound is a pressure wave produced by the vibrations of material objects. Any elastic substance, whether solid, liquid, or gas, can transmit sound; however, most sounds that we hear are transmitted through the air. These sound waves travel at a certain speed that varies with temperature. The transmission of sound requires a medium; if there is nothing to compress and expand, there can be no sound. Therefore, due to the lack of a medium (air) there can be no sound in the vacuum of space.

THE REAL ATMOSPHERE

It would be fortunate if the Earth's real atmosphere corresponded to a standard atmospheric model, but thermal effects of the Sun, the presence of continents and oceans, and the Earth's rotation all combine to stir up the atmosphere into a nonuniform, nonstandard mass of gases in motion. Although a standard atmosphere provides the criteria necessary for design of an aircraft, it is essential that "nonstandard" performance in the real atmosphere be anticipated also. This nonstandard performance shows up in numerous ways, some of which are discussed in the following paragraphs.

Unquestionably, the most important real atmospheric effect is the relative motion of the atmosphere. Although in the standard atmosphere the air is motionless with respect to the Earth, it is known that the air mass through which an airplane flies is constantly in a state of motion with respect to the surface of the Earth. Its motion is variable both in time and space and is exceedingly complex. The motion can be divided into two classes:

- Large-scale motions
- Small-scale motions

Large-scale motions of the atmosphere (winds) affect the navigation and the performance of an aircraft. The airplane flies in the sea of air, and the sea of air is in motion relative to the Earth; therefore, the pilot must compensate for crosswinds, headwinds, and tailwinds when navigating from one airport to another. Winds also affect the range of an airplane and the fuel required to fly from one ground point to another.

The small-scale motion of the atmosphere is called *turbulence* (or *gustiness*). The response of an aircraft to turbulence is an important matter. In passenger aircraft, light turbulence might cause minor discomfort and in severe turbulence, injuries if seat belts are not fastened. In cases of precision flying such as air-to-air refueling, bombing, gunnery, or aerial photography, turbulence-induced motions of the aircraft are a nuisance.

Turbulence-induced stresses and strains over a long period might cause fatigue in the airframe and in cases of extremely severe turbulence, the loss of control of an aircraft or even immediate structural failure is possible. The structural design of an airplane is subject to *gust load criteria* prescribed by various civil agencies (FAA in the United States) and the military services.

There are several causes of turbulence. The unequal heating of the Earth's surface by the Sun will cause convective currents to rise and make the plane's motion through such unequal currents rough. On a clear day, the turbulence is not visible but will be felt; hence, the name *clear air turbulence* (CAT). Turbulence also occurs because of winds blowing over irregular terrain or, by different magnitude or direction, winds blowing side by side and producing a *wind shear* effect.

The thunderstorm is one of the most violent of all cases of turbulence where strong updrafts and downdrafts exist side by side. The severity of the aircraft motion caused by turbulence will depend upon the magnitude of the updrafts and downdrafts and their directions. Many aircraft have been lost to thunderstorm turbulence because of structural failure or loss of control. Commercial airliners generally fly around such storms for the comfort and safety of their passengers, using radar to determine a clear flight path.

Another real atmospheric effect is that of moisture. Water in the air, in either its liquid or vapor form, is not accounted for in the pure dry standard atmosphere and will affect an aircraft in varying degrees. Everyone is familiar with the forms of precipitation that can adversely affect aircraft performance: icing on wings, zero visibility in fog or snow, and physical damage caused by hail. Water vapor is less dense than dry air. Because of this, an aircraft requires a longer takeoff distance in humid air than in the more dense dry air.

Air density is a very important factor in the lift, drag, and engine power output of an aircraft and depends upon the temperature and pressure locally. Because the standard atmosphere does not indicate true conditions at a particular time and place, it is important for a pilot to obtain the local temperature and pressure readings. This is especially important to high-performance aircraft such as jet airliners where takeoff distance and engine power output are affected by air density. Takeoff from a high-altitude airport when the temperature is high creates a low-density situation requiring a longer takeoff run and possible off-loading of fuel and passengers.

The local pressure is important in aircraft using pressure altimeters. A pilot must set the aircraft pressure altimeter to local measured sea-level pressure rather than to standard sea-level pressure to obtain accurate altitude readings above sea level.

Although the preceding discussion considers only a few of the many effects of a nonstandard atmosphere on aircraft design and performance, the standard atmosphere still remains as a primary reference in the preliminary design stage of an aircraft; however, nonstandard conditions are incorporated into detailed operating handbooks, especially for high-performance airplanes.

AIR AS FLUID

Viscosity. There are basically three states of matter: solid, liquid, and gas. H_2O is commonly called ice in the solid state, water in the liquid state, and water vapor in the gaseous state. Assume a piece of ice with side forces applied to it (*shearing forces*). Very large forces are needed to deform or break it. The solid has a very high internal friction or resistance to shearing. The word for internal friction is *viscosity* and for a solid its value is generally very large.

Liquids and gases are considered to be fluids because they behave differently from a solid. Consider two layers of water or air. If shear forces are applied to these layers, a substantial and sustained relative motion of the layers is observed with the air layers sliding faster over one another than the water layers; however, the fact that a shear force must be applied to deform the fluids indicates that they also possess internal friction.

Water, under normal temperatures, is about 50 times more viscous than air. Ice is 5×10^{16} times more viscous than air. In general, solids have extremely high viscosities whereas fluids have low viscosities. Under the category of fluids, liquids generally possess higher viscosities than gases. Air, of primary interest in aerodynamics, has a relatively small viscosity and in some theories it is described as a *perfect fluid*, one that has zero viscosity or is *inviscid*. But it will be shown that even this small viscosity of air (or internal friction) has important effects on an airplane in terms of lift and drag, especially at high speeds.

Compressibility. All fluids are compressible to some extent—density increases under increasing pressure—but liquids are generally highly incompressible compared with gases. Even gases can be treated as incompressible provided the flow speeds involved are not great. For subsonic flow over an airplane that is flying slower than 350 mph, air may be treated as *incompressible*: no change in density throughout the flow. At faster speeds, the effects of compressibility must be taken into account.

AIRSPEED MEASUREMENT

Bernoulli's equation states that in a streamline fluid flow, the greater the speed of the flow, the less the static pressure; and the less the speed of the flow, the greater the static pressure. There exists a simple exchange between the dynamic and static pressures such that their total remains the same. As one increases, the other must decrease. (See the glossary.) Bernoulli's equation reduces to:

$$\text{Dynamic pressure} + \text{static pressure} = \text{total pressure}$$

Pressure measurement

Let us now examine how total, static, and dynamic pressures in a flow are measured. Figure 2-3(A) shows the fluid flow around a simple hollow bent tube, called a *pitot tube,* which is connected to a pressure measurement readout instrument. The fluid dams up immediately at the tube entrance and comes to rest at the *stagnation point* while the rest of the fluid divides up to flow around the tube. By Bernoulli's equation, the static pressure at the stagnation point is the total pressure because the dynamic pressure reduces to zero when the flow stagnates. The pitot tube is, therefore, a total-pressure measuring device.

Figure 2-3(B) shows the fluid flow around another hollow tube, except now the end facing the flow is closed, and a number of holes have been drilled into the tube's side. This *static tube* may be connected to a pressure measuring readout instrument as before. Except at the stagnation point, the fluid is parallel to the tube everywhere. The static pressure of the fluid acts normal to the tube's surface. Because pressure must be continuous, the static pressure normal to the holes is communicated into the interior of the tube; therefore, the static tube with the holes parallel to the flow direction is a static-pressure measuring device.

Figure 2-3(C) shows a combined *pitot-static tube.* When properly connected to opposite ends of a pressure-measuring readout instrument, the difference between total pressure and static pressure is measured. By Bernoulli's equation this difference is the dynamic pressure, defined as $\frac{1}{2}\rho V^2$. If the fluid density is known, the fluid flow speed can be calculated. In actual use on aircraft, the pitot-static tube is connected directly to an airspeed indicator that, by proper gearing, will automatically display the aircraft airspeed to the pilot. The pitot-static tube is sometimes mounted forward on a boom extending from the airplane nose to ensure the closest possible measurement of the undisturbed approaching flow, also called the *free-stream condition.*

Altitude measurement

Altitude of an airplane is determined by a pressure instrument called an *altimeter,* calibrated in units of height (feet or meters) using the standard atmosphere from Table 2-1. Basically, the altimeter measures the static pressure at a particular altitude. An altimeter is connected to the static tube or port as shown in Fig. 2-3(D).

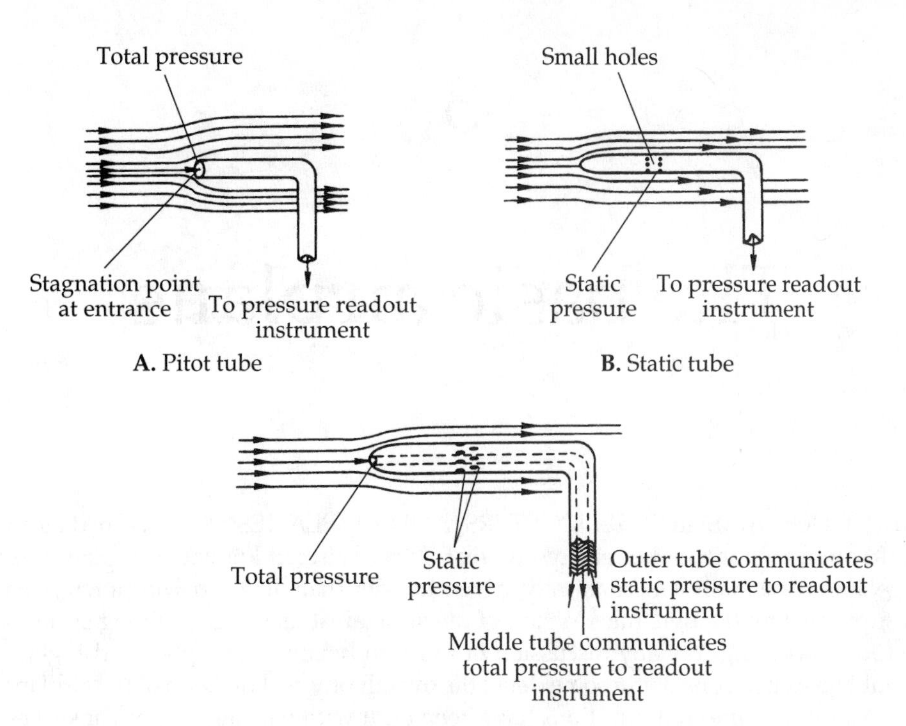

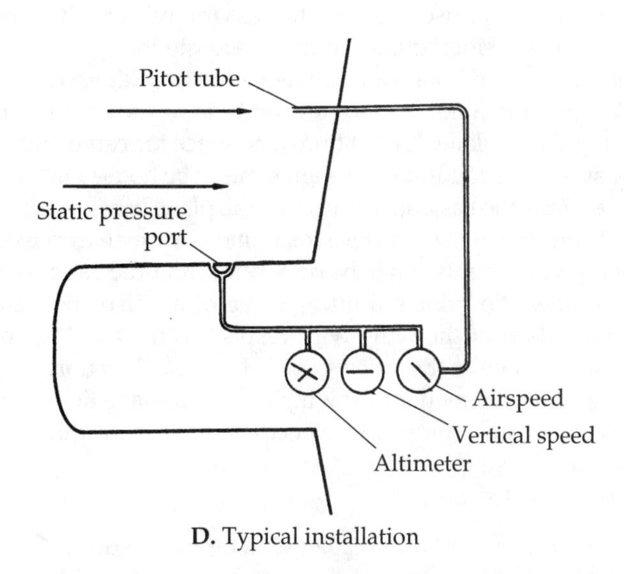

Fig. 2-3. Pressure measuring devices of the pitot-static system. Some installations combine the pitot and static tubes as in Figure 2-3(C), others separate the pitot tube from the static port as in Figure 2-3(D).

3

The basic airplane

ALTHOUGH AIRSHIPS, HELICOPTERS, AND AIRPLANES are all defined as aircraft, our attention will be centered on that class of aircraft known as airplanes. An airplane is essentially a mechanically driven, heavier-than-air, fixed-wing aircraft that is supported by the dynamic reaction of the air against its wings or lifting surfaces. Before proceeding into any discussion of the aerodynamics of high-speed flight, it would be well to consider in some detail the overall physical makeup of the airplane.

Many airplane configurations have been built with varying degrees of success (flying wing, tailless, canard, biplane, etc.); however, the basic airplane consists of a monoplane with a wing, fuselage, and tail assembly. Aerodynamic aspects of these components are considered later in this discussion.

Figure 3-1 is a high-speed, subsonic, three-engine jet airliner. The body of an airplane is called the *fuselage*. It houses the crew and the controls necessary for operating and controlling the airplane. It might provide space for cargo and passengers or various military systems. In addition, an engine might be housed in the fuselage. The fuselage is, in one sense, the basic structure of the airplane because many of the other large components are attached to it. It is streamlined as much as possible to reduce drag. Designs vary with the mission to be performed, and the variations are endless.

The *wing* provides the principal lifting force of an airplane. Lift is obtained from the dynamic action of the wing with respect to the air. The cross-sectional shape of the wing is known as the *airfoil section*. The airfoil section shape, planform shape of the wing, and placement of the wing on the fuselage depend upon the airplane's mission and the best compromise necessary in the overall airplane design. Figure 3-2 illustrates the shapes most often used.

The tail assembly consists of:

- The *vertical stabilizer* (fin) and *rudder*, which provide directional control and stability in yaw.
- The *horizontal stabilizer* and *elevator*, which provide control and stability in pitch.

Control surfaces are all those moving surfaces of an airplane that are used for attitude, lift, and drag control. *Yaw control* (turning the airplane to the left or right) is

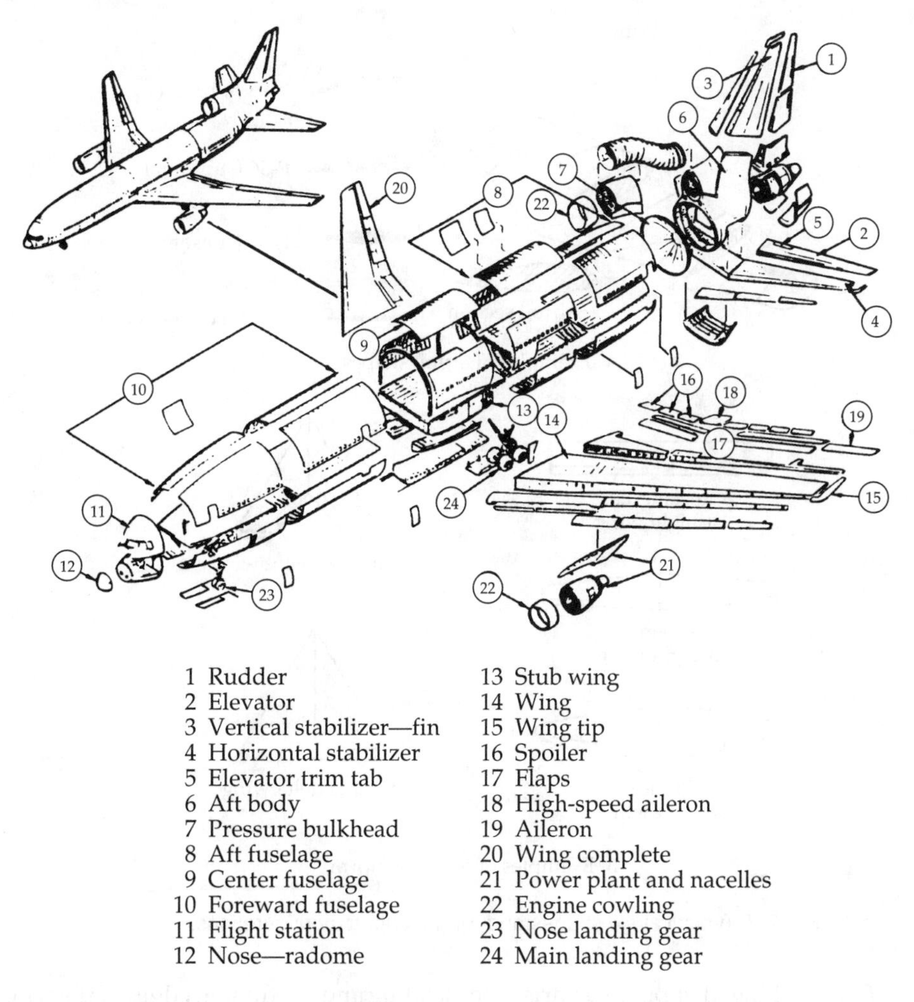

Fig. 3-1. Exploded view of a Lockheed L-1011 airliner that is capable of flying at Mach 0.85.

provided by the rudder, which is generally attached to the fin. *Pitch control* (nosing the airplane up or down) is provided by the elevators, which are generally attached to the horizontal stabilizer. *Roll control* (rolling the wing to the right or left) is provided by the ailerons located generally near the outer trailing edge of the wing.

Trim tabs are small, auxiliary, hinged control surface inserts on the elevator, rudder, and aileron surfaces. The tabs:

- Balance the airplane if it is too nose heavy, tail heavy, or wing heavy to fly in stable cruise condition.
- Maintain the elevator, rudder, or ailerons at whatever particular setting the pilot wishes without the pilot maintaining pressures on the controls.
- Help move the elevators, rudder, and ailerons, which relieves the pilot of the effort necessary to move the surfaces.

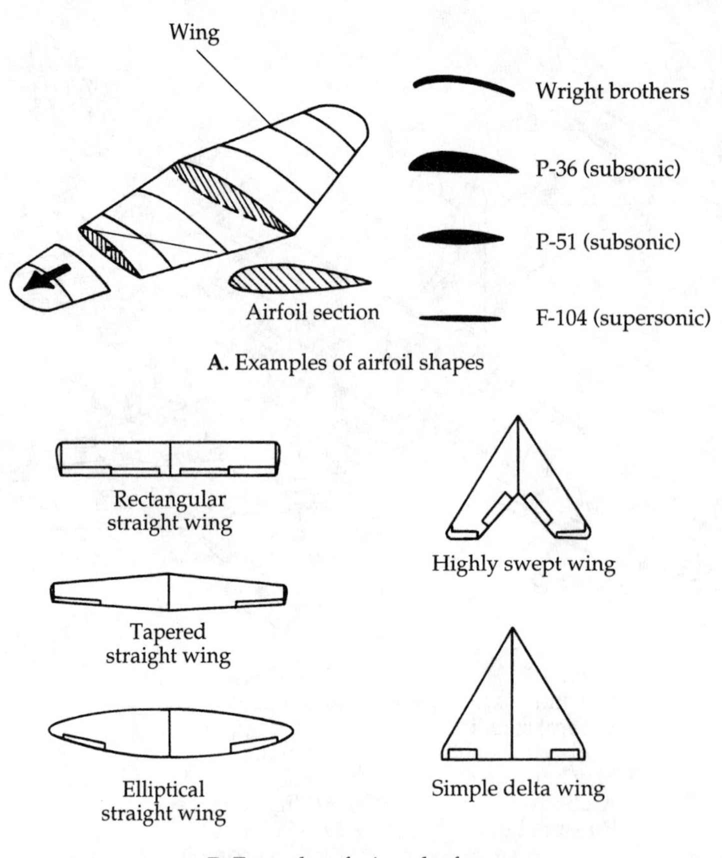

Fig. 3-2. Typical airfoil shapes and wing planform configurations.

Flaps are hinged or pivoted parts of the leading and/or trailing edges of the wing used to increase lift at reduced airspeeds. They are used primarily for landing and takeoff. *Spoilers* reduce the lift on an airplane wing quickly. By operating independently on both sides of the wing, they can provide an alternate form of roll control. Figure 3-3 illustrates the attitude control surfaces. Figure 3-4 shows a simple aileron and flap installation and a more complicated arrangement used on a large jet airliner.

An airplane must possess a thrust-producing device, a *powerplant*, to sustain flight. The powerplant consists of the engine (and propeller, if present,) and the related accessories. The main engine types are:

- Reciprocating (piston)
- Reaction
  - Ram jet
  - Pulse jet
  - Turbojet
  - Turboprop (only partial reaction engine)
  - Rocket

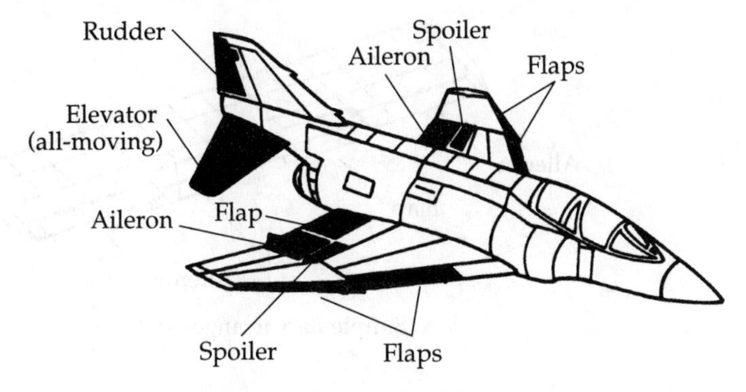

A. Control surfaces on F-4 Phantom

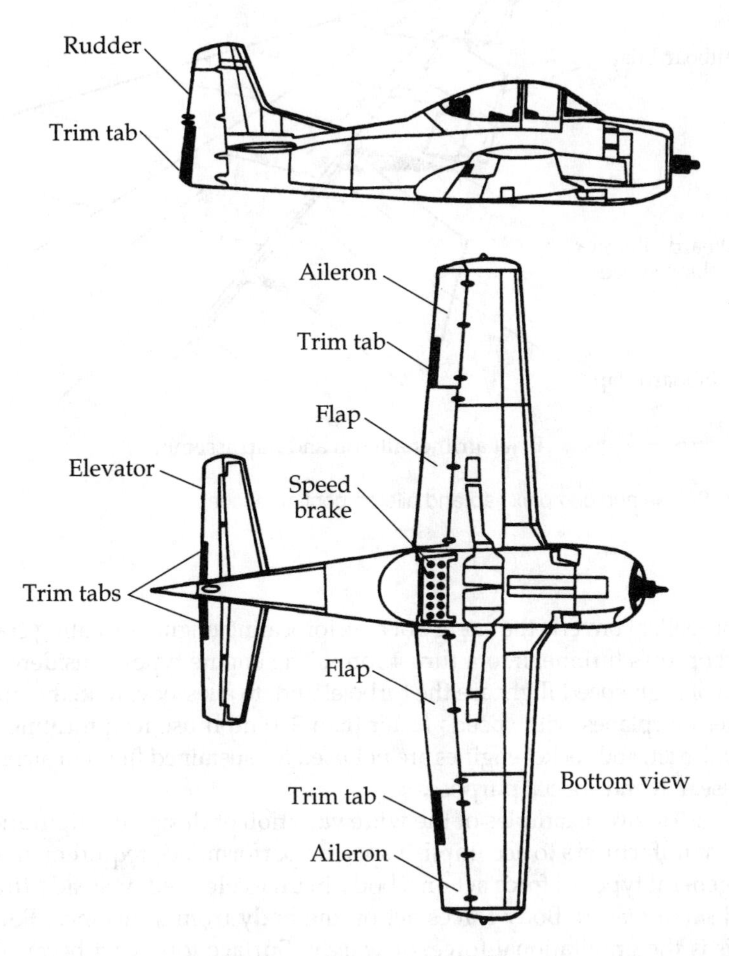

B. Control surfaces on T-28

Fig. 3-3. Control surfaces on the supersonic F-4 Phantom jet fighter and the subsonic T-28 trainer.

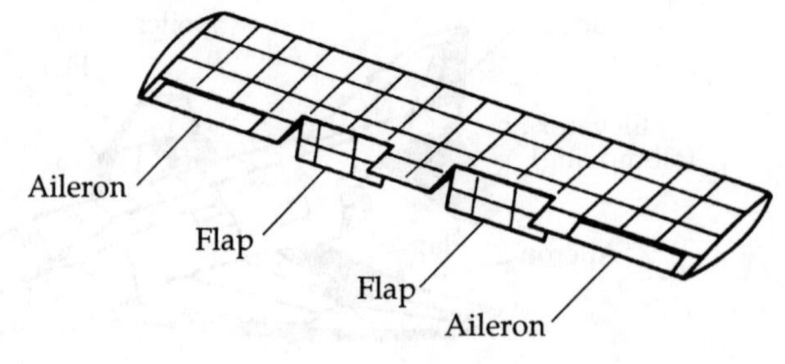

A. Simple flap arrangement

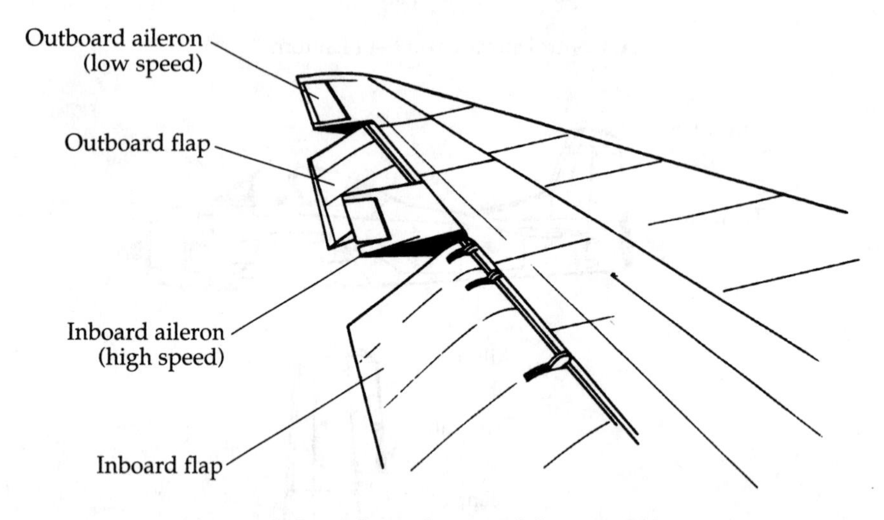

B. Jet airliner aileron and flap assembly

Fig. 3-4. Simple and complex flap and aileron configurations.

The propeller converts the energy of a reciprocating engine's rotating crankshaft, or the turboprop's turbine, into a thrust force. The engine types considered for our discussion of high-speed flight are the turbojet and, to a lesser extent, the turboprop. Few modern airplanes with speeds faster than 350 mph use reciprocating engines. Ram jet, pulse jet, and rocket engines are not used for sustained flight in airplanes, except for research and special purposes.

Figure 3-5 shows examples of the wide variation of design configurations provided by manufacturers to accomplish specific performance requirements.

Two general types of force act on a body in unaccelerated or steady flight: *body forces* and *surface forces*. Body forces act on the body from a distance. For the airplane, this is the gravitational force, or weight. Surface forces act because of contact between the medium and the body, that is, between the air and the airplane surface. Surface forces are *lift*, *drag*, and *thrust*; therefore, the four forces acting on an airplane are weight, thrust, lift, and drag.

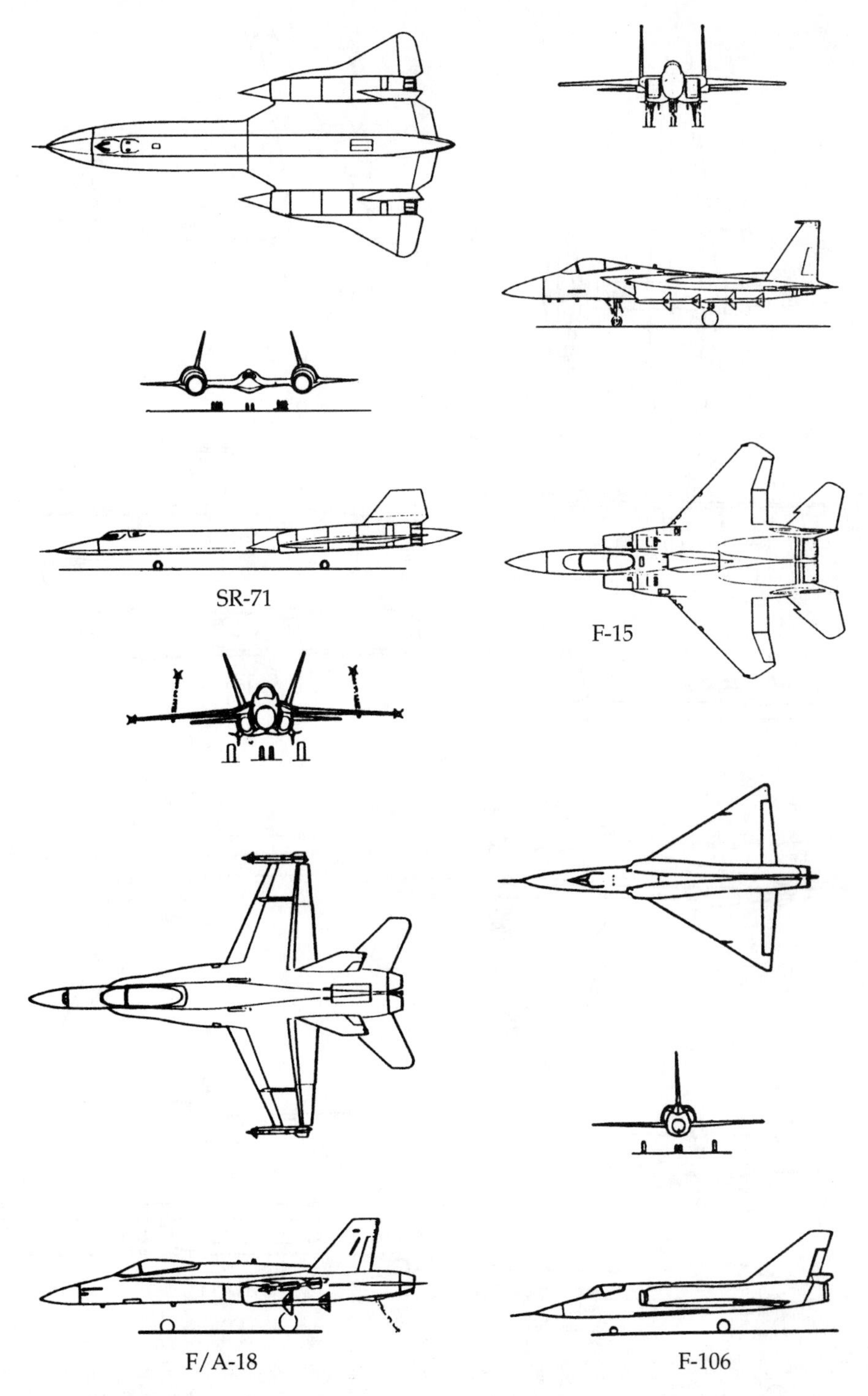

Fig. 3-5. Various high-speed airplane configurations based upon performance within the confines of their missions.

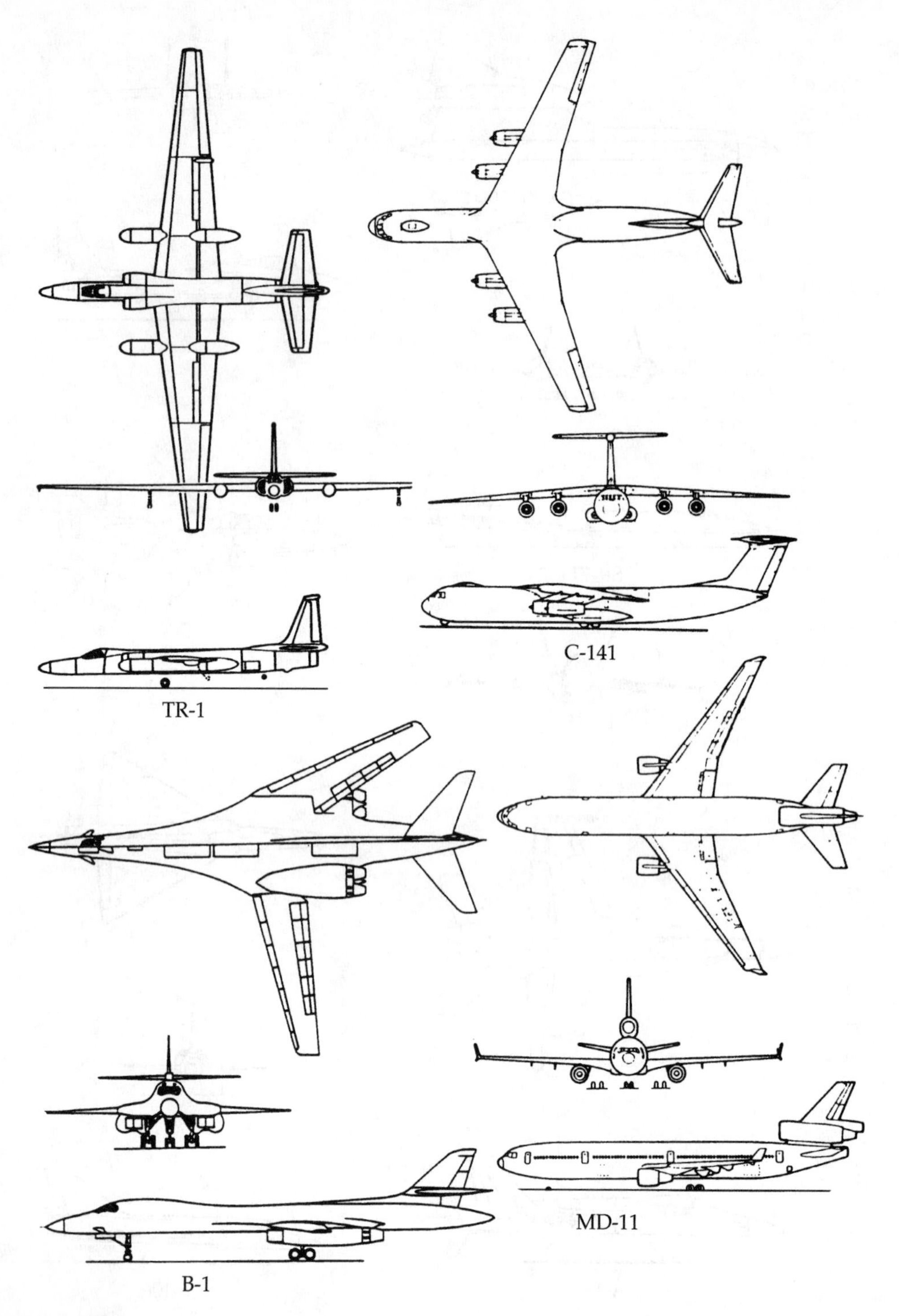

Fig. 3-5. Continued.

Weight. The weight includes the airplane itself, the payload, and the fuel. Since fuel is consumed as the airplane flies, the weight decreases. Weight acts in a direction toward the center of the Earth.

Thrust. The driving force of whatever propulsive system is used, engine-driven propeller, jet engine, rocket engine, and so forth, is the thrust. Thrust acts along the longitudinal axis of the airplane, except for vertical takeoff airplanes. Thrust is required to overcome drag.

Lift. This force is generated by the flow of air around the airplane. The wing produces most of the lifting force. Lift represents the component of the resultant aerodynamic force normal to the line of flight.

Drag. This force is also caused by the flow of air around the airplane, but is the component of the resultant aerodynamic force along the line of flight. A pound of drag requires a pound of thrust.

In the simplest flight situation, an airplane will travel in straight-and-level flight at a uniform velocity. Figure 3-6 shows the disposition of the four forces under these conditions. To maintain this basic flight situation, the lift equals the weight, and the thrust equals the drag. Weight and thrust are physical attributes of an airplane; they generally are known or can be easily determined and controlled. But lift and drag are caused by the dynamic movement of the airplane through the air.

The major concern of aerodynamics is the manner in which the lift and drag forces are created, which is considered in more detail in chapter 4.

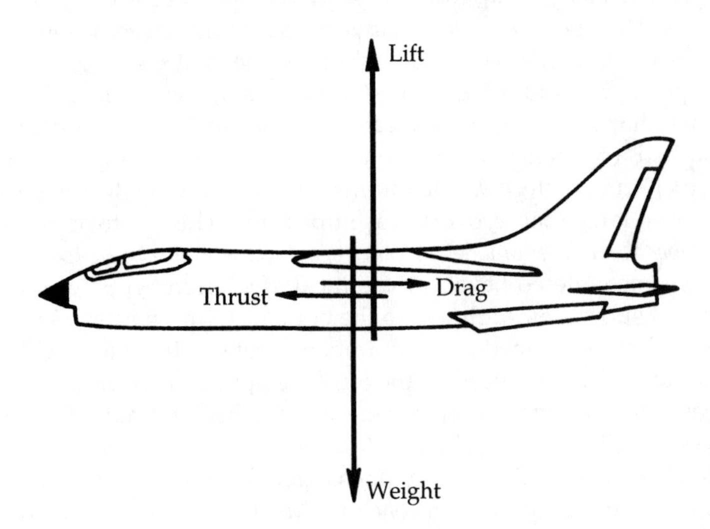

Fig. 3-6. Forces on an airplane in normal flight.

4

Aerodynamic concepts for high-speed flight

THE LOCKHEED F-80 of the middle 1940s, the first operational U.S. jet fighter, pushed the science of aerodynamics to the edge of the known, subsonic regime and allowed engineers a peek into the unknown, supersonic realm. The F-80's speed of over 500 mph was a large jump from the 400-plus mph speeds of the latest piston-engine fighters. This speed was also relatively close to the speed of sound, which is 760 mph at sea level, standard conditions, but is somewhat slower at high altitudes.

As the speed of sound, Mach 1, was approached, usually in a dive, peculiar things began to happen to the high-speed subsonic airplane of the 1940s. It began to *buffet* and had a tendency to "tuck under" or nose down into a more extreme dive. If this tuck-under effect was not immediately noticed by the pilot, the control forces became so large that recovery was impossible. This led to assigning a *critical Mach number* to all subsonic jet airplanes, which was not to be exceeded. Also, the difficulties encountered beyond the critical Mach number as well as some of the mysteries of supersonic flight led to the popular term *sound barrier*.

The sound barrier seemed to be an impenetrable wall beyond which the airplane could not fly. As the speed of the airplane approached Mach 1, the drag—and consequently the power required—approached infinity. Control forces became excessive. Stability deteriorated to a dangerous value. None of the design concepts that had remained relatively unchanged for almost 40 years could solve the problems of exceeding or even reaching Mach 1. German wartime research changed all that.

Supersonic flight was a barrier only for a short time. It was merely a challenge to the engineers in the aircraft industry, the military services, and especially the National Advisory Committee for Aeronautics (NACA)—a forerunner of the National Aeronautics and Space Administration (NASA)—who were encouraged by the captured German research data. Entirely new research techniques evolved. New sonic and supersonic wind tunnels were constructed, and old tunnels were modified to provide higher speeds.

Theoretical studies substantiated by exhaustive wind-tunnel tests indicated that the drag of a properly designed airplane form increased sharply at Mach 1, but then leveled off again in the supersonic regime. Control forces also increased, but they could be determined close enough for design purposes. Stability changed radically from subsonic to supersonic; however, satisfactory stability could be obtained by proper design configurations.

Engineers were now faced with a dilemma. Supersonic airplanes could be designed with all available data. But they had to design a combination subsonic and supersonic airplane. The supersonic airplane still had to take off and fly subsonically, then fly through Mach 1 and above. It then had to reverse the procedure and come in for a definitely subsonic landing.

During the late 1940s and early 1950s, however, most of the early jet airplanes' stability and control problems associated with flight around Mach 1 had been satisfactorily solved without unduly compromising low-speed flight characteristics, for instance in the North American F-86 and Republic F-84F. Their swept wings reduced some of the high-speed drag and improved stability close to Mach 1. Hydraulically powered control systems, replacing the former manual systems, provided satisfactory control forces. These airplanes could reach and slightly exceed Mach 1 in a dive; therefore, the airplanes were called *transonic*.

Except for a few rocket-powered research airplanes in this time period, sufficiently powerful engines were not yet available to exceed Mach 1 in level flight. Supersonic flight had to await the development of the Pratt & Whitney J-57 and General Electric J-79 turbojet engines under test during the early 1950s.

The advent of the J-57 engine just about doubled the thrust previously available from the J-65 powerplant of the F-84F. With afterburner, the 15,000-pound thrust of the J-57 tripled that of the F-86's J-47 engine. At last, the airplane designer had sufficient power to overcome the drag rise at Mach 1 and try for supersonic speed!

Rocket-powered research aircraft such as the Bell X-1 had penetrated the sound barrier in the late 1940s, mainly by a brute-force methodology. The North American F-100 Super Sabre, which was powered by the J-57 engine, became the first operational supersonic airplane. The F-100 entered service in 1954 and flew at approximately Mach 1.25 (825 mph) at 35,000 feet. The F-100 proved that the so-called sound barrier was not a problem for operational aircraft. Coincidentally, the F-100 was the first of the "Century series" of supersonic fighters.

When the sound barrier had been proven to be surmountable by the operational F-100 airplane, rapid ingress into the supersonic regime was made by following Century-series fighter airplanes. McDonnell's F-101, with two J-57 powerplants, pushed its way to Mach 1.5, or 1,000 mph. Convair's F-102, with one J-57 engine, used its strange looking delta wing and "Coke-bottle" fuselage to advantage in attaining supersonic flight, as well as exceptional high-altitude performance of well over 50,000 feet.

By the end of the 1950s, several airplanes had flown at twice the speed of sound, Mach 2.0, faster than 1,300 mph: the Republic F-105 and Convair F-106 (both powered by a Pratt & Whitney J-75 engine); the Lockheed F-104 (powered by a General Electric J-79 engine); and the McDonnell F-4 (powered by two J-79s). It had taken 50 years to attain Mach 1, but fewer than 10 years to double it.

So much for history. What's so different about a high-speed jet-powered airplane? Without considering the jet powerplant, let's take a look at the basic difference between a high-performance prop job and a jet.

A good place to start is the wing. Wings on a jet-powered airplane, at close look, are thinner. The thickest point is farther back from the leading edge. The higher speed jet airplanes have swept wings. To illustrate the difference high speed makes, let's go back to the basics—"How does a wing provide lift?"—and go on from there. But first, we should emphasize the limitations and purpose of this discussion.

This chapter provides a highly simplified discussion of high-speed flight mainly in terms of drag and drag reduction to achieve high speeds. Chapter 5 presents a more detailed discussion of wing lift, aspect ratio, induced drag, stall characteristics, and low-speed flight problems of high-speed airplanes. Then again, in chapter 8, supersonic flow is presented in more detail and the supersonic airplane's characteristics are summarized.

In order to avoid excessive footnotes and supplementary explanations in the text, definitions of the more specialized terms used in this chapter are presented in the glossary.

LIFT

Lift is the force exerted primarily by the wings, although the fuselage and horizontal stabilizer under certain conditions will produce lift. Lift is created by the action of the air moving past the airfoil.

Using the classic principle devised by a scientist named Bernoulli when airplanes were just a gleam in a few dreamers' eyes, we'll show how a typical low-speed airfoil develops lift. Without getting into technical details, Bernoulli proved that as air flows through a constriction in a tube, the pressure is less at the constriction. Also, in order to get across the narrow point, the air must move faster. As shown in Fig. 4-1, an airplane wing is merely an extension of Bernoulli's principle. Due to the curvature (*camber*) of the wing, the air molecules must travel faster along the upper surface than the lower surface. The faster airspeed along the upper surface then produces a lower pressure, providing a major part of the lift just as Bernoulli said in 1738.

All of the above is based on slower subsonic flight and also slower subsonic airflow, whether we use Bernoulli's tube or a wing section.

The greatest pressure differences occur near the maximum camber of the wing. Because this pressure differential results in the net lifting force, the greatest portion of the wing's lift is produced forward of this maximum camber point, toward the leading edge.

As the angle of attack is increased, the lift continues to increase because of changes in the airflow and pressure distribution; however, when the stall angle of attack is reached, the airflow over the wing can no longer remain attached to the upper surface of the airfoil, and separation occurs. This separation and resulting turbulence result in large loss of lift, and a great increase in drag; in other words, the wing is *stalled* (Fig. 4-2).

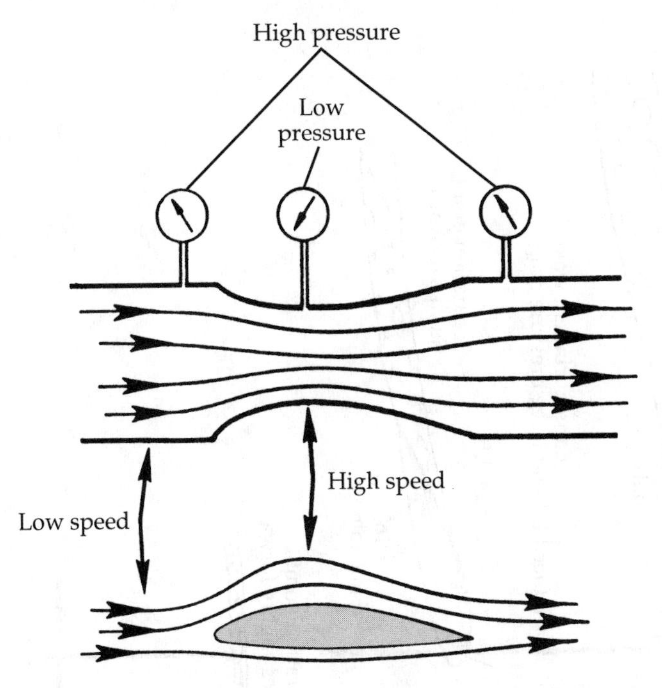

Fig. 4-1. The air flows faster in the constriction as well as over the upper surface of the wing.

DRAG

Drag is the force acting on a body to resist forward motion. Whenever a body is moved through a fluid, such as air, drag is produced. Airplane aerodynamic drag is composed of three parts:

- *Induced drag* caused by lift being created.
- *Parasite drag,* which is also called *form drag, skin drag,* or *interference drag.*
- *Compressibility* or *wave drag* caused by shock wave formation.

Induced drag

Induced drag is a byproduct of lift. It is an undesirable, but unavoidable, consequence of developing or generating lift. The induced drag is related to the wing span and aspect ratio. Wings with a high aspect ratio are desirable; however, high aspect ratios are not always feasible in high-speed, swept-wing airplanes.

The choice of aspect ratio and other features becomes one of compromise between structural and aerodynamic considerations, and the purpose for which the airplane is designed. Induced drag is approximately inversely proportional to the square of the airspeed. It is at a minimum during high-speed flight and at a maximum during low-speed operation.

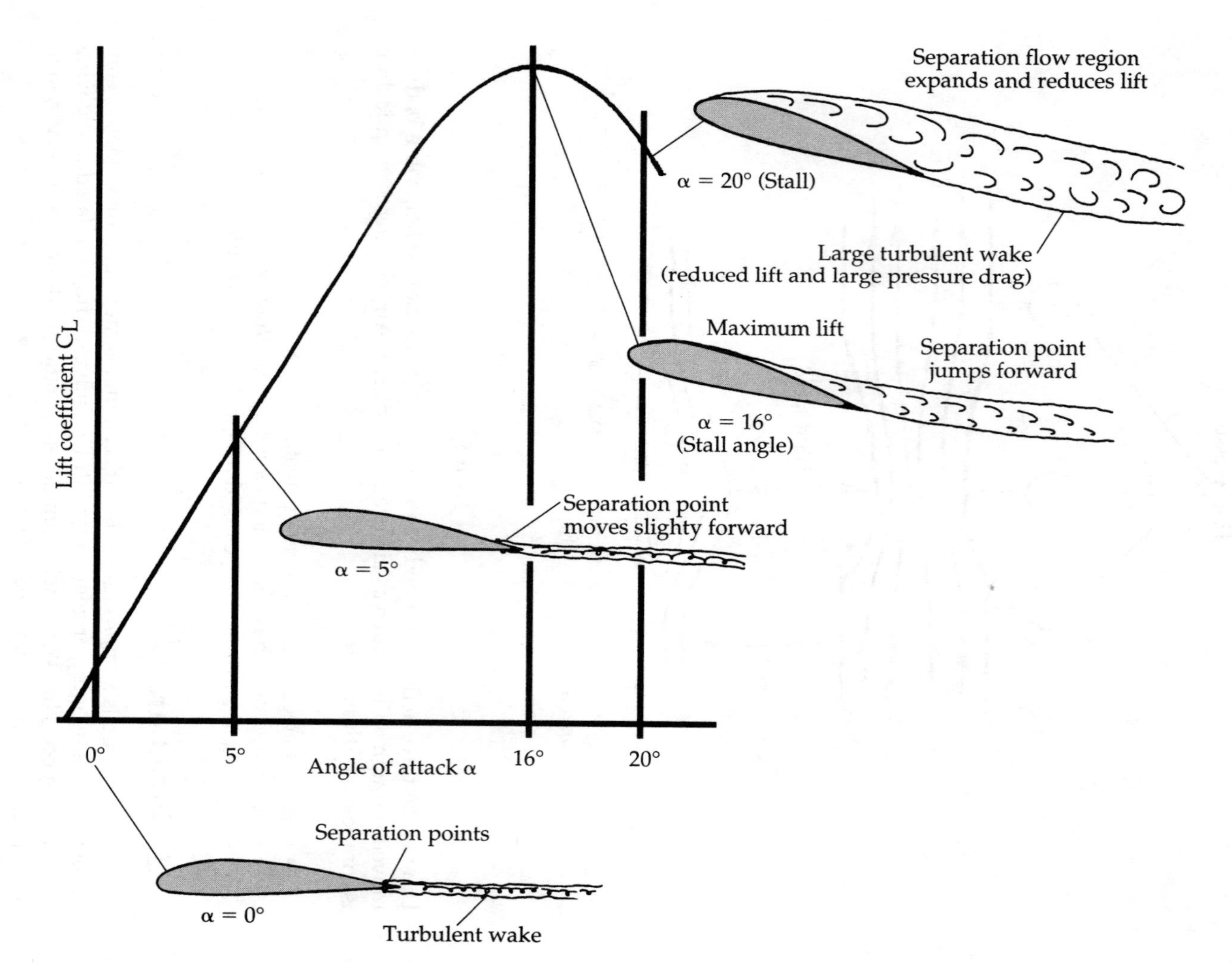

Fig. 4-2. Variation of lift with angle of attack. Lift increases with angle of attack until stall angle is reached.

Winglets or vertical airfoils at the wingtips reduce induced drag by increasing the effective aspect ratio without the greater structural problems of increased wing span. Figure 4-3 shows winglets on the Learjet Model 55C business jet. Induced drag and aspect ratio are discussed in more detail in chapter 5.

Fig. 4-3. A Learjet model 55C business jet with winglets. Winglets increase the effective aspect ratio, which reduces induced drag without increasing wing bending moments.

Parasite drag

A wing surface, even at zero lift, will have profile drag, as a result of skin friction, and form drag, which is due to the pressure distribution on the surface. Form drag is the greatest contributor to profile drag. Form drag can be minimized by streamlining. Skin friction drag also contributes to profile drag. Skin friction drag can be minimized by keeping surface roughness and surface area to a minimum, commensurate with design requirements. All drag not due to lift is considered parasite drag.

The relative proportion of form and skin friction drag depends upon the shape and aerodynamic cleanliness of the aircraft wing and any other components. The other components of the airplane, such as fuselage, tail, and nacelles, contribute to drag because of their own form and skin friction. Any loss of momentum of the airstream from powerplant cooling, air conditioning, or leakage through construction or access gaps is, in effect, an additional drag. When the various components of the airplane are put together, the total drag will be greater than the sum of the individual components because of "interference" of one surface on another.

The greatest amount of interference drag usually occurs at the wing-body intersection, where the growth of the boundary layer on the fuselage reduces the boundary layer velocities on the wing root surface. This reduction in energy allows the wing-root boundary layer to be more easily separated in the presence of an adverse pressure gradient. Because the upper wing surface has the more critical pressure gradients, a low-wing position on a circular fuselage would create greater interference drag than a high-wing position. Adequate filleting and control of local pressure gradients are necessary to minimize such additional interference drag. Figure 4-4 shows an example of wing fillets.

Induced drag is inversely proportional to the square of airspeed, but parasite drag is directly proportional to the square of the airspeed.

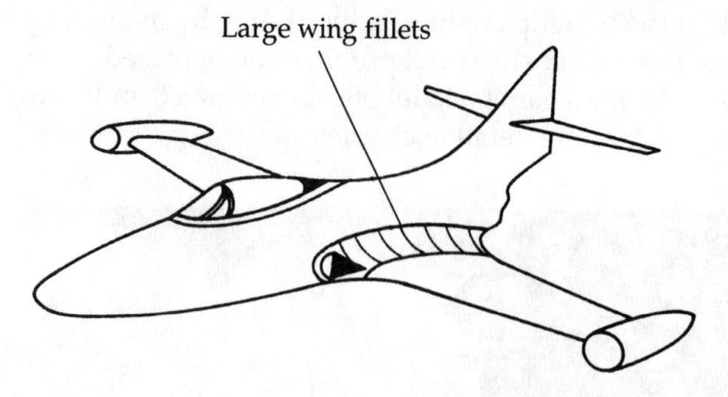

Fig. 4-4. Wing fillets reduce interference drag between the wing and fuselage.

EFFECTS OF HIGH-SPEED FLIGHT

For low-speed flight, as previously stated, the study of aerodynamics is greatly simplified by the fact that air can experience relatively small changes in pressure with only negligible changes in density. The airflow is termed incompressible in this slow-speed regime; however, at high speeds, the pressure changes that take place are quite large, and significant changes in air density occur. The study of airflow at high speeds must account for these changes in air density and must consider that the air is compressible, and that there will be "compressibility effects."

Now, let's take Bernoulli's tube in Fig. 4-5 and gradually increase the airflow rate, which increases the airspeed through the tube, and the constriction.

We see that the air goes faster through the constriction as well as along the upper surface of the airfoil until it "dams up." In other words, even though the entire wing is moving at subsonic speed, part of the local airflow is actually at sonic speed.

Now, this "shock wave" has all kinds of effects on the wing, as well as the airplane to which it is attached. At this point, we need to define some common terms applicable to jet airplanes.

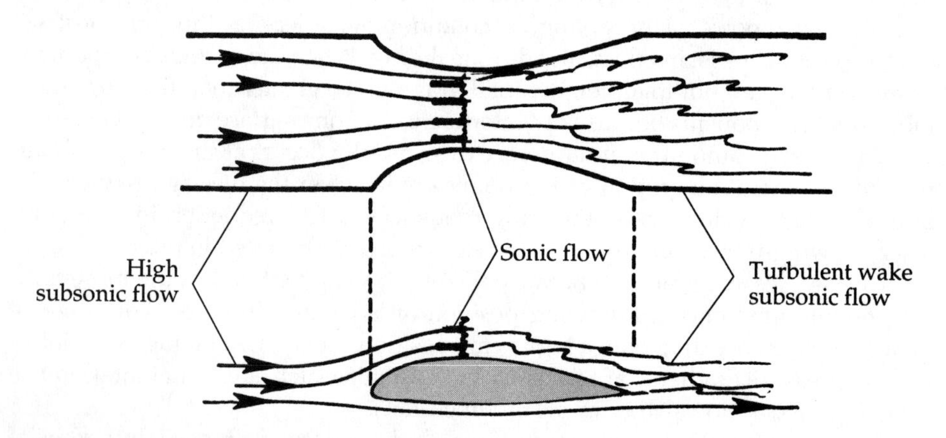

Fig. 4-5. As airflow is increased, a shock wave forms in the constriction as well as on the upper surface of the wing due to sonic flow.

SPEED OF SOUND, AIRCRAFT VELOCITY, AND MACH NUMBER

A parameter of great importance in high-speed flight is the speed of sound, the rate at which small pressure disturbances will spread through the air. As an aircraft moves through the air, velocity and pressure changes create pressure disturbances in the airflow surrounding the aircraft. If the aircraft is traveling at low speed, the pressure disturbances spread through the air ahead of the aircraft, much like the bow wave of a boat.

The airflow immediately ahead of the airplane is influenced by the pressure field on the aircraft. If the object is traveling at some speed approaching the speed of sound, the airflow ahead of the aircraft will not be influenced by the pressure field on the aircraft because pressure disturbances cannot be spread ahead of the object fast enough. A *compression wave* will form at the leading edge, and all changes in velocity and pressure will take place quite sharply and suddenly. The airflow ahead of the aircraft is not influenced until the air particles are suddenly forced out of the way by the concentrated pressure wave set up by the aircraft.

The speed of sound in air depends only on the static temperature of the air, decreasing with lower temperature and increasing with higher temperature (Fig. 4-6). The speed of sound is 661.48 knots at 59.0°F (15°C), slowing to 602.0 knots at –30.0°F (–34.4°C), and slowing even more to 573.6 knots at –69.7°F (–56.5°C).

Mach number is defined as *the ratio of true airspeed (TAS) to the speed of sound*. If the aircraft TAS is used to compute the Mach number, it is called the *airplane Mach number*. If a velocity at some point on the airplane is used, the resultant value is referred to as a *local Mach number*.

Thus, the airplane will experience compressibility effects at flight speeds below the speed of sound because subsonic and supersonic flow might exist over different portions of the airplane at various times during high-speed flight. The local areas of supersonic flow that occur on an airplane flying at speeds above the critical Mach number are accompanied by the formation of shock waves.

The airflow accelerates easily from subsonic to supersonic over a smooth airflow surface; however, when the flow is decelerated from supersonic back to subsonic without any direction change, a shock wave will form. The shock wave forms a boundary between the two regions of flow. Sharp increases in both static pressure and density, and a loss in energy to the airstream, occur across the shock wave. This energy loss might be accompanied by separation of the flow. The strength of a given shock wave and the magnitude of the changes across it increase with increasing Mach number.

COMPRESSIBILITY DRAG, WAVE DRAG, AND TRANSONIC FLIGHT

The major portion of *compressibility drag* is a result of the formation of shock waves on the wings, and the drag becomes more significant as flight speed increases. Above critical Mach number, the airflow behind the shock wave might separate (Fig. 4-7), which results in additional increases in drag. The associated pressure pattern changes might result in loss of lift or aerodynamic buffet.

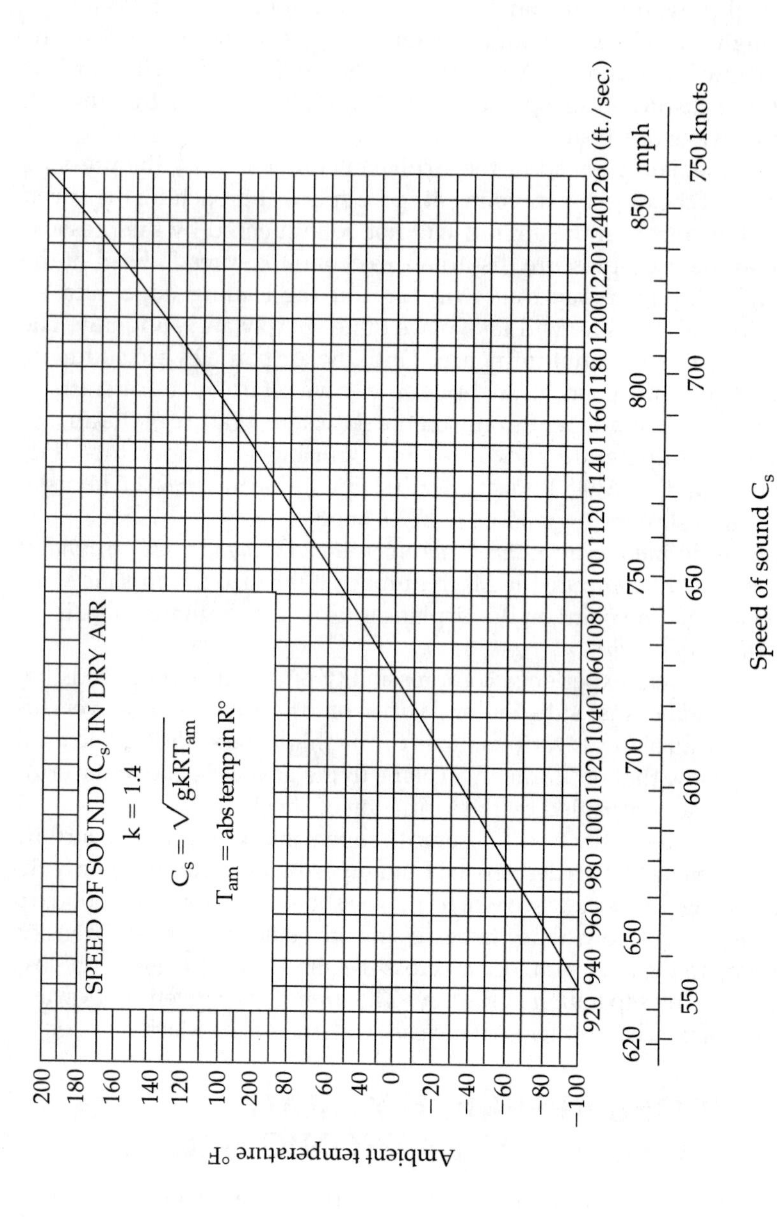

Fig. 4-6. The speed of sound varies with air temperature.

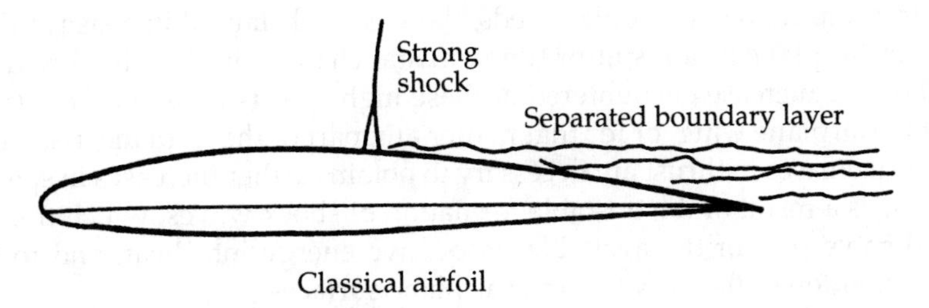

Fig. 4-7. Above critical Mach number, the airflow behind the shock wave might separate, resulting in increased drag.

Supersonic flow is more well behaved than transonic flow, and there are adequate theories to predict the aerodynamic forces and moments present. As previously discussed, in transonic flow the flow is unsteady and the shock waves on the body surface might jump back and forth along the surface, disrupting and separating the flow over the wing surface and sending pulsing unsteady flow back to the tail surfaces of the airplane. The result is that the pilot feels a buffeting and vibration of both wing and tail controls.

With proper design, however, airplane configurations have gradually evolved to the point where flying through the transonic region has posed little or no difficulty in terms of airframe buffeting or loss of lift. Figure 4-8 shows the transonic flow range.

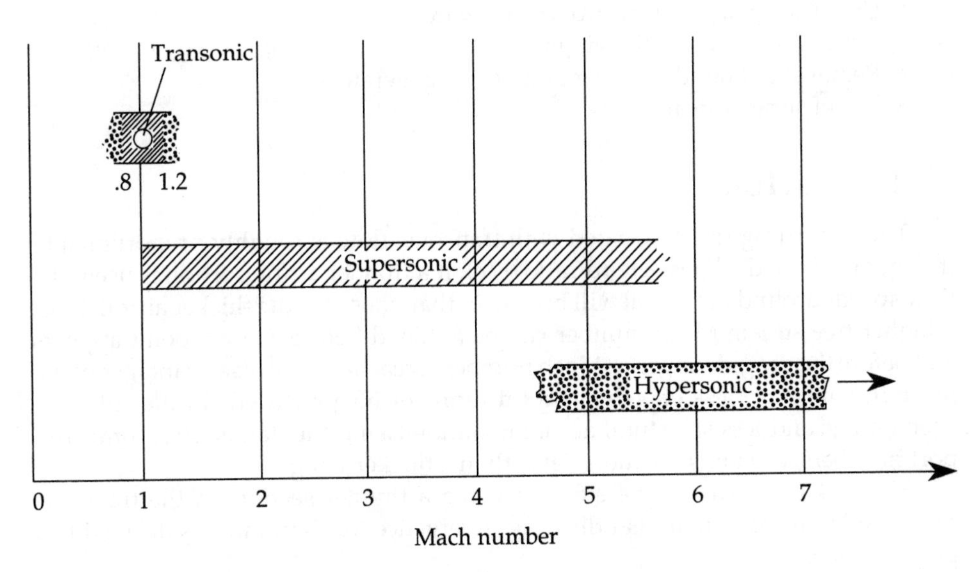

Fig. 4-8. Flight regime terminology showing the transonic speed range.

At transonic and supersonic speeds, there is a substantial increase in the total drag of the airplane as a result of fundamental changes in the pressure distribution. The drag increase encountered at these high speeds is called *wave drag*. The drag of the airplane wing, or for that matter any part of the airplane, rises sharply, and large increases in thrust are necessary to obtain further increases in speed. The wave drag is a result of the unstable formation of shock waves, which transforms a considerable part of the available propulsive energy into heat, and to the induced separation of the flow from the airplane surfaces.

Figure 4-9 shows the variations of airplane drag with Mach number. The infamous sound barrier shows up rather clearly in Fig. 4-9 because to fly supersonically the airplane must provide enough thrust to exceed the maximum *transonic drag* encountered. In the early days of transonic flight, the sound barrier represented a real barrier to faster speeds. The dotted-line drag curve in Fig. 4-8 shows that an excessive amount of thrust is needed to propel a World War II fighter even close to Mach 1.

The question of whether to delay the *drag-divergence Mach number* to a value closer to 1.0 is a fascinating subject of novel aerodynamic designs. (The drag-divergence Mach number is the point where airfoil drag rises sharply (Fig. 4-10).) What this really suggests is the ability to fly at near-sonic velocities with the same available engine thrust before encountering large wave drag. There are a number of ways of delaying the transonic wave drag rise (or equivalently, increasing the drag-divergence Mach number closer to 1.0):

- Use of thin airfoils.
- Use of wing sweep forward or backward.
- Use of low-aspect-ratio wings.
- Removal of boundary layer and vortex generators.
- Use of supercritical wings.

Thin airfoils

The wave drag rise associated with transonic flow is roughly proportional to the square of the thickness-chord ratio (t/c). If a thinner airfoil section is used, the flow speeds around the airfoil will be slower than those for the thicker airfoil; thus, a higher free-stream Mach number can be attained before a sonic point appears and before the drag-divergence Mach number is reached. The disadvantages of using thin wings are less effectiveness (in terms of lift produced) in the subsonic speed range and less structural accommodation (wing fuel tanks, structural support members, armament stations, etc.) than a thicker wing.

Figure 4-10 illustrates the effect of using a thinner section on the transonic drag. Notice, in particular, that the drag-divergence Mach number is delayed to a greater value.

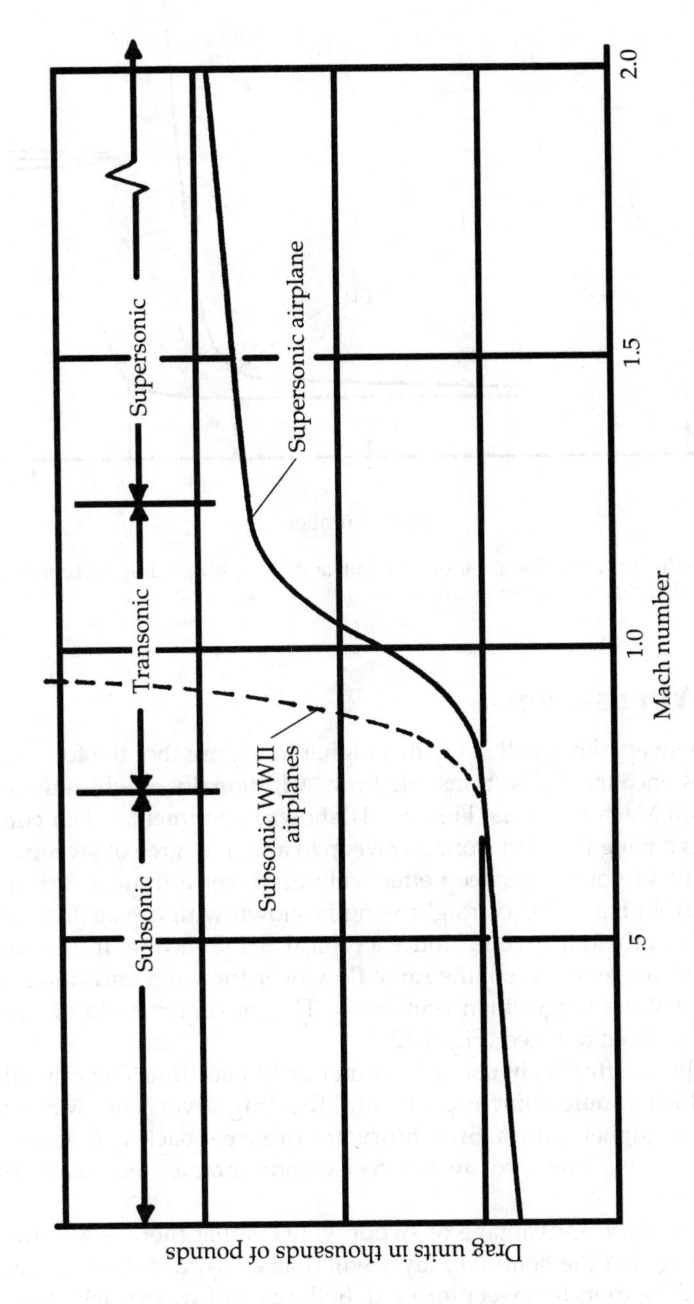

Fig. 4-9. Typical variation of airplane drag with Mach number showing the transonic drag rise, or "sound barrier." A World War II fighter cannot reach supersonic speeds.

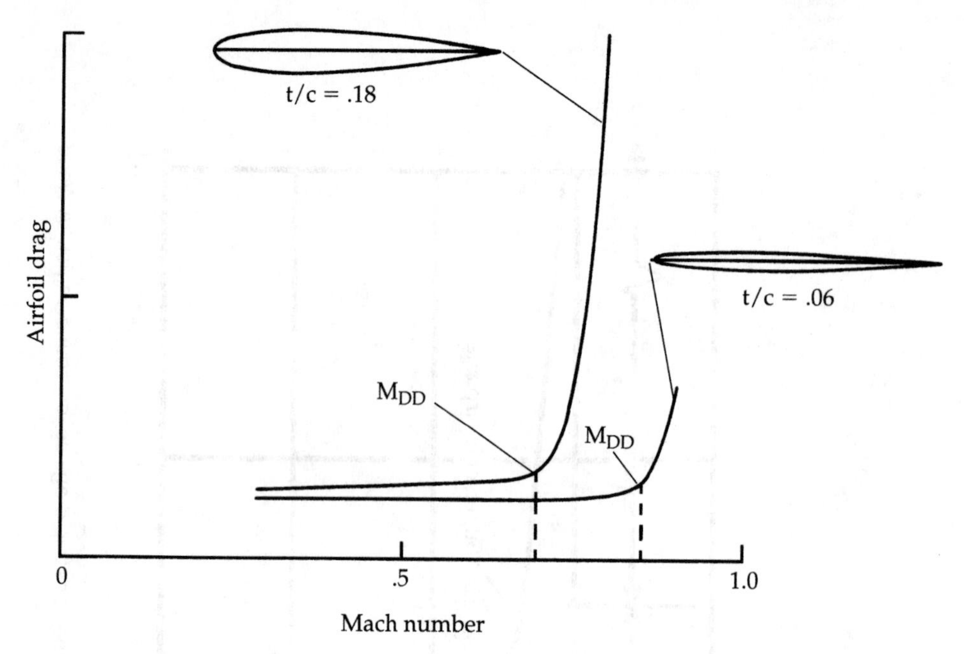

Fig. 4-10. Effect of airfoil thickness on transonic drag. M_{DD} = drag divergence Mach number; t/c is airfoil thickness divided by chord.

Wing sweep

A swept wing will delay to a higher Mach number the formation of the shock waves encountered in transonic flow. Additionally, it will reduce the wave drag over all Mach numbers. Figure 4-11 shows experimental data confirming this result as a wing is swept from no sweep to a high degree of sweep.

The use of wing sweep effectively produces a thinner airfoil section (t/c reduced). In Fig. 4-12, a straight wing is shown with the airflow approaching perpendicularly to the wing. Notice a typical airfoil section. If the wing is now swept to some angle of sweep, the same flow over the wing encounters new airfoil sections that are longer than previously. The maximum ratio of thickness to chord (t/c) has been reduced (Fig. 4-12).

This is effectively using a thinner airfoil section. The critical Mach number (at which a sonic point appears) and the drag-divergence Mach number are delayed to higher values. Sweepforward or sweepback will accomplish these desired results. Forward sweep has disadvantages, however, as discussed in chapter 5.

A major disadvantage of swept wings is that there is a spanwise flow along the wing, and the boundary layer will thicken toward the tips for sweepback and toward the roots for sweepforward. In the case of sweepback, there is an early separation and stall of the wingtip sections and the ailerons lose their roll-control effectiveness.

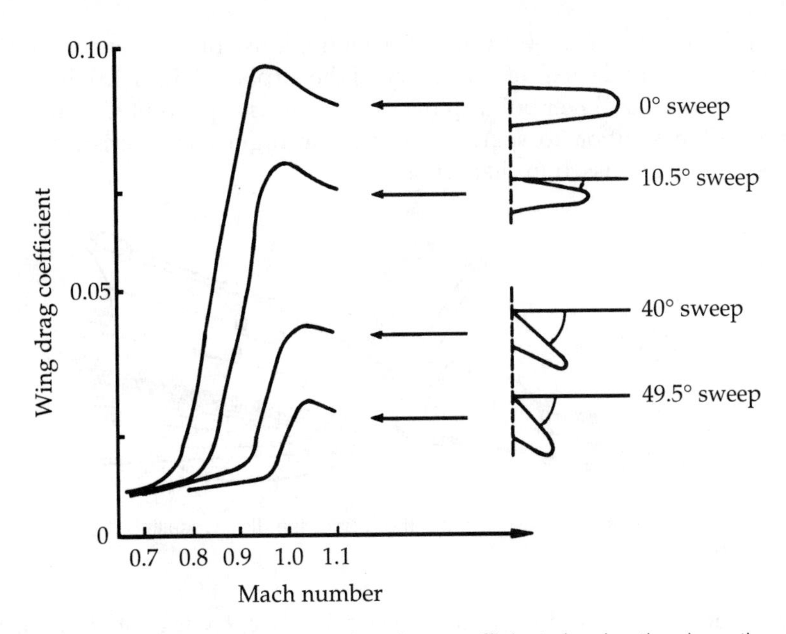

Fig. 4-11. Effect of sweep on wing drag coefficient showing the dramatic reduction in drag for swept wings in transonic flow.

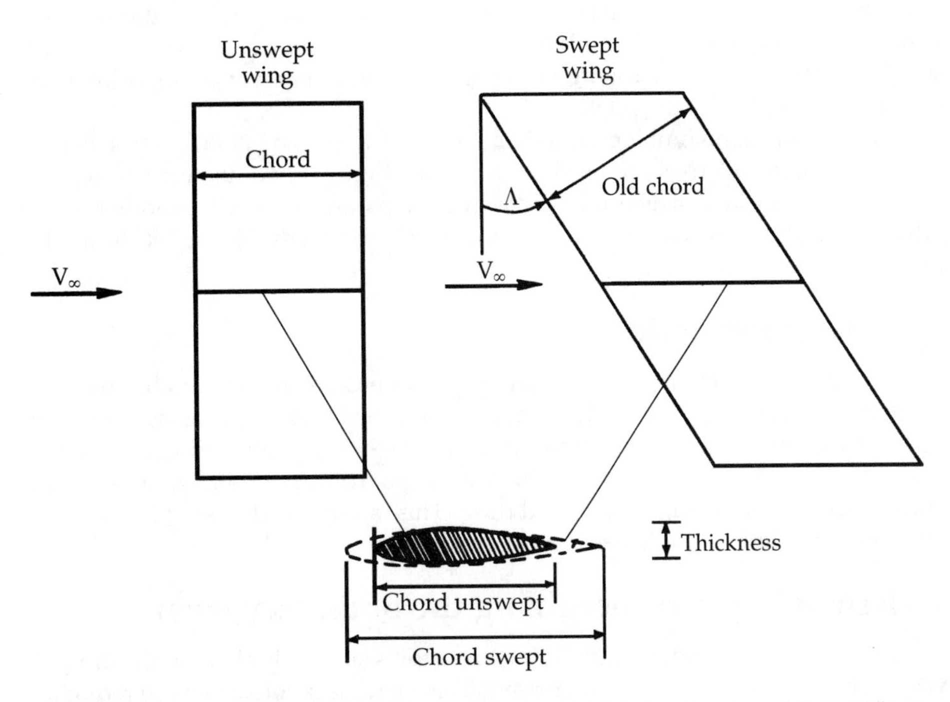

Fig. 4-12. Wingsweep effectively produces a thinner airfoil in the direction of flight while retaining a thicker airfoil for structural purposes.

The spanwise flow can be reduced through the use of stall fences, which are thin plates parallel to the axis of symmetry of the airplane (Fig. 4-13). In this manner, a strong boundary-layer buildup over the ailerons is prevented. Wing twist is another possible solution to spanwise flow. The flight characteristics of swept wings are further discussed in chapter 5.

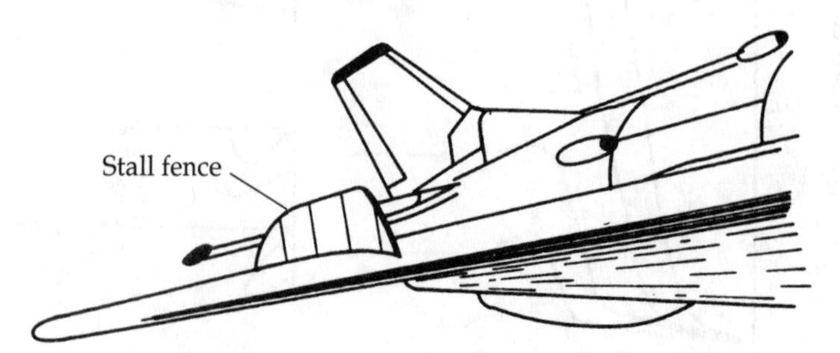

Fig. 4-13. Stall plates reduce the spanwise flow resulting from wingsweep.

A problem encountered on high-aspect ratio swept wings such as used on commercial transports is *aileron reversal*. If conventional ailerons are installed near the tip of thin flexible wings, extreme forces can occur as they are deflected at high speeds. Their deflection can cause twisting of the entire wing, affecting the rolling performance of the aircraft. At some high speed, the twisting deformation is great enough to nullify the effect of aileron deflection. At speeds above this point, this wing deflection can create rolling moments called aileron reversal opposite to the direction controlled by the pilot.

Most jet airliners combine a number of devices to obtain lateral control. For example, outboard ailerons are used for low speed only; the outboard ailerons are automatically locked in neutral when wing flaps are retracted. Another set of ailerons is installed inboard for lateral control at high speeds. Figure 3-4 shows this aileron combination.

Low-aspect-ratio

The wing's aspect ratio (ratio of wing span to mean chord) is another parameter that influences the critical Mach number and the transonic drag rise. Substantial increases in the critical Mach number occur when an aspect ratio of less than 4.0 is used; however, low-aspect-ratio wings are at a disadvantage at subsonic speeds because of the higher induced drag. This is further discussed in chapter 5 on low-speed flight of high-speed airplanes.

Removing or reenergizing the boundary layer

If some of the boundary layer along an airfoil's surface is bled off, the drag-divergence Mach number can be increased. This increase results from the reduction or elimination of shock interactions between the subsonic boundary layer and the supersonic flow outside it.

Vortex generators are small plates mounted along the surface of a wing and protruding perpendicularly to the surface as shown in Fig. 4-14. They are in reality small wings. By creating a strong tip vortex, the generators feed high-energy air from outside the boundary layer into the slow-moving air inside the boundary layer. This condition reduces the adverse pressure gradients and prevents the boundary layer from stalling; thus, small increases in the drag-divergence Mach number can be achieved. This method is economically beneficial for airplanes designed for cruise at the highest possible drag-divergence Mach number.

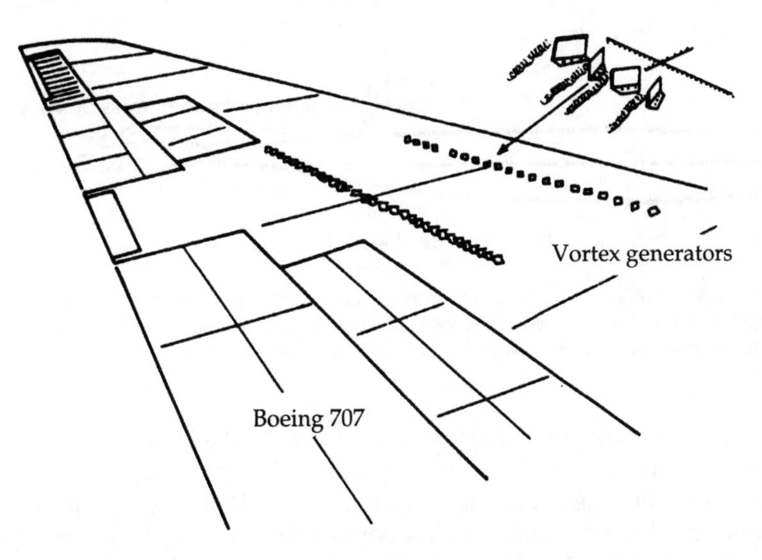

Fig. 4-14. Vortex generators feed high-energy air from outside the boundary layer into the slow-moving air inside the boundary layer.

Supercritical wing

One of the more recent developments in transonic technology is the NASA supercritical wing developed by Dr. Richard T. Whitcomb of the NASA Langley Research Center. His efforts are destined to be an important influence of future wing design. A substantial rise in the drag-divergence Mach number is realized with the use of this wing.

Figure 4-15 shows a classic airfoil operating near the Mach 1 region (supercritical beyond the critical Mach number) with its associated shocks and separated boundary layer. Figure 4-15 also shows the supercritical airfoil operating at the same Mach number. The airfoil has a flattened upper surface, which delays the formation and strength of the shocks to a point close to the trailing edge. Additionally, the shock-induced separation is greatly decreased. The critical Mach number is delayed even up to 0.99, which represents a major increase in commercial airplane performance.

The curvature of an airfoil gives the wing its lift. Because of the flattened upper surface of the supercritical airfoil, lift is reduced. To counteract this problem, the new supercritical wing has increased camber at the trailing edge.

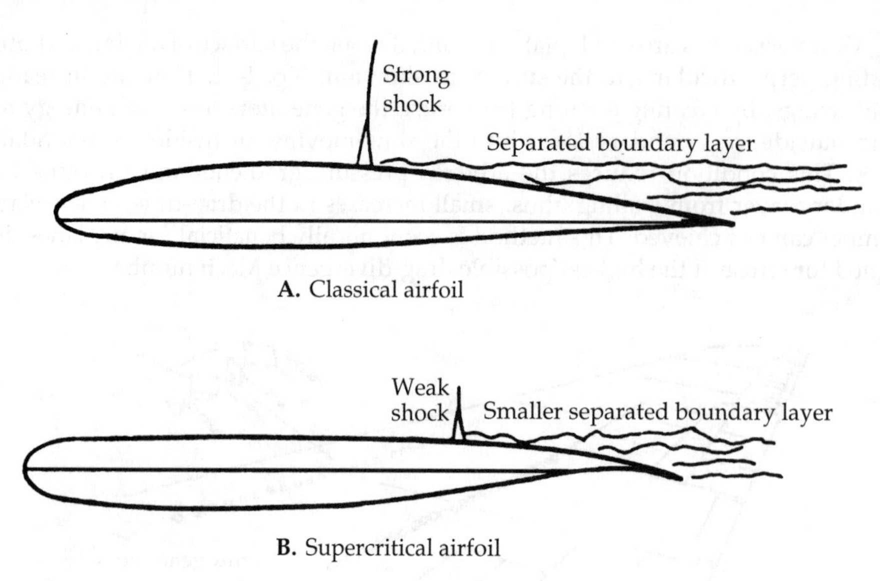

Fig. 4-15. The NASA-developed supercritical airfoil compared to a classical airfoil at critical Mach number. The weaker shock and smaller separated boundary layer for the supercritical airfoil provides lower drag at a higher Mach number.

There are two main advantages of the supercritical airfoil. First, by using the same thickness-chord ratio, the supercritical airfoil permits high subsonic cruise near Mach 1 before the transonic drag rise. Second, at lower drag-divergence Mach numbers, the supercritical airfoil permits the use of a thicker wing section without a drag penalty. This airfoil reduces structural weight and permits higher lift at lower speeds.

SUPERSONIC FLIGHT

The previous discussion centers mainly on transonic drag rise and the way it can be delayed through proper design. Many of the techniques also are directly applicable in designing the airplane to fly with minimum wave drag in the supersonic regime.

Returning to the discussion of shock formation, it was shown that a bow shock wave will exist for free-stream Mach numbers above 1.0 (Fig. 4-16). In three dimensions, the bow shock is in reality a cone shape (a *Mach cone*) as it extends back from the nose of the airplane. The Mach cone becomes increasingly swept back with increasing Mach numbers. As long as the wing is swept back behind the Mach cone, there is subsonic flow over most of the wing and relatively low drag.

A delta wing has the advantage of a large sweep angle, but also greater wing area than a simple swept wing to compensate for the loss of lift usually experienced in sweepback. At still higher supersonic Mach numbers, however, the Mach cone might approach the leading edge of even a highly swept delta wing. This con-

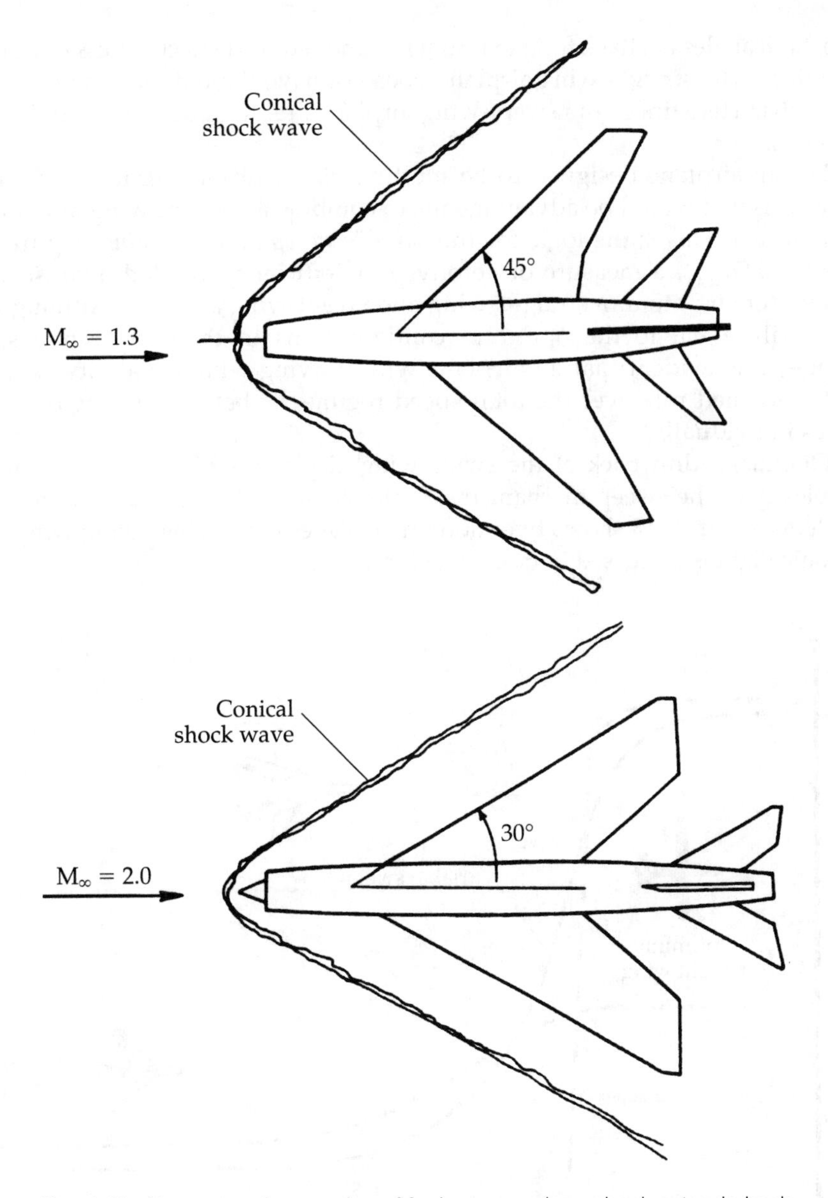

Fig. 4-16. At supersonic speeds, a Mach cone or bow shock extends back from the nose of the airplane. As long as the wing is swept back behind the Mach cone, there is subsonic flow over most of the wing and relatively low drag.

dition causes the total drag to increase rapidly and, in fact, a straight wing (no sweep) becomes preferable.

Sweepback has been used primarily to minimize transonic and supersonic wave drag. At subsonic Mach numbers, however, the disadvantages are dominant. They include high induced drag as a result of small wing span or low aspect ratio, high angles of attack for maximum lift, and reduced effectiveness of trailing-edge flaps. The straight-wing airplane does not have these disadvantages. Low-speed characteristics of the swept-wing airplane are discussed in more detail in chapter 5.

For an airplane designed to be multimission (subsonic cruise and supersonic cruise), it would be advantageous to combine a straight-wing and swept-wing design. This is the logic for the *variable sweep* or *swing-wing*. Figure 4-17 shows $(L/D)_{MAX}$, a measure of aerodynamic efficiency, plotted against Mach number for an optimum straight-wing and swept-wing airplane. Although not necessarily equal to the optimum configurations in their respective speed regimes, it is evident that an airplane with a swing-wing capability can, in a multimissioned role over the total speed regime, be better than the other airplanes individually.

One major drawback of the swing-wing airplane is the added weight and complexity of the sweep mechanisms. Technological advances are solving these problems. Figure 4-18 shows two modern airplanes employing swing wings. Supersonic flight is discussed in more detail in chapter 8.

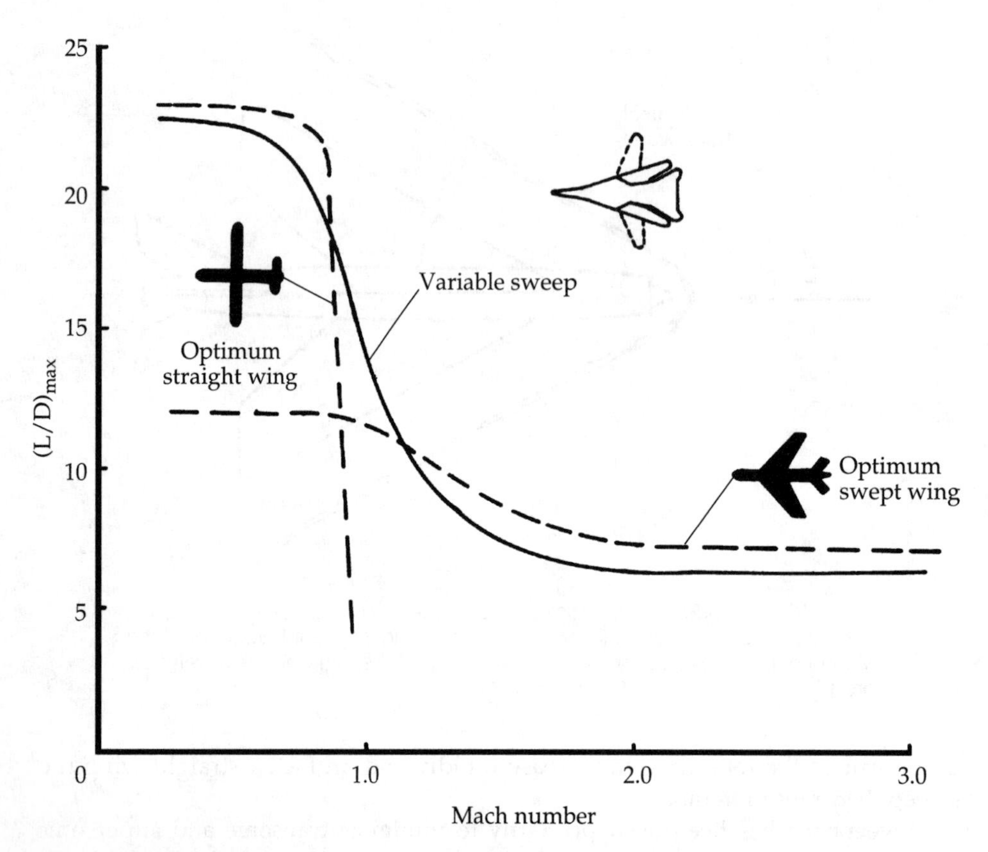

Fig. 4-17. Wing efficiency (L/D_{MAX} or lift divided by drag) for straight and swept wings. A variable-sweep wing combines characteristics of both with accompanying structural problems.

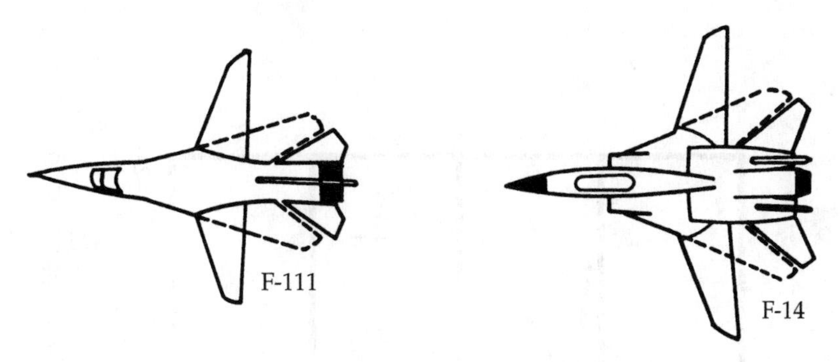

Fig. 4-18. The General Dynamics F-111 and the Grumman F-14 are examples of airplanes incorporating variable sweep.

PERFORMANCE

Recall that the four basic forces acting on an airplane are lift, drag, weight, and thrust; early portions of this chapter concentrate on drag and, to a lesser extent, lift. Performance is basically the effects that the application of these forces have on the flight path of the airplane. Stability and control, considered in chapter 7, is the effect that these forces have on the attitude of the airplane for a short term. For performance purposes, the airplane is assumed to possess stability and a usable flight control system; however, as discussed in chapter 7, the high-speed airplane possesses unique and sometimes complex stability and control systems.

The various items of airplane performance result from the combination of airplane and powerplant characteristics. The aerodynamic characteristics of the airplane generally define the thrust requirements at various conditions of flight; the powerplant characteristics generally define the thrust available at various conditions of flight. The matching of the aerodynamic configuration with the powerplant will be accomplished to provide maximum performance at the specific design condition: range, endurance, climb, etc.

STRAIGHT AND LEVEL FLIGHT

When the airplane is in steady, level flight, the condition of equilibrium must prevail. The unaccelerated condition of flight is achieved with the airplane trimmed for lift equal to weight and the powerplant set for a thrust to equal the airplane drag. In certain conditions of airplane performance, it is convenient to consider the airplane requirements by the thrust required (or drag). In the case of a propeller-driven airplane, it is more applicable to consider the power required. Since our discussion is based on jet airplanes, the airplane in steady level flight will require:

- Lift equal to weight (and)
- Thrust available that is equal to thrust required (drag).

The variation of thrust required with velocity is illustrated in Fig. 4-19. Each specific curve of thrust required is valid for a particular aerodynamic configuration

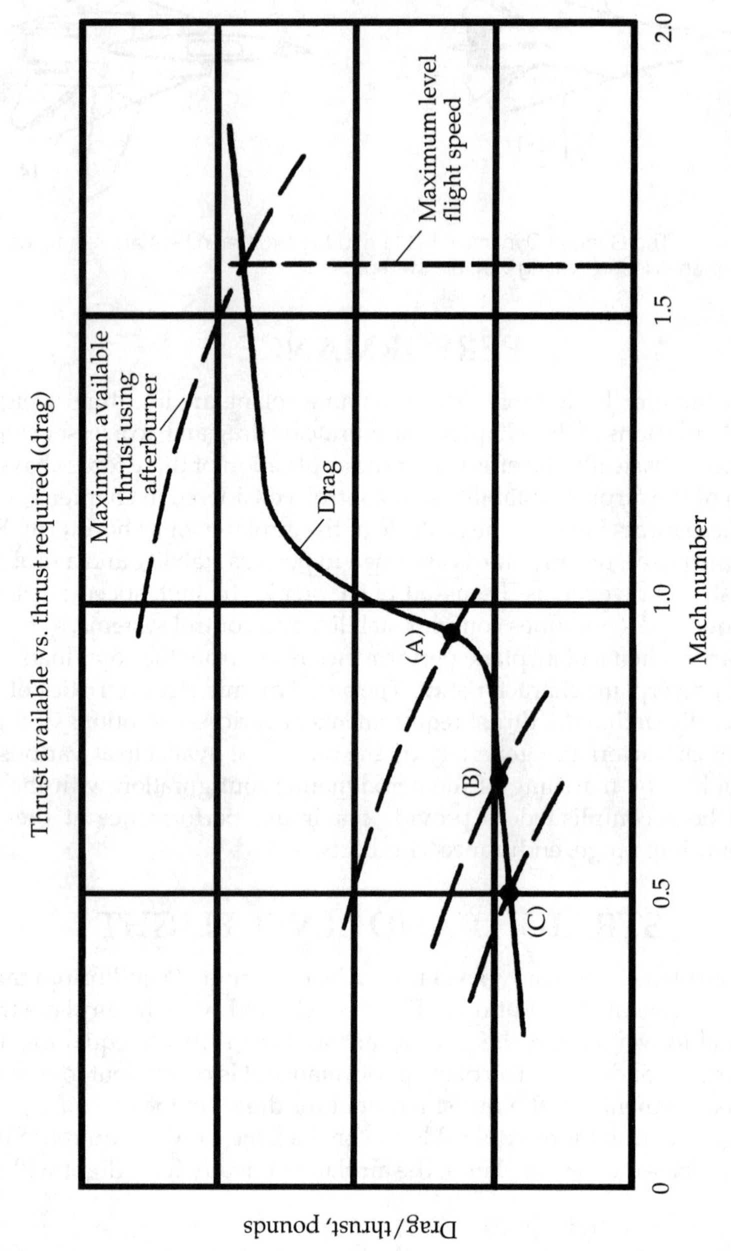

Fig. 4-19. The intersection of the thrust-available curve and the thrust-required (drag) curve for a particular engine setting determines the speed for that condition at a particular altitude.

at a given weight and altitude. These curves define the thrust required to achieve equilibrium, lift-equal-weight, and constant-altitude flight at various airspeeds.

As shown by the curves of Fig. 4-19, if it is desired to operate the airplane at the airspeed corresponding to point A, the thrust-required curves define a particular value of thrust that must be made available from the powerplant to achieve equilibrium. A different airspeed, such as that corresponding to point B, changes the value of thrust or power required to achieve equilibrium. Of course, the change of airspeed to point B also would require a change in angle of attack to maintain a constant lift equal to the airplane weight.

Similarly, to establish airspeed and achieve equilibrium at point C will require a particular angle of attack and powerplant thrust. In this case, flight at point C would be in the vicinity of the minimum flying speed and a major portion of the thrust required would be due to induced drag.

The maximum level-flight speed for the airplane will be obtained when the thrust required equals the maximum thrust available from the powerplant. The minimum level-flight airspeed is not usually defined by thrust requirement because conditions of stall, or stability and control problems, generally predominate.

A plot of airplane drag verses thrust required at various altitudes reveals a typical level-flight speed profile as shown in Fig. 4-20. This figure shows the per-

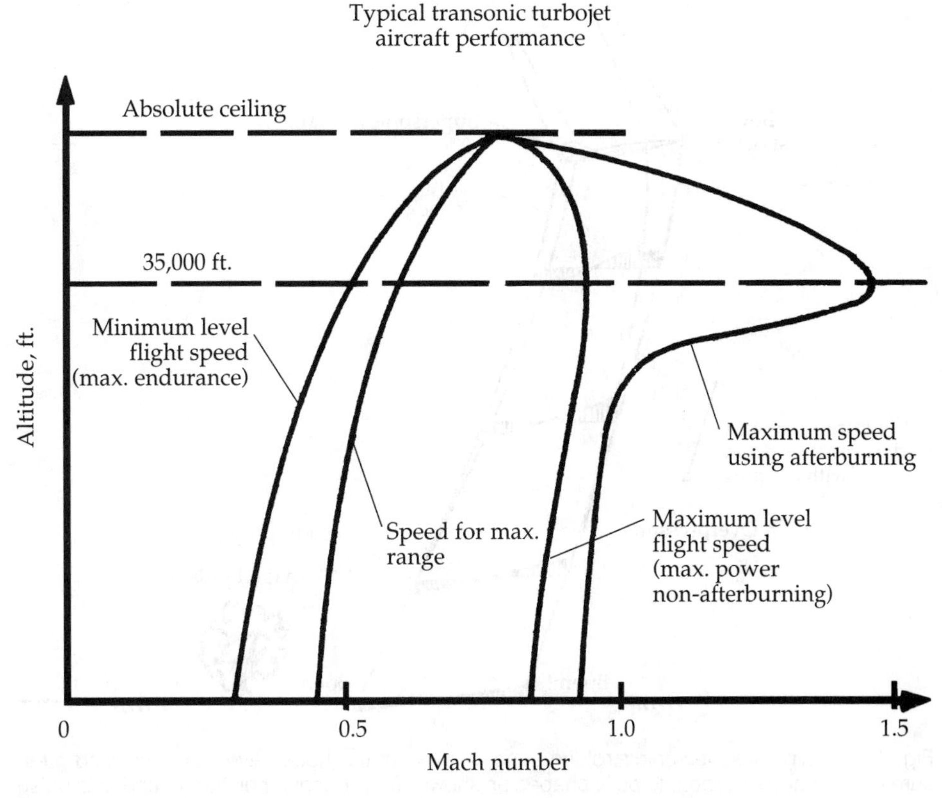

Fig. 4-20. Typical transonic turbojet aircraft performance envelope.

formance of a typical supersonic jet fighter powered by an afterburning turbojet engine. Chapter 6, regarding the jet engine, reveals that afterburning provides a thrust increase of around 50 percent. Afterburning thrust is required to overcome the transonic drag rise as shown in Fig. 4-19 and produce the maximum speed bulge at 35,000 feet altitude shown if Fig. 4-20.

All performance curves meet at the airplane's ceiling (Fig. 4-20); the airplane is flying at maximum angle of attack and stall speed, maximum endurance, maximum range, and maximum top speed are all the same. Even a slight turn at this altitude would provide enough load factor increase to cause the airplane to stall because lift exactly equals weight for all steady-state flight conditions.

SONIC BOOM

One of the more objectionable problems facing supersonic flight is commonly referred to as the sonic boom. A typical airplane generates two main shock waves: one at the nose (*bow shock*) and one off the tail (*tail shock*). Shock waves coming off the canopy, wing leading edges, engine nacelles, and the like, tend to merge with the main shocks some distance from the airplane (Fig. 4-21). The resulting pressure pulse changes appear to be N-shaped as shown.

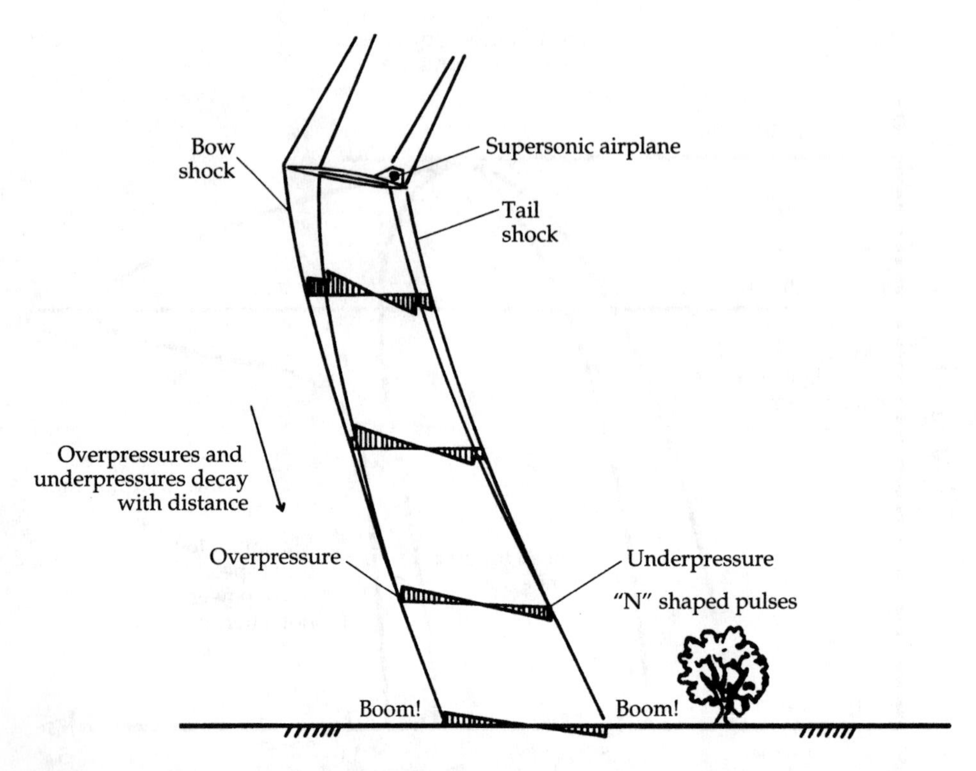

Fig. 4-21. A typical supersonic airplane generates two main shock waves. The resulting pressure pulse changes appear to be N-shaped as shown. To an observer on the ground, this pulse is felt as an abrupt compression and final recompression and is felt as a double jolt or boom.

To an observer on the ground, this pulse is felt as an abrupt compression above atmospheric pressure, followed by a rapid decompression below atmospheric pressure, and a final recompression to atmospheric pressure. The total change takes place in 1/10 second or less and is felt and heard as a double jolt or boom.

The sonic boom—the overpressures that cause the boom—is controlled by factors such as airplane angle of attack, altitude, cross-sectional area, Mach number, atmospheric turbulence, atmospheric conditions, and terrain. The overpressures: increase with increasing airplane angle of attack and cross-sectional area; decrease with increasing altitude; first increase and then decrease with increasing Mach number.

HYPERSONIC FLIGHT

Hypersonic flight is arbitrarily defined as flight at speeds beyond Mach 5.0 (Fig. 4-8), although no drastic flow changes are evident at this limit. To date, speeds of this magnitude have been achieved only by rockets, spacecraft, and the NASA X-15 research airplane. Hypersonic flight is discussed in more detail in chapter 10.

Two formidable problems are encountered at these speeds. First, the shock waves generated by a body trail back at such a high angle that they can seriously interact with the boundary layers about the body. For the most part, these boundary layers are highly turbulent in nature. Second, the air undergoes a drastic temperature increase across the strong shocks.

Aerodynamic heating of the body is a major problem (Fig. 1-4). For sustained hypersonic flight, most normal metals used in today's airplanes would quickly melt; therefore, new materials or methods that can withstand the high-temperature effects are required. The temperature of the leading edge of the airplane wing can be reduced by using a high degree of sweepback. Additionally, to obtain a good lift/drag ratio, a flat-plate wing design is used.

Control surfaces for hypersonic flight must be placed strategically so that they encounter sufficient dynamic pressure around them to operate. Otherwise—if shielded from the approaching flow by the fuselage, for example—they will be ineffective.

Although commercial hypersonic flight is a long way from being realized, NASA is conducting studies to obtain the basic knowledge necessary for design.

Propulsion is another major problem at hypersonic speeds. Economically, the most promising prospect is the *ramjet* engine. This engine works on the principle that at high Mach numbers the shock waves compress the air for combustion in the engine; therefore, this engine does away with many moving parts and represents an efficient propulsion method. NASA research is also continuing in this field.

5

The high-speed airplane at low speeds

AERODYNAMIC DESIGN FOR HIGH-SPEED FLIGHT, primarily in terms of drag, is a primary concern for an aerodynamicist; however, the high-speed airplane must take off and land at low speeds. Unfortunately, the aerodynamic design concepts required for high-speed flight are detrimental to low-speed flight.

Chapter 4, regarding aerodynamic concepts, discusses airplane configuration in general terms with an emphasis on high-speed flight characteristics. This chapter examines these concepts in more detail, especially in regard to low-speed flight.

LIFT OF A WING

The lift of a wing is a function of:

- The type of airfoil used (NASA has developed hundreds of airfoil shapes with different characteristics).
- The camber of the airfoil. The more camber, the higher lift as well as higher drag.
- The thickness of the airfoil. A thick airfoil such as 12 percent develops more lift, as well as more drag, than a 4-percent airfoil.
- Angle of attack. Lift increases as angle of attack increases up to the stall. Drag also increases with angle of attack.
- Speed of the airplane (and airfoil). The higher the speed, the greater its lift at a constant angle of attack. At high speed, the angle of attack must be lower than at slow speeds to develop the same lift.
- Wing area. The greater the area (wing span x mean chord), the more lift.
- Aspect ratio, the ratio of span to mean chord of the wing, or, the ratio of the square of the span to the total area of the wing.

Airfoil terminology

Because the shape of an airfoil and the inclination to the airstream (angle of attack) are so important in determining the pressure distribution, it is necessary to properly define the airfoil terminology. Figure 5-1 shows typical airfoils and illustrates the various items of airfoil terminology:

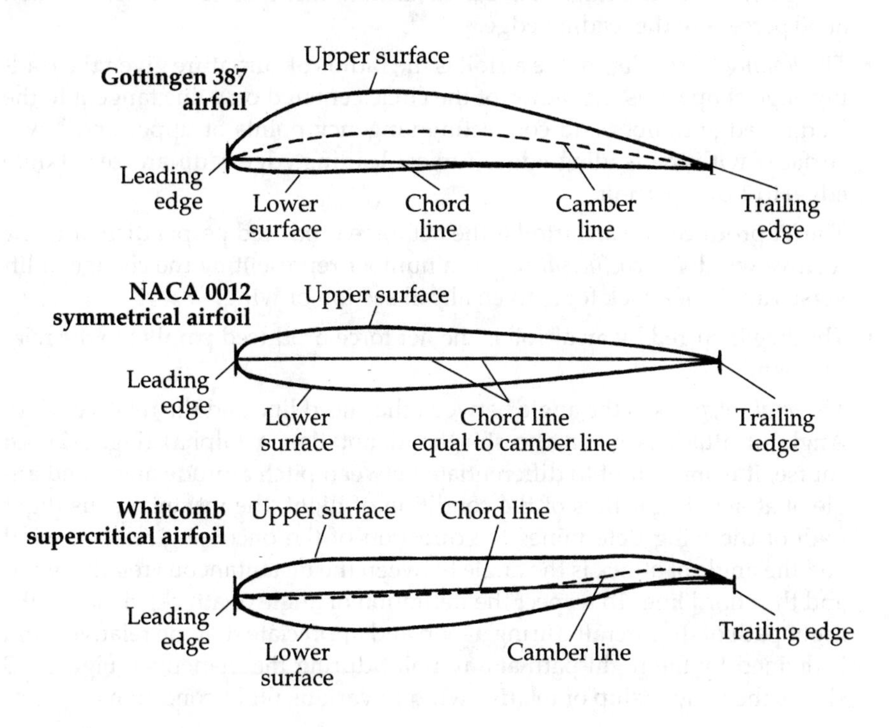

Fig. 5-1. Airfoil terminology.

- The *chord line* is a straight line connecting the leading and trailing edges of the airfoil.
- The *chord* is the characteristic dimension of the airfoil.
- The *mean-camber line* is a line drawn halfway between the upper and lower surfaces. Actually the chord line connects the ends of the mean-camber line.
- The shape of the mean-camber line is very important in determining the aerodynamic characteristics of an airfoil section. The *maximum camber* (displacement of the mean line from the chord line) and the location of the maximum camber help to define the shape of the mean-camber line. These quantities are expressed as fractions or percent of the basic chord dimension. A typical low-speed airfoil might have a maximum camber of 4 percent located 40 percent aft of the leading edge.

- The thickness and thickness distribution of the profile are important properties of a section. The maximum thickness and location of maximum thickness define thickness and distribution of thickness; they are expressed as fractions or percent of the chord. A typical low-speed airfoil might have a maximum thickness of 12 percent located 30 percent aft of the leading edge. A high-speed airfoil might have a maximum thickness of 4 percent located at 50 percent of the leading edge.
- The *leading-edge radius* of the airfoil is the radius of curvature given the leading edge shape. It is the radius of the circle centered on a line tangent to the leading-edge camber and connecting tangency points of upper and lower surfaces with the leading edge. Typical leading-edge radii are zero (knife edge) to 1 or 2 percent.
- The *lift* produced by an airfoil is the net force produced perpendicular to the relative wind. *Lift coefficient* (C_L) is a number representing the change of lift versus angle of attack for a given airfoil section or wing.
- The *drag* incurred by an airfoil is the net force produced parallel to the relative wind.
- The *angle of attack* is the angle between the chord line and the relative wind. Angle of attack is given the shorthand notation α (alpha) (Fig. 5-2). Of course, it is important to differentiate between pitch attitude angle and angle of attack. Regardless of the condition of flight, the instantaneous flight path of the wing determines the direction of the oncoming relative wind and the angle of attack is the angle between the instantaneous relative wind and the chord line. To respect the definition of angle of attack, visualize the flight path of the aircraft during a loop and appreciate that the relative wind is defined by the flight path at any point during the maneuver. Figure 5-3 shows the relationship of relative wind to various flight conditions.

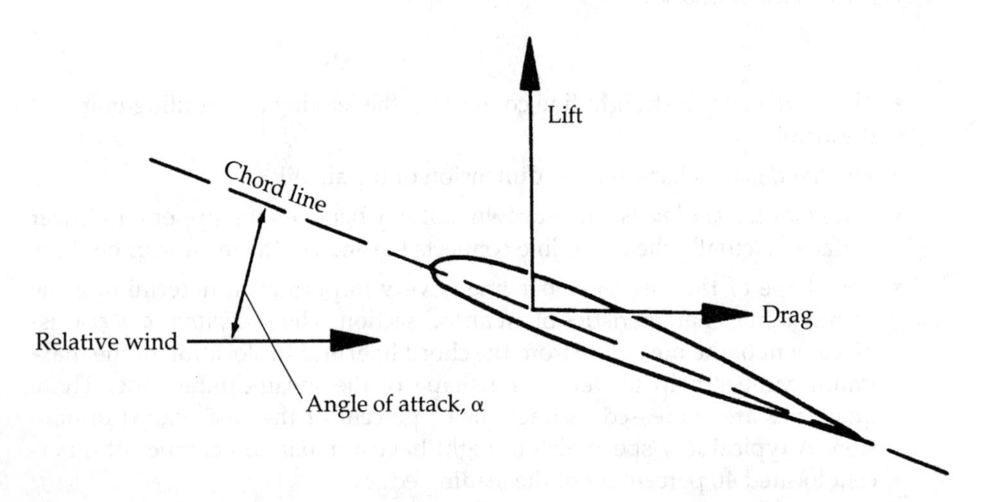

Fig. 5-2. Angle of attack is the angle between the chord line and the relative wind.

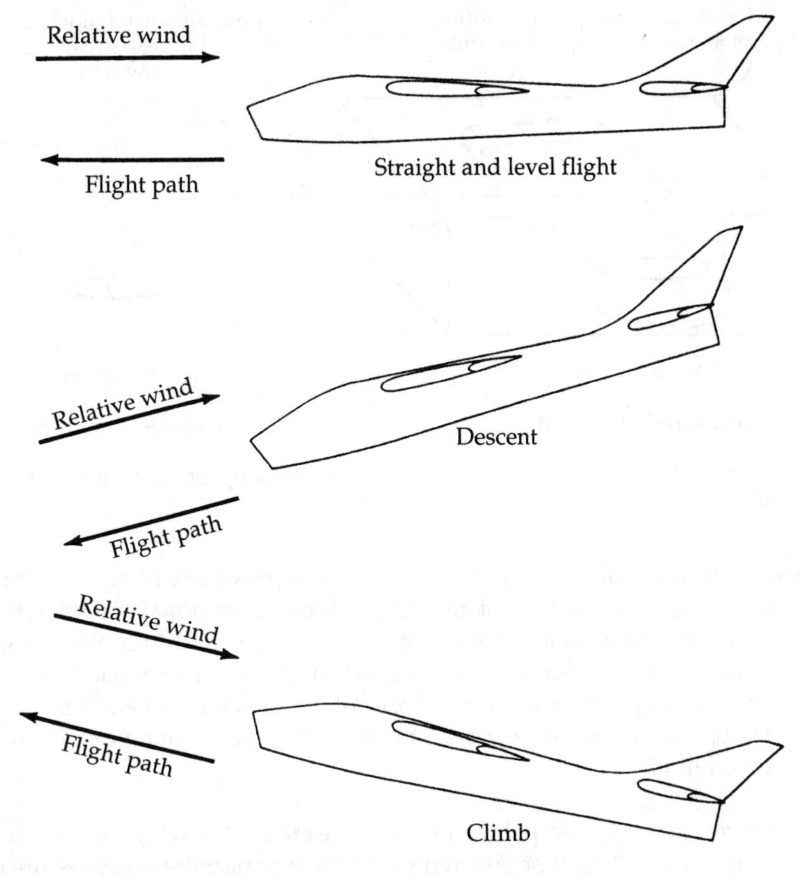

Fig. 5-3. Relationship of relative wind to various flight conditions.

Level flight—relation between angle of attack and airspeed

The high-speed airplane has a very wide level-flight speed range. A jet airliner can maintain level flight at speeds from approximately 130 knots to faster than 500 knots. The carrier-based Navy jet fighter can also fly level as slow as 120 knots to as fast as Mach 2.0.

Recall that lift of a fixed-area wing is a function of angle of attack and airspeed; thus, at high airspeeds, angle of attack is low and, conversely, at slow airspeeds, angle of attack is high. For every airspeed, there is a corresponding angle of attack to maintain level flight, assuming no weight change.

Wing planform and stall patterns

NASA has obtained and documented aerodynamic characteristics on hundreds of airfoil shapes. These data were obtained in a wind tunnel that provided two-dimensional flow (Fig. 5-4). The two-dimensional wing tested in the wind tunnel

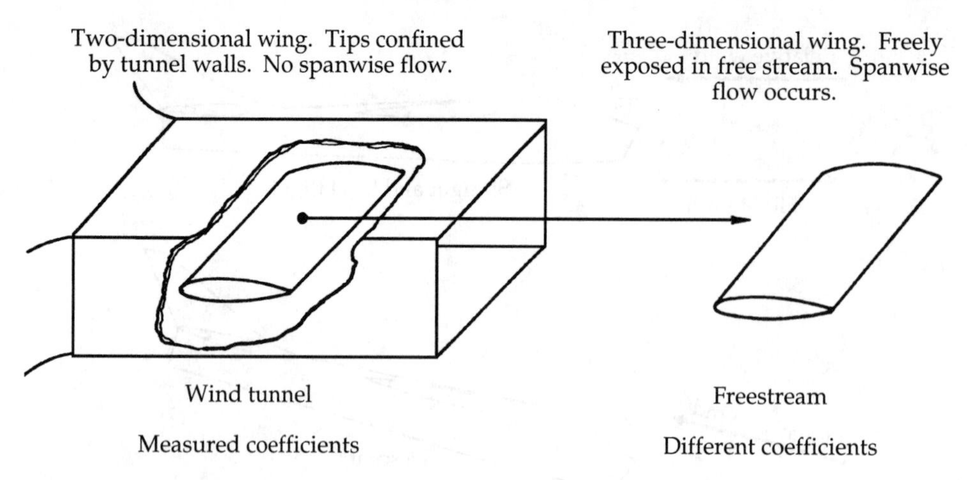

Fig. 5-4. Airfoil characteristics as tested in the wind tunnel must be corrected for the actual three-dimensional wing.

spanned the tunnel walls and did not allow for the possibility of airflow about the wing tips, that is, spanwise flow of air. But the three-dimensional real wing is freely exposed in the free stream, and spanwise flow can occur. The two-dimensional results must be modified to account for the effects of three-dimensional flow.

In order to fully describe the planform of a real wing, several terms are required. The terms having the greatest influence on the aerodynamic characteristics are illustrated in Fig. 5-5:

- The *wing area* (S) is simply the plan surface area of the wing. Although a portion of the area might be covered by fuselage or nacelles, the pressure carryover on these surfaces allows legitimate consideration of the entire plan area.
- The *wing span* (b) is measured tip to tip.
- The *average chord* (c) is the geometric average. The product of the span and the average chord is the wing area ($b \times c = S$).
- The *aspect ratio* (AR) is the proportion of the span and the average chord.

  $$AR = b \div c$$

  The aspect ratio is a *fineness ratio* of the wing, and this quantity is very powerful in determining the aerodynamic characteristics and structural weight. Typical aspect ratios vary from 35.0 for a high-performance sailplane to 2.5 for a jet fighter.
- The root chord, C_R, is the chord at the wing centerline and the tip chord, C_T, is measured at the tip.
- Considering the wing planform to have straight lines for the leading and trailing edges, the taper ratio is the ratio of the tip chord to the root chord. The taper ratio affects the lift distribution and the structural weight of the wing. A rectangular wing has a taper ratio of 1.0; the pointed tip delta wing has a taper ratio of 0.0.

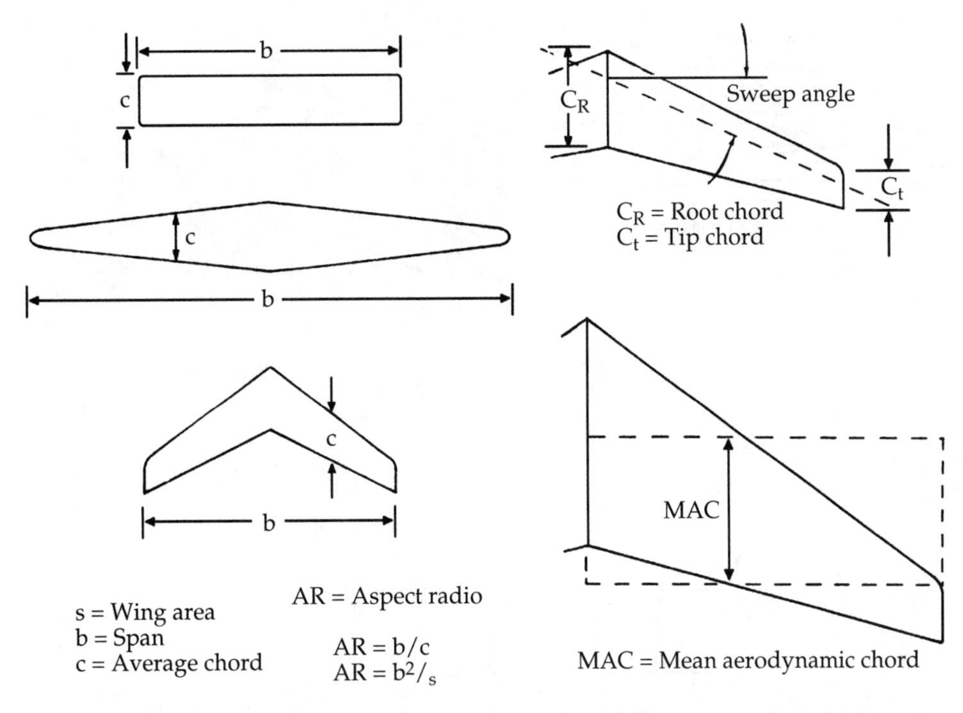

Fig. 5-5. Aspect ratio, taper ratio, and degree of sweepback influence aerodynamic characteristics.

- The *sweep angle* is usually measured as the angle between the line of 25-percent chord and a perpendicular to the root chord. The sweep of a wing causes definite changes in compressibility, maximum lift, and stall characteristics as discussed in chapter 4.
- The *mean aerodynamic chord* (MAC) is the chord drawn through the centroid (geographical center) of plan area. A rectangular wing of this chord and the same span would have identical pitching moment characteristics. The MAC is located on the reference axis of the airplane and is a primary reference for longitudinal stability considerations. Note that the MAC is not the average chord through the centroid of the area. As an example, the pointed-tip delta wing with a taper ratio of zero would have an average chord equal to one-half the root chord, but an MAC equal to two-thirds of the root chord.

The aspect ratio, taper ratio, and sweepback of a planform are the principal factors that determine the aerodynamic characteristics of a wing. These same quantities also have a definite influence on the structural weight and stiffness of a wing.

DEVELOPMENT OF LIFT BY A WING

In order to appreciate the effect of the planform on the aerodynamic characteristics, it is necessary to study the manner in which a wing produces lift. Figure 5-6

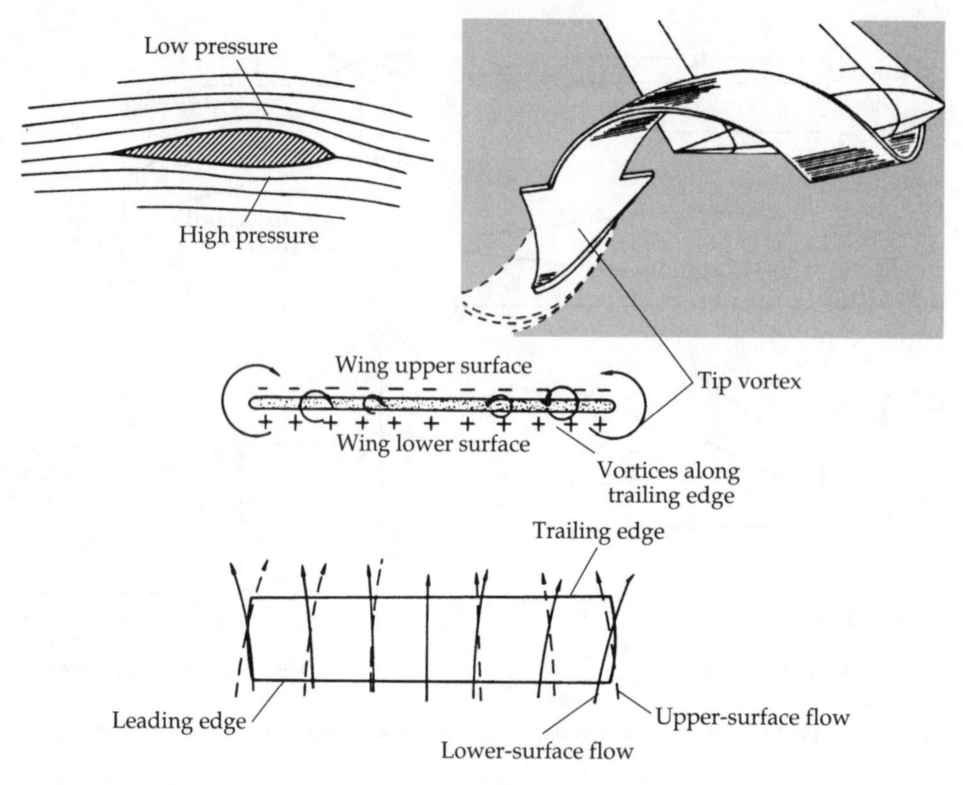

Fig. 5-6. In order to develop lift, the upper surface of the wing is at a lower pressure than the lower surface; therefore, some amount of spanwise flow exists for the three-dimensional wing.

illustrates the three-dimensional flow pattern that results when the rectangular wing creates lift.

If a wing is producing lift, a pressure differential will exist between the upper and lower surfaces; that is, for positive lift, the static pressure on the upper surface will be less than on the lower surface. At the tips of the wing, the existence of this pressure differential creates the spanwise flow components shown in Fig. 5-6. For the rectangular wing, the lateral flow developed at the tip is quite strong, and a strong vortex is created at the tip. The lateral flow and consequent vortex strength reduces inboard from the tip until it is zero at the centerline.

As shown in Fig. 5-6, there is an upward flow of air outside this span of the wing and a downward flow behind the trailing edge of the wing. This downward flow is separate and distinct from the ordinary downwash along the entire wing. In the case of wingtip vortices, the corresponding upward flow is outside the wing span, not in front of it; therefore, the lift that is at right angles to the airflow is slightly rearward, contributing to its drag.

In other words, the real wing, due to the tip vortices, flies at a slightly higher angle of attack to produce the same lift as an ideal, two-dimensional wing (no tip losses). Because drag as well as lift increases with angle of attack, the real wing has greater drag. The increased drag is called *induced drag*. For a real wing, lift is accompanied by induced drag; therefore, a higher aspect ratio means less violent wingtip vortices and lower induced drag.

The higher the aspect ratio, the lower the induced drag and therefore the higher the efficiency of the airplane. High-aspect ratio wings present structural problems; therefore, aspect ratios for flight at subsonic speed range from 6-to-1 to about 10-to-1. Figure 5-7 shows the effect of aspect ratio on the lift coefficient of the three-dimensional wing compared to the two-dimensional wing.

Figure 5-8 shows the difference in wing lift for a two-dimensional wing as tested in a wind tunnel compared to the wing lift of an actual three-dimensional wing.

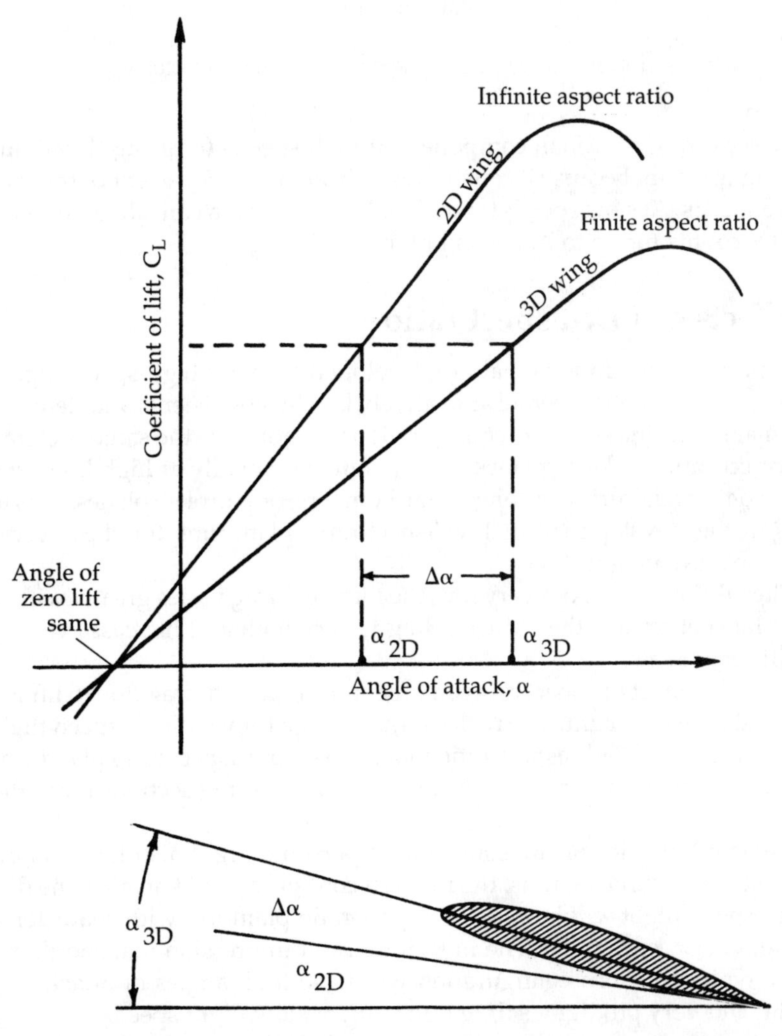

Fig. 5-7. The actual three-dimensional wing must fly at a higher angle of attack due to tip losses and higher drag than a two-dimensional wing as tested in the wind tunnel. This increased drag is called *induced drag*.

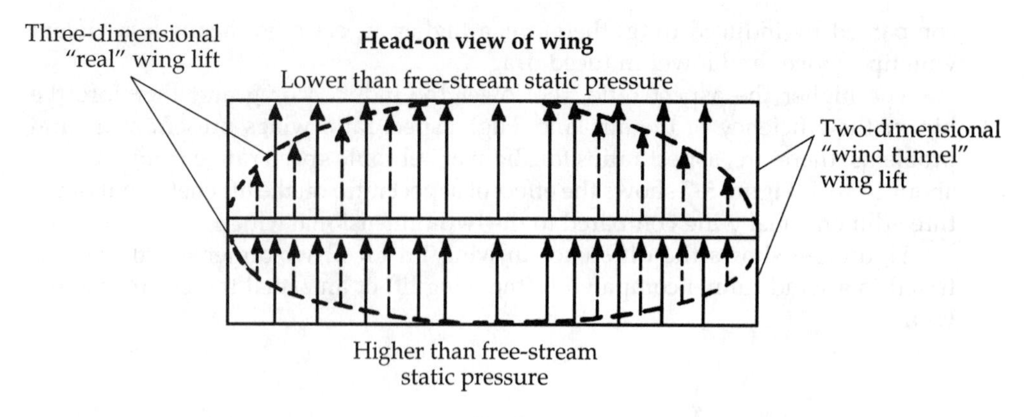

Fig. 5-8. Difference in lift of a real wing compared to a wind-tunnel-tested wing.

Induced drag is a small component at high speeds (cruising flight) and relatively unimportant because it constitutes only about 5–15 percent of the total drag at those speeds. At slow speeds (takeoff or landing), it is a considerable component since it accounts for up to 70 percent of the total drag.

Effects of low-aspect ratio

Airplane configurations that are developed for very high-speed flight (especially supersonic flight) operate at relatively low lift coefficients and demand great aerodynamic cleanness. These configurations do not have the same preference for high-aspect ratio as the airplanes that operate continually at high lift coefficients such as commercial airliners, long-range bombers, or patrol airplanes. This usually results in the development of low-aspect-ratio planforms for these very high-speed airplane configurations.

When the aspect ratio is very low, the induced drag varies greatly with lift and at high lift coefficients, the induced drag is very high and increases very rapidly with lift coefficient.

While the effect of aspect ratio on lift curve slope and drag due to lift is an important relationship, it must be realized that design for very high-speed flight does not favor the use of high-aspect-ratio planforms. Low aspect ratio planforms have structural advantages and allow the use of thin, low-drag sections for high-speed flight.

The aerodynamics of transonic and supersonic flight also favor short-span, low-aspect-ratio surfaces; thus, the modern configuration of an airplane designed for high-speed flight will have a low-aspect-ratio planform with characteristic aspect ratios between 2 and 4. The most important impression that should result is that the typical modern configuration will have high angles of attack for maximum lift and very prodigious drag due to lift at slow flight speeds.

The modern configuration of a high-speed airplane usually has a low-aspect-ratio planform with high wing loading. When wing sweepback is coupled with low-aspect ratio, the wing lift curve has distinct curvature and is very flat at high angles of attack; that is, at high C_L, the C_L increases very slowly with an increase in angle of attack (Fig. 5-9). In addition, the drag curve shows an extremely rapid rise at high lift coefficients because the drag due to lift is so very large. These effects produce flying qualities that are distinctly different from a more "conventional" high-aspect-ratio airplane configuration.

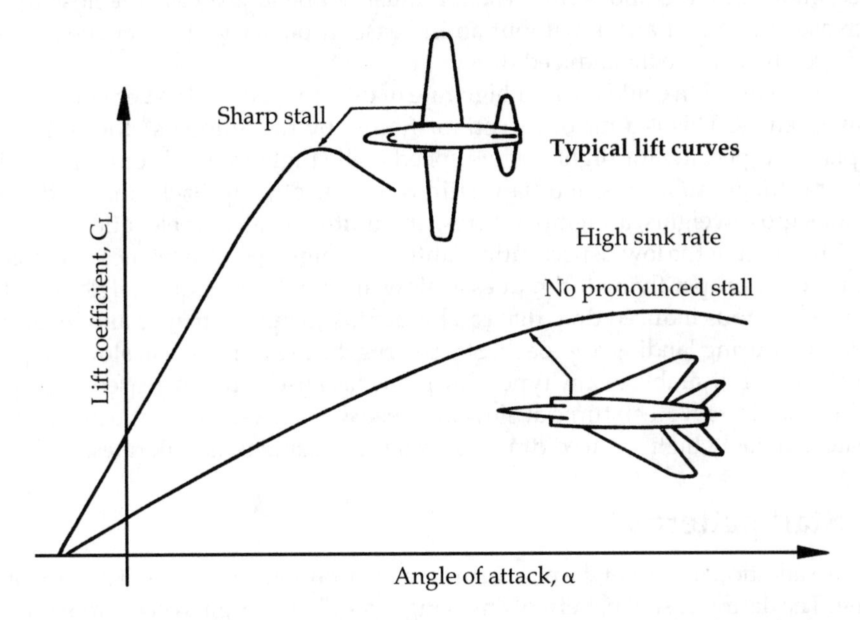

Fig. 5-9. Sweepback tends to flatten the lift curve. At high angle of attack, the swept wing develops an excessive amount of induced drag.

Some of the most important ramifications of the modern high-speed configuration are during takeoff and approach for landing.

During takeoff, the airplane must not be overrotated to an excessive angle of attack. Any given airplane will have some fixed angle of attack (and C_L) that produces the best takeoff performance, and this angle of attack will vary with weight, density altitude, or temperature. An excessive angle of attack produces additional induced drag and might have an undesirable effect on takeoff performance. Takeoff acceleration might be seriously reduced, and a large increase in takeoff distance might occur. Also, the initial climb performance might be marginal at an excessively slow airspeed.

There are modern configurations of airplanes of very low-aspect ratio (plus sweepback) that cannot fly out of ground effect if overrotated during a high-altitude, high-gross-weight takeoff. With the more conventional airplane configuration, an excess angle of attack produces a well-defined stall; however, the high-speed airplane configuration at an excessive angle of attack has no sharply defined stall, as shown in Fig. 5-9, but develops an excessive amount of induced drag.

During approach to landing, the pilot must exercise proper technique to control the flight path. ("Attitude plus power equals performance.") The modern high-speed aerodynamic configuration at slow speeds will have a low lift-drag ratio due to the high induced drag and can require relatively high power settings during the power approach. If the pilot interprets that the airplane is below the desired glide path, the pilot's first reaction must not be to just ease the nose up. An increase in angle of attack without an increase in power will lower the airspeed and greatly increase the induced drag.

Such a reaction could create a high rate of descent and lead to very undesirable consequences. This is a major reason for flight "by the numbers" for high-speed airplanes, especially for flight at slow speeds. The flight manual contains detailed power settings, airspeeds, and flap settings for takeoff, approach, and landing for various gross weights, air temperatures, and altitudes (airport elevations).

The effect of the low-aspect-ratio planform of high-speed airplanes emphasizes the need for proper flying techniques at slow airspeeds. Excessive angles of attack create enormous induced drag that can hinder takeoff performance and incur high sink rates during landing approach. Steep turns during approach at slow airspeeds are always undesirable in any type of airplane because of the increased stall speed and induced drag. Steep turns at slow airspeeds in a low-aspect-ratio airplane can create extremely high induced drag and can incur dangerous sink rates.

Stall patterns

An additional effect of the planform area distribution is on the stall pattern of the wing. The desirable stall pattern of any wing is a stall that begins on the root sections first. The advantages of root stall first are: ailerons remain effective at high angles of attack; favorable stall warning results from the buffet on the tail surfaces and aft portion of the fuselage; and the loss of downwash behind the root usually provides a stable nose-down moment to the airplane. Such a stall pattern is desirable, but might be difficult to obtain with certain wing configurations. Stall patterns inherent with various planforms are illustrated in Fig. 5-10. The various planforms are:

- Elliptical
- Rectangular
- Moderately tapered
- Highly tapered
- Pointed wingtip
- Sweepback

Elliptical. The elliptical planform has constant local lift coefficients throughout the span from root to tip. Such a lift distribution means that all sections will reach stall at essentially the same wing angle of attack, and stall will begin and progress uniformly throughout the span. While the elliptical wing would reach high lift coefficients before incipient stall, there would be little advance warning of complete stall. Also, the ailerons might lack effectiveness when the wing operates near the stall and lateral control might be difficult.

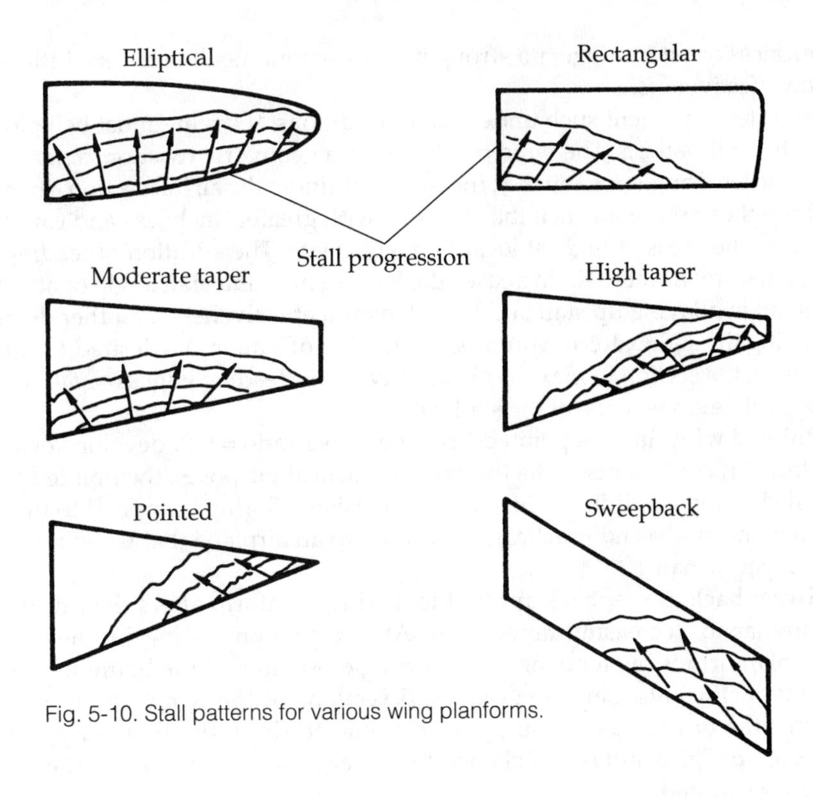

Fig. 5-10. Stall patterns for various wing planforms.

Rectangular. The lift distribution of the rectangular wing exhibits low local lift coefficients at the tip and high local lift coefficients at the root. Since the wing will initiate stall in the area of highest local lift coefficients, the rectangular wing is characterized by a strong root stall tendency. Of course, this stall pattern is favorable because there is adequate stall warning buffet, adequate aileron effectiveness, and usually strong stable moment changes on the airplane. Because of the great aerodynamic and structural inefficiency of this planform, the rectangular wing finds limited application only to low-cost, low-speed lightplanes. The simplicity of construction and favorable stall characteristics are predominating requirements of such an airplane.

Moderately tapered. The wing of moderate taper (taper ratio = 0.5) has a lift distribution that closely approximates that of the elliptical wing; hence, the stall pattern is much the same as the elliptical wing.

Highly tapered. The highly tapered wing (with a taper ratio of = 0.25) shows the stall tendency inherent with high taper. The lift distribution of such a wing has distinct peaks just inboard from the tip. Since the wing stall is started in the vicinity of the highest local lift coefficient, this planform has a strong *tip stall* tendency; the initial stall does not start at the exact tip, but at the station inboard from the tip where highest local lift coefficients prevail.

If an actual wing were allowed to stall in this fashion, the occurrence of stall would be typified by aileron buffet and wing drop. There would be no buffet of the tail surfaces or aft fuselage, no strong nose-down moment, and very little, if any, aileron effectiveness.

In order to prevent such undesirable conditions, the wing must be tailored to favor the stall pattern. The wing can be given a geometric twist or *washout* to decrease the local angles of attack at the tip. In addition, the airfoil section can be varied throughout the span such that sections with greater thickness and camber are located in the areas of highest local lift coefficients. The addition of leading-edge slots or slats toward the tip increases the local $C_L$max and stall angle of attack and are useful in allaying tip stall and loss of aileron effectiveness. Another device for improving the stall pattern would be the forcing of stall in the desired location by decreasing the section $C_L$max in this vicinity. Sharp leading edges or *stall strips* are a powerful device to control the stall pattern.

Pointed wingtip. The pointed tip wing (taper ratio = 0.0) develops extremely high local lift coefficients at the tip. For all practical purposes, the pointed tip will be stalled at any condition of lift unless extensive tailoring is applied to the wing. Such a planform has no practical application to an airplane that is definitely subsonic in performance.

Sweepback. Sweepback applied to a wing planform alters the lift distribution similar to decreasing taper ratio. Also, a predominating influence of the swept planform is the tendency for a strong cross flow of the boundary layer at high lift coefficients. Since the outboard sections of the wing trail the inboard sections, the outboard suction pressures tend to draw the boundary layer toward the tip. The result is a thickened low-energy boundary layer at the tips that is easily separated.

A strong spanwise crossflow develops on the upper surface of the swept wing at high angles of attack. Slots, slats, and flow fences help to allay the strong tendency for spanwise flow.

When sweepback and taper are combined in a planform, the inherent-tip stall tendency is considerable. If tip stall of any significance is allowed to occur on the swept wing, an additional complication results: The forward shift in the wing center of pressure creates an unstable nose-up pitching moment.

An additional effect of sweepback is the reduction in the slope of the lift curve and maximum lift coefficient (Fig. 5-9). When the sweepback is large and combined with low-aspect ratio, the lift curve is very shallow, and maximum lift coefficient can occur at tremendous angles of attack. The lift curve of one typical low-aspect-ratio, highly tapered, swept-wing airplane depicts a maximum lift coefficient at approximately 45° angle of attack.

Such drastic angles of attack are impractical in many respects. If the airplane is operated at such high angles of attack, an extreme landing gear configuration is required, induced drag is extremely high, and the stability of the airplane might seriously deteriorate; thus, the modern configuration of an airplane might have *minimum control speeds* set by these factors rather than simple stall speeds based on $C_L$max.

When a wing of a given planform has various high-lift devices added, the lift distribution and stall pattern can be greatly affected. Deflection of trailing-edge flaps increases the local lift coefficients in the flapped areas, and since the stall angle of the flapped section is decreased, initial stall usually begins in the flapped area. The extension of slats simply allows the slatted areas to go to higher lift coefficients and angles of attack, and generally delays stall in that vicinity.

Why not sweepforward?

Aerodynamically, sweepforward has the same transonic drag-reducing characteristics as sweepback. In addition, sweepforward has other aerodynamic advantages over sweepback. For example it is more efficient than sweepback. The lift generated by a swept-back wing, or a straight wing for that matter, is usually limited by tip stall. Due to taper, vortex effects and spanwise flow of a swept-back wing, the tip sections are normally more highly loaded than the inner portions; thus, the tips will stall before the rest of the wing, preventing the wing from reaching its maximum angle of attack. Since spanwise flow is at a minimum, the tips of the forward-swept wing remain unstalled past the point where the inner portion stalls; thus, more of the wing's lifting potential is usable.

The span loading is also better distributed. Swept-back wings are normally designed for an elliptical span loading for cruising flight because this generates the lowest induced drag (drag due to lift); however, at high angles of attack, the loads on the outer sections build up rapidly due to spanwise flow; therefore, induced drag also increases. A forward-swept wing produces the desirable elliptical loading at maximum lift. This provides a better lift-to-drag ratio (an indication of wing efficiency) in high-G maneuvers.

Spanwise flow on a swept-back wing reduces the effectiveness of the ailerons especially at slow speeds and high angles of attack. Recall from chapter 4 that wing fences and vortex generators are sometimes used to minimize spanwise flow effects. Forward-swept wings, however, do not require these aerodynamic "fixes," and the ailerons remain responsive even after the rest of the wing has stalled.

Forward-swept wings then have the following advantages:

- More lift than a similar size swept-back wing, or a smaller wing for the same lift.
- Less induced drag. Shorter takeoffs and landings due to improved effectiveness of high-lift devices (slots and flaps).
- Increased aileron effectiveness at high angles of attack.
- The same lower transonic and supersonic drag as swept-back wings.

If forward-swept wings are so great, why do all high-speed aircraft use only swept-back wings? Obviously, there must be a major disadvantage, and there is. The forward-swept wing is structurally unstable.

A wing, especially with an aspect ratio of around 6 (the ideal for a high-speed jet airliner), bends under load along its flexural axis. For a straight-wing (no sweep), the tip merely moves up causing no appreciable increase in tip angle of attack and therefore no increase in tip loading or wing bending load (Fig. 5-11). A

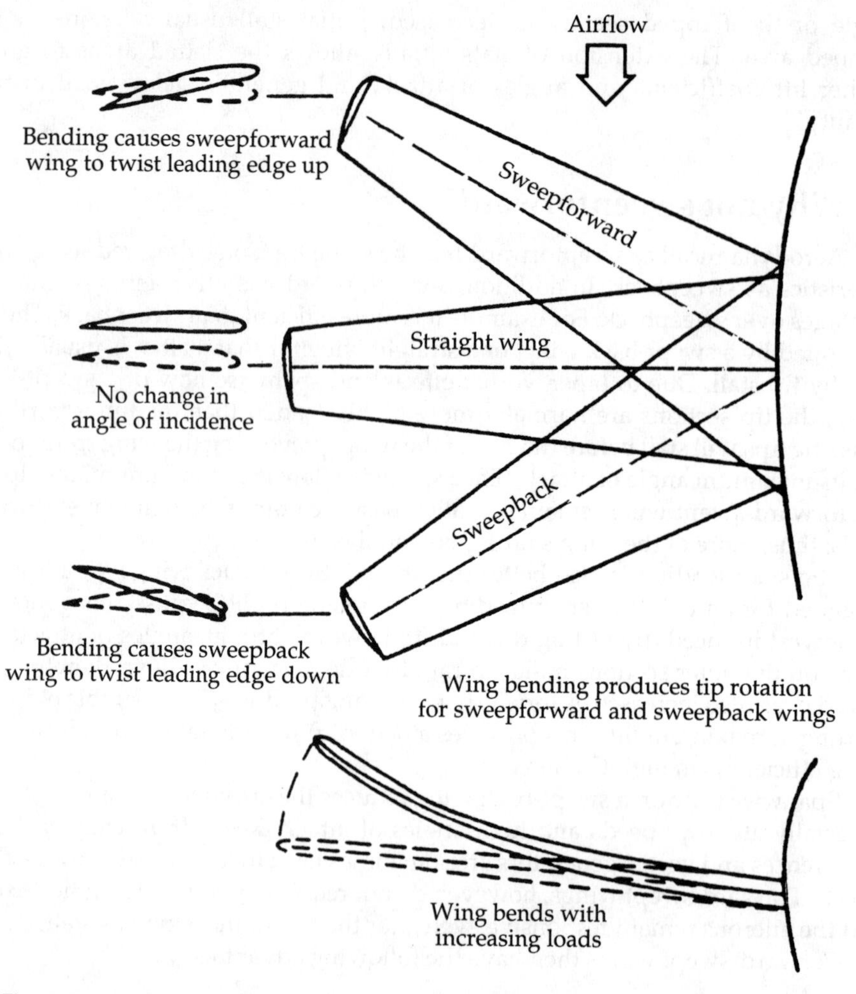

Fig. 5-11. Wings flex with increasing air loads. The straight wing merely flexes upwards with no appreciable change in angle of attack. The sweepback wing twists leading edge down, which is a stable condition because this decreases the load on the wing; however, a sweepforward wing twists leading-edge up, which is an unstable condition that increases wing loading that might lead to structural failure.

swept-back wing, with an increase in angle of attack, twists leading-edge down, which reduces the tip load; thus a swept-back wing is stable, tending to reduce the wing-bending load.

With a forward-swept wing, however, the leading edge twists upward (Fig. 5-11), tending to increase the angle of attack as well as the loading. This structural instability, or divergence, increases as the degree of forward sweep increases. To overcome this instability, a structurally stiff, heavy wing is required. The weight penalty for forward sweep is considered just too great a price to pay for its advantages.

During World War II, the German Junkers 287 jet bomber was designed with a swept-forward wing; however, only a prototype was constructed by the war's end. Recently the Defense Advanced Research Projects Agency (DARPA) sponsored a research program into forward-swept wings, resulting in the X-29. The use of composite structure instead of metal can overcome divergence by varying the direction and thickness of the carbon-ply wing skins to control the direction of the flexural axis and therefore the twisting under high G loads. The use of composites for major structural assemblies of production airplanes has not been sufficiently developed; therefore, the X-29 is a pure research aircraft.

SLOWING DOWN THE HIGH-SPEED AIRPLANE

Recall that lift of an airfoil (or wing) is a function of curvature (camber and thickness) of the airfoil section: the greater the camber and the greater the thickness, therefore, the greater the lift. A thick wing with a lot of camber develops much more lift than a thin wing with subsequently little camber.

It is in the interest of safety to perform takeoff and landing maneuvers at a speed that is as slow as possible; therefore, the airplane should take off and land at speeds close to *stall airspeed*. This occurs at $C_L$max. (Recall that C_L is the lift coefficient, an indicator of the lift produced by an airfoil vs. angle of attack (Fig. 5-12).) Because the airplane wing is designed for high-speed flight, some changes must take place in order to provide lift at slow speeds.

One of the most fascinating subjects regarding the aerodynamics of flight is the vast number of "aerodynamic devices"—for want of a better term—affixed to a simple wing to achieve increases or decreases in lift and drag: *slats, slots, flaps,*

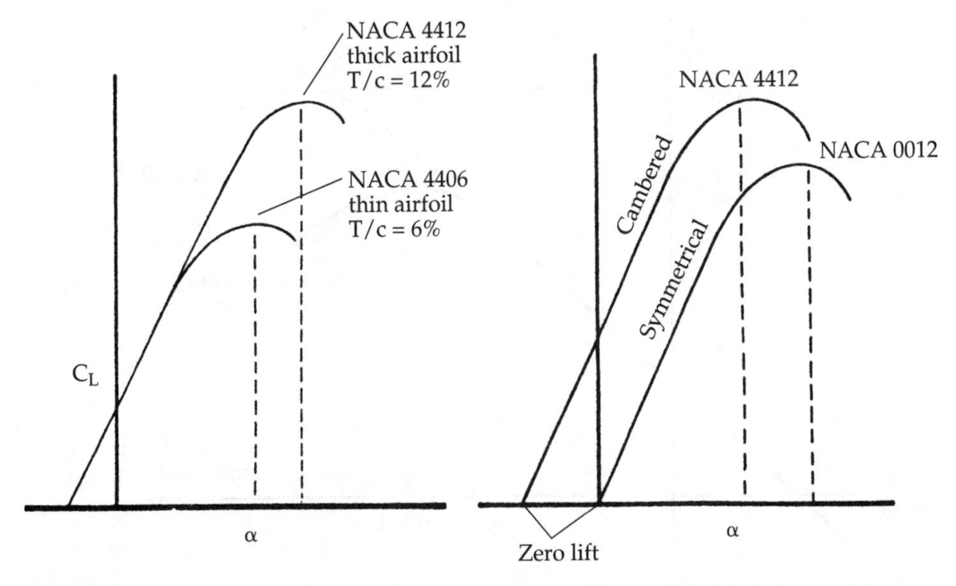

Fig. 5-12. Greater thickness (left) plus greater camber (right) equal greater lift.

spoilers, and *dive brakes*. The only way to reduce speed to a minimum is to increase $C_L$max and/or the wing area. Slots and flaps are used for this purpose.

Slots

The maximum coefficient of lift can be increased through the use of a slot formed by a leading-edge auxiliary airfoil called a *slat*. Figure 5-13 illustrates the operating principle. When the slot is open, the air flows through the slot and over the airfoil. The slot is a *boundary-layer control* device, and the air thus channeled energizes the boundary layer about the wing and retards the separation. The airfoil can then be flown at a higher angle of attack before stall occurs and thus get a higher $C_L$max value. Figure 5-13 is a curve showing C_L as a function of angle of attack for the normal and the slotted airfoil. Notice particularly that for angles of attack less than the stall angle, the airfoil lift curve is relatively unaffected whether the slot is opened or closed.

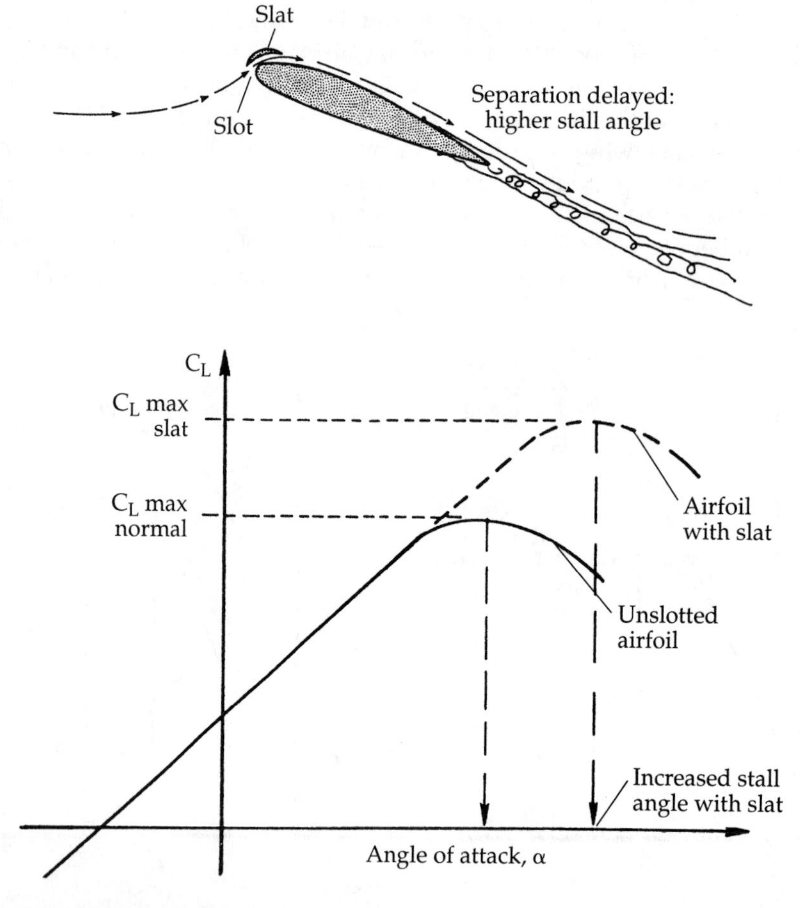

Fig. 5-13. Maximum lift of a wing can be increased with a slot that energizes the boundary layer and retards separation.

Slots are either fixed (in position) or automatic. The fixed slot is self-explanatory; the leading-edge slat is mounted a fixed distance from the airfoil. Its main disadvantage is that it creates excessive drag at high speeds. The automatic slot depends on air pressure lifting the slat away from the wing at high angles of attack to open the slot. At reduced angles of attack, the slat is flush against the wing leading edge and reduces drag at high speeds, compared with the fixed slot. The automatic slot's disadvantages are added weight, complexity, and cost.

One main disadvantage of both types of slots is the excessive stall angle that is created. The airplane must approach for a landing in an extreme nose-up attitude, which promotes reduced visibility and complicates landing gear design.

Flaps

Flaps can be used to increase the maximum lift coefficient or increase the wing area or both. A change in the maximum lift coefficient can be realized by a change in the shape of the airfoil section or by increased camber. The *trailing-edge flap* is one method of accomplishing this. Figure 5-14 shows a normal airfoil and the

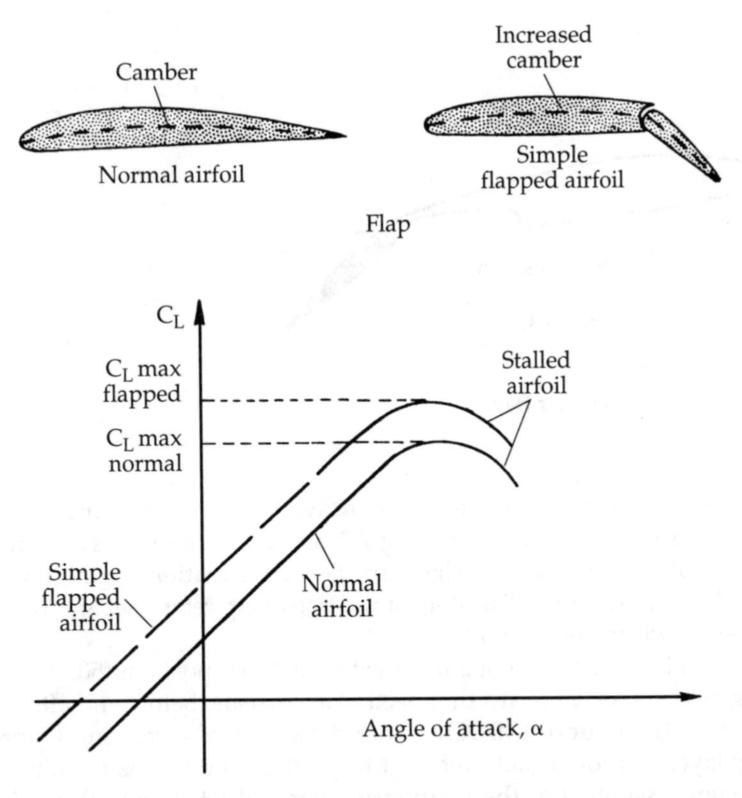

Fig. 5-14. The maximum lift coefficient for an airfoil with a simple flap is greater than the maximum lift coefficient for the unflapped airfoil.

same shaped airfoil with a simple flap in the down position for increased camber. The maximum lift coefficient for the airfoil with the simple flap is greater than that for the unflapped airfoil. Also, the coefficients of lift are increased over the entire angle-of-attack range (Fig. 5-14). Note also that the stall angle is essentially unchanged from that of the unflapped airfoil. This is opposed to the slot operation where a higher stall angle was obtained. The flapped airfoil reduces the disadvantage that the slot has in excessive landing angles.

Figure 5-15 shows a *Fowler flap* that is hinged such that it can move back and increase the wing area. Also, the Fowler flap can be rotated down to increase the camber. A very large increase in maximum lift coefficient is realized at the expense of design complexity.

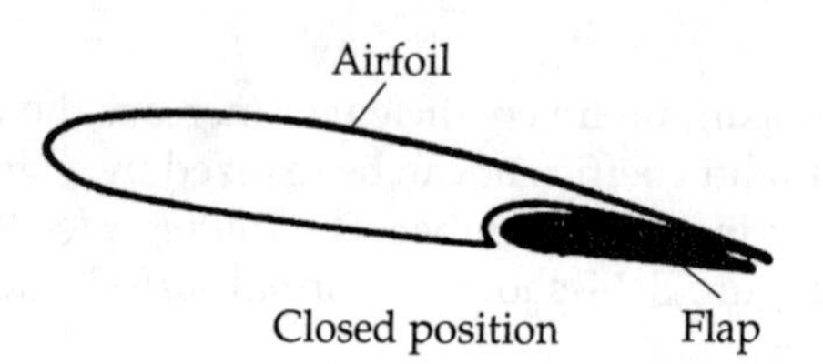

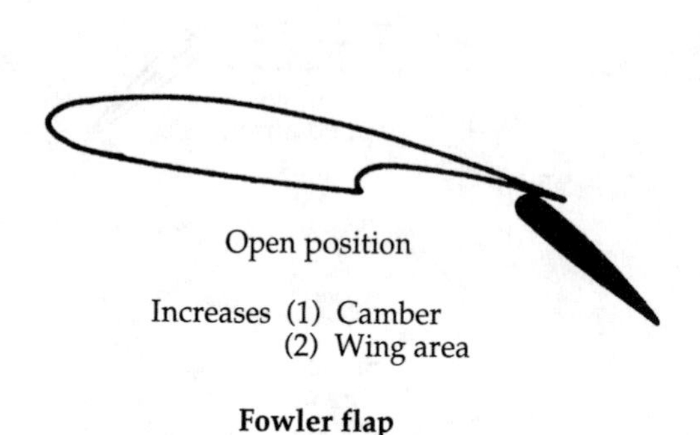

Fig. 5-15. A Fowler flap increases both the camber and the wing area.

Many combinations of slots and flaps are available for use on airplanes. Figure 5-16 shows the arrangement on a Boeing 737 airplane that utilizes a leading-edge slat and a triple-slotted trailing-edge flap. This combination is a highly efficient lift-increasing arrangement. The slots in the flaps help retard separation over the flap segments, which enhances lift.

It can also be noted that flaps in an extreme down position (50°–90°) act as a high-drag device and can retard the speed of an airplane before and after landing.

Boundary-layer control (BLC). Another method of increasing $C_L$max is by boundary-layer control, usually referred to as BLC. The idea is to either remove the low-energy segment of the boundary layer and let it be replaced by high-energy flow from above or by adding kinetic energy to the boundary layer directly.

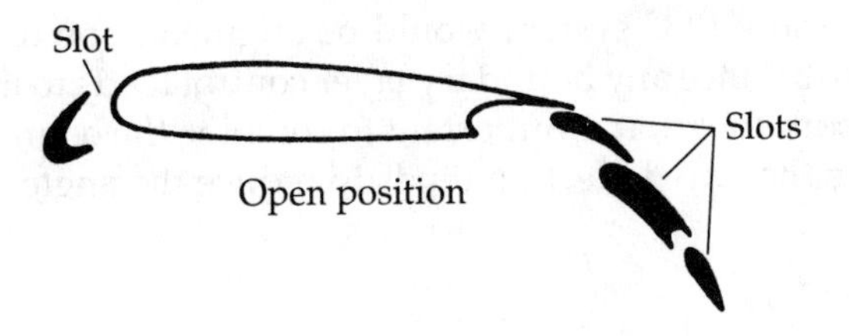

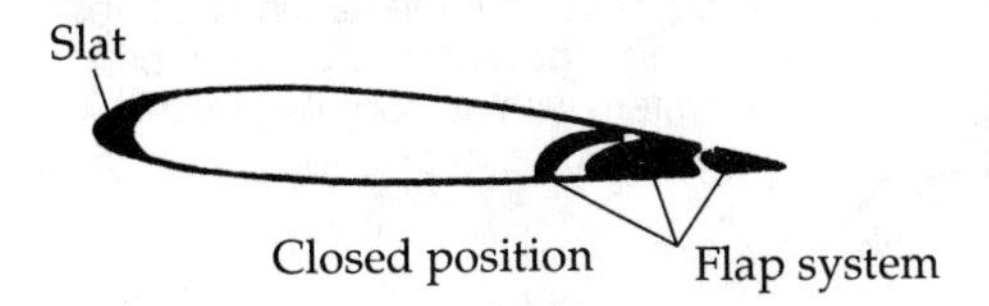

Fig. 5-16. The complex slotted flap system of the Boeing 737.

Both of these methods maintain a laminar flow for a longer distance over the airfoil, delay separation, and provide a higher angle of attack before stall occurs, and thus a higher $C_L$max. The slot was shown to be one means of passing high-energy flow over the top surface of a wing.

The low-energy boundary layer can be sucked through slots or holes in the wing; high-energy air can be blown into the boundary layer through backward-facing holes or slots as shown in Fig. 5-17.

A suction BLC system requires the installation of a separate pump; a "blown" BLC system can utilize high pressure from a jet-engine compressor. The

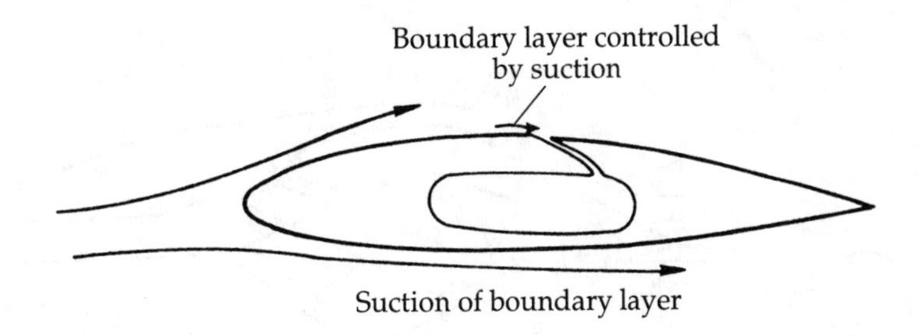

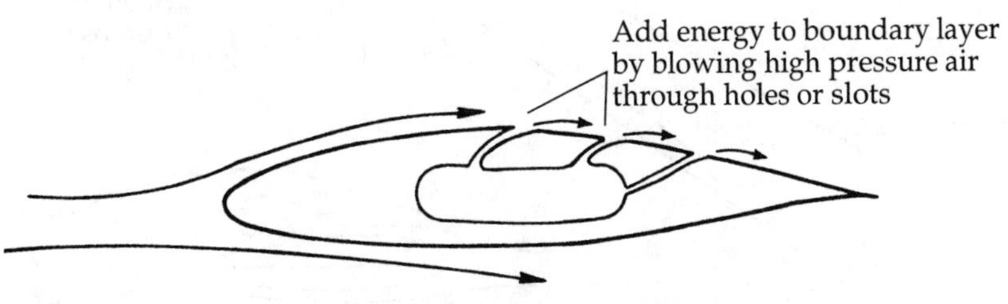

Fig. 5-17. Another method of increasing the maximum lift coefficient is by boundary layer control, which has not proven practical for an entire wing.

typical installation of a high-pressure BLC system would be augmentation of a deflected flap as shown in Fig. 5-18. Since any boundary layer control tends to increase the angle of attack for maximum lift, it is important to combine the boundary layer control with flaps since the flap deflection tends to reduce the angle of attack for maximum lift.

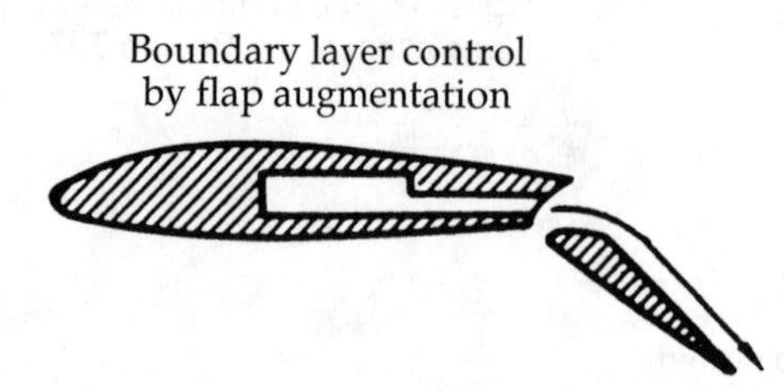

Fig. 5-18. Augmentation of a flap using a boundary layer control system with air supplied by the jet engine compressor.

Spoilers

Spoilers are devices used to reduce the lift on the airplane wing. They might serve the purpose, as on gliders, to vary the total lift and control the glide angle. Or on large commercial jets, spoilers might be used to help aileron control by "dumping" lift on one wing and help to roll the airplane. Also, on landing, with all spoilers up, the lift is quickly destroyed, and the airplane will settle on its landing gear without bouncing. Figure 5-19 shows the spoiler arrangement on a Boeing 707 wing.

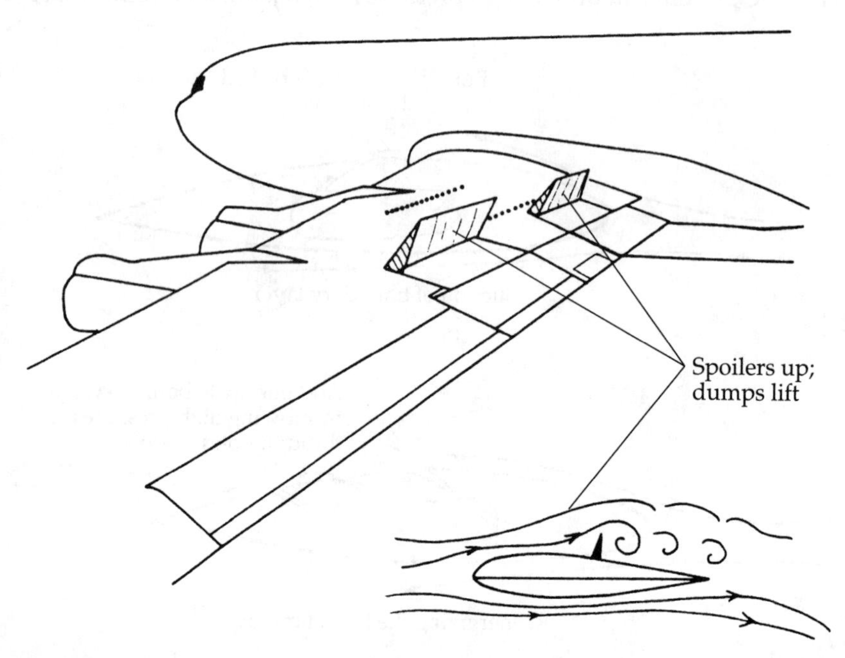

Fig. 5-19. Spoilers reduce the lift of a wing. Differential spoiler deployment can be used for roll control. Lift is quickly dumped with all spoilers up during the landing rollout.

Dive brakes

Dive brakes, sometimes called *speed brakes*, are used in airplanes to control descent speed. Whether slowing down quickly when approaching for a landing, after landing, or in a dive, these aerodynamic brakes are helpful. Essentially, they promote a large separation wake and increase the pressure drag. Figure 5-20 shows two military aircraft dive brake arrangements.

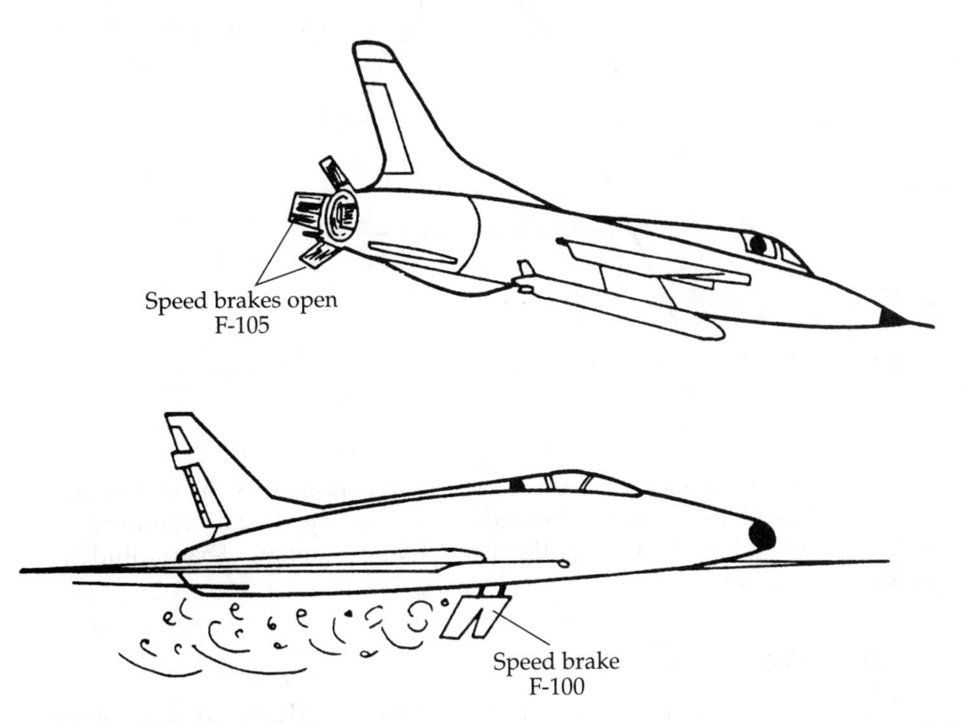

Fig. 5-20. Dive or speed brakes are strictly drag producing devices to slow down a clean airplane in a dive or during an approach to landing.

Stall characteristics

The present discussion has concentrated on operating near or at the stall condition $C_L$max. A further word about stalling is in order. A wing should possess favorable stall characteristics so that the pilot has adequate warning of the stall, the stall is gradual, and there is little tendency for a wing to drop and initiate a spin after a stall. This can be achieved by "forcing" the stall to occur at the wing-root section first and let it progress toward the wingtips. The outboard wingtip stations should be the last to stall so that the ailerons remain effective, not immersed in a turbulent "dead-air" wake. Use of twist, namely *washout*, is often employed so that the wing-root section reaches the stall angle first. Also, airfoil sections with gradual stall characteristics are more favorable than ones with sharp stall characteristics (Fig. 5-21).

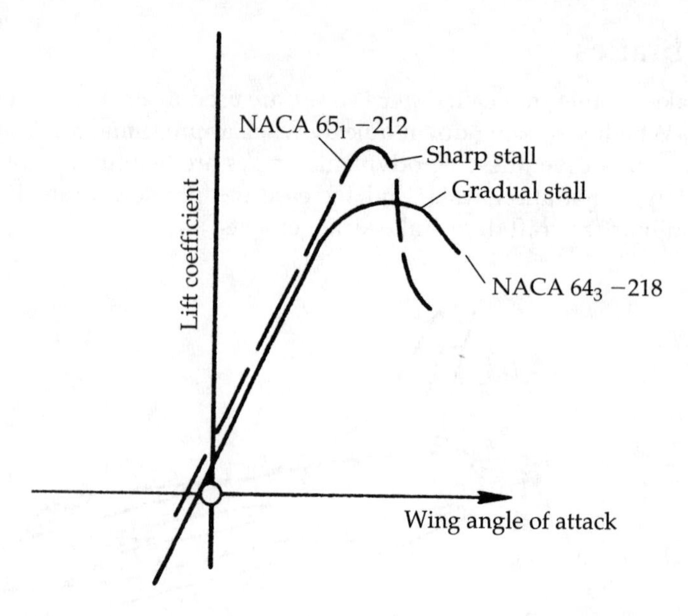

Fig. 5-21. Airfoil sections with gradual stall characteristics are more favorable than airfoils with sharp stall characteristics.

As the inboard root stations stall, turbulent flow from the wing strikes the tail and buffets the pilot's controls. This condition is an adequate stall warning device. With a gradual stall on both wings, the plane should maintain a level attitude with few spin tendencies.

Landing speeds

Even when all available aerodynamic devices are deployed, a jet airplane touchdown speed normally exceeds 125 mph. Reverse engine thrust may be applied to supplement wheel braking to reduce landing roll. (Chapter 6 has more information about engines and reverse thrust.) Some military airplanes deploy a parachute from an aft compartment to provide added drag upon landing.

The commercial jet airliner employs extensive slot and flap airfoil systems to achieve safe, slow-speed flight. After touchdown, spoilers are raised to reduce wing lift and reverse thrust is engaged to supplement wheel brakes, to control and reduce the landing roll, and reduce tire loads.

6

The jet engine (gas turbine engine)

ALL JET ENGINES, including rocket motors, pulsejets, and ramjets, belong to a class of powerplants called *reaction engines*; however, this book is confined to the *gas turbine engine*, previously and hereafter referred to as a jet engine, which is the basic propulsion system for high-speed flight. Pulsejets, ramjets, and rocket motors have powered certain research aircraft; however, they are mainly used in missiles and space vehicles.

Jet *propulsion* is the propelling force produced in the direction opposite to the flow of a mass of gas or liquid under pressure that is escaping through an opening called a *jet nozzle*. In other words, jet propulsion is the practical application of Newton's Third Law (remember high school science?):

To every action force there is an equal and opposite reaction force.

The force that makes the nozzle of a fire hose spraying high-pressure water difficult to hold and causes the nozzle to flop around on the ground, if dropped, is a simple example of Newton's law, as well as an example of jet propulsion. The recoil effect when firing a rifle is another example.

DEVELOPMENT OF THE GAS TURBINE ENGINE

Jet propulsion for aircraft did not become practical until the mid-1940s; however, various examples of jet propulsion preceded 1940. In 1908, Rene Lorin, a French engineer, proposed using a piston engine to compress air that would then be mixed with fuel and burned to produce pulses of hot gases that would be expelled through a nozzle to generate a propelling force.

During the early 1930s, the Italian Caproni monoplane actually flew with a similar system. A piston engine powered a compressor. The compressed air was mixed with fuel and burned, producing hot gases that expanded through a jet nozzle, producing thrust. Although the Caproni actually flew, it was woefully inefficient due mainly to the inefficient compressor and its source of power.

It remained for Sir Frank Whittle of Great Britain and Hans von Ohain of Germany to successfully integrate the compressor driven by a turbine principle for a jet engine that could be used to power an airplane. Whittle filed his initial patent for a turbojet engine using a centrifugal compressor in 1929. The first Whittle engine flew in the British Gloster G.40 on May 15, 1941.

Independently of Whittle, von Ohain in Germany also was working on a jet engine for aircraft. His first demonstration engine was run in 1937. His first flight engine was the Hes 3B, which flew in the Heinkel HE178 on August 27, 1939.

The Whittle and von Ohain engines led to successful jet-powered aircraft by the end of World War II. The British Gloster Meteor, powered by two centrifugal-flow compressor turbojets, first flew in March 1943 and saw limited service in World War II; however, Germany developed the more advanced turbojet engine using axial flow compressors.

By the end of the war in Europe in April 1945, Germany had turbojet engines in production by three different manufacturers: BMW, Junkers, and Heinkel-Hirth. These engines powered the operational Messerschmitt 262 fighter, Heinkel HE-126 fighter, and the Arado AR-234 reconnaissance bomber. These early production turbojets had a static sea-level thrust of around 1,800 pounds; however, turbojets in the 5,500-pound to 7,000-pound thrust region were under development.

In the United States, the General Electric Company built and tested America's first jet engine based on the English Whittle design in 1942. The General Electric I-A produced 1,300 pounds of thrust. Two I-As were installed in the Bell P-59 Aerocomet, which made its first flight in October 1942. Later Aerocomets were powered by General Electric J-31 engines of 2,000 pounds thrust. In June 1944, the Lockheed P-80 made its first flight with a General Electric J-33. After the end of World War II, jet engine development advanced at a rapid pace. General Electric, Allison, Pratt & Whitney, Curtiss-Wright, and Westinghouse were all involved in jet engine development.

Note: Most of the following text and all of the photographs and illustrations were supplied by General Electric Aircraft Engine Groups and Pratt & Whitney, division of United Technologies. Other illustrations were supplied by NASA and McDonnell Douglas.

THE BASIC JET ENGINE

The piston engine and the gas turbine develop power or thrust by burning a combustible mixture of fuel and air. Both engines convert the energy of the expanding gases into propulsive force. The piston engine does this by changing the energy of combustion into mechanical energy, which is used to turn a propeller. Aircraft propulsion is obtained as the propeller imparts a relatively small acceleration to a large mass of air. The gas turbine, in its basic turbojet configuration, imparts a relatively large acceleration to a smaller mass of air, and thus produces thrust or propulsive force directly. Here the similarity between the two types of engines ceases.

HOW A TURBOJET OPERATES

The simplest gas turbine engine for aircraft is a turbojet, which is essentially a machine designed for the sole purpose of producing high-velocity gases at the jet nozzle. The engine is started by rotating the compressor with a starter, and then igniting a mixture of fuel and air in the combustion chamber with one or more *igniters* that resemble automobile spark plugs. When the engine has started and its compressor is rotating at sufficient speed, the starter and the igniters are turned off. The engine will then run without further assistance as long as fuel and air in the proper proportions continue to enter the combustion chamber and burn.

The reason why a turbojet will run as it does lies in the compressor. The gases created by a fuel-and-air mixture burning under normal atmospheric pressure do not expand enough to do useful work. Air under pressure must be mixed with fuel before gases produced by combustion can be used to make a piston engine or turbojet engine operate. Power and thrust produced by an engine is proportional to the amount of air that can be compressed and used by the engine; more air means more power and thrust.

Finding a way to accomplish the compressing of air was the biggest challenge for designers during the early years of turbojet engine development. Recall that Frank Whittle solved the problem by using a centrifugal compressor similar to those employed in superchargers for aircraft piston engines. Whittle provided the power required to turn the compressor by mounting a gas-driven turbine immediately to the rear of an engine's combustion chamber in approximately the same manner used today.

High power is necessary to drive the compressor in a turbojet engine. Workable gas turbine engines would have been developed sooner if anyone had known how to build a turbine that would produce sufficient power to turn the compressor and yet leave enough energy in the exhaust gases to push an airplane. Superior compressor and turbine combinations eventually led to successful engines.

To indicate how much power is absorbed by a compressor of a moderately large turbojet, let us assume that we have an engine with about a 12-to-1 compression ratio that produces 10,000 pounds of thrust for takeoff. In this engine, the turbine has to produce approximately 35,000 shaft horsepower just to drive the compressor when the engine is operating at full thrust. Only the power left over is available to produce the thrust needed to propel the airplane.

TYPES OF GAS TURBINE ENGINES

A *turbojet* is a gas turbine in which no excess power (above that required by the compressor) is supplied by the turbine. The available energy in the exhaust gases is converted to kinetic energy of the jet, which supplies the propulsion force.

A *turboprop* is a gas turbine engine in which the turbine provides power in excess of that required to drive the compressor, which is used to drive a propeller. Ordinarily, it also has an appreciable jet thrust.

A *turbofan* or *fanjet* is similar to a turboprop, except that the excess power of the turbine is used to drive a fan or low-pressure compressor in an auxiliary duct, usually annular around the primary duct.

Afterburning can be used with turbojet and turbofan engines to provide additional thrust at the expense of fuel economy. In afterburning, additional fuel is added to the exhaust gases and burned, thereby increasing the temperature, the jet velocity, and the thrust. (Afterburning is subsequently discussed in detail in this chapter.)

The jet engines of high-speed flight are the turbojet, turbofan (also called fanjet), and to a lesser extent, the turboprop. The afterburning turbojet and fanjet are also widely used, mainly in military aircraft.

THE GAS TURBINE

The basic gas turbine, whether applied to the turbojet, turboshaft, turboprop, or turbofan principle, is referred to as the *gas generator* or *core engine*.

The gas turbine is comprised of three primary components: the compressor, the combustor, and the turbine (Fig. 6-1). Recall that the compressor is driven by the turbine through an interconnecting shaft. Compressed air flows to the combustor, where it is mixed with injected fuel, and the fuel-air mixture is ignited. The hot combustion gases flow through the turbine. The turbine extracts energy from the hot gases, converting it to power to drive the compressor and any mechanical load connected to the drive.

The gas turbine and the reciprocating (piston) engine are internal-combustion engines. The piston engine operates on the principle of the *Otto cycle*; the gas turbine operates on the principle of the *Brayton cycle*. Figure 6-2 shows the events of the gas turbine engine (Brayton cycle). The piston engine operates with intermittent power pulses; the gas turbine operates with a steady, continuous power flow.

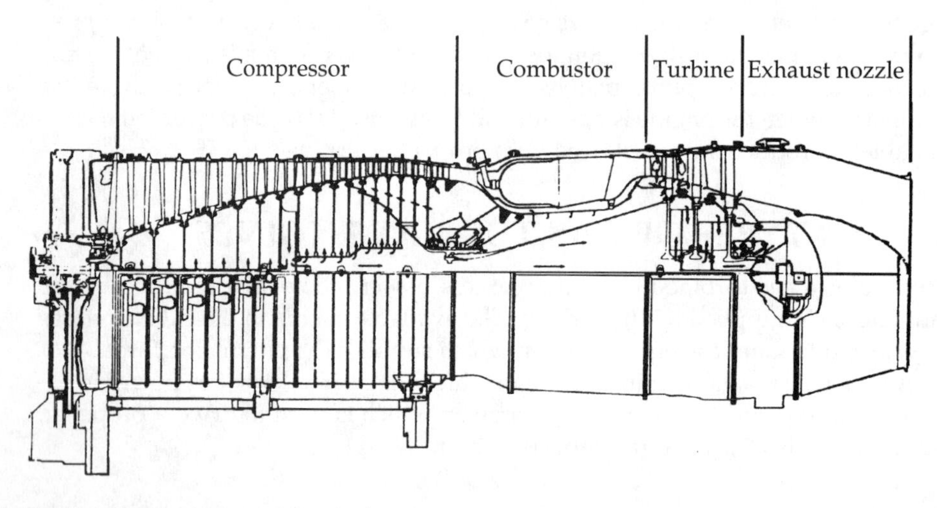

Fig. 6-1. The basic gas turbine engine.

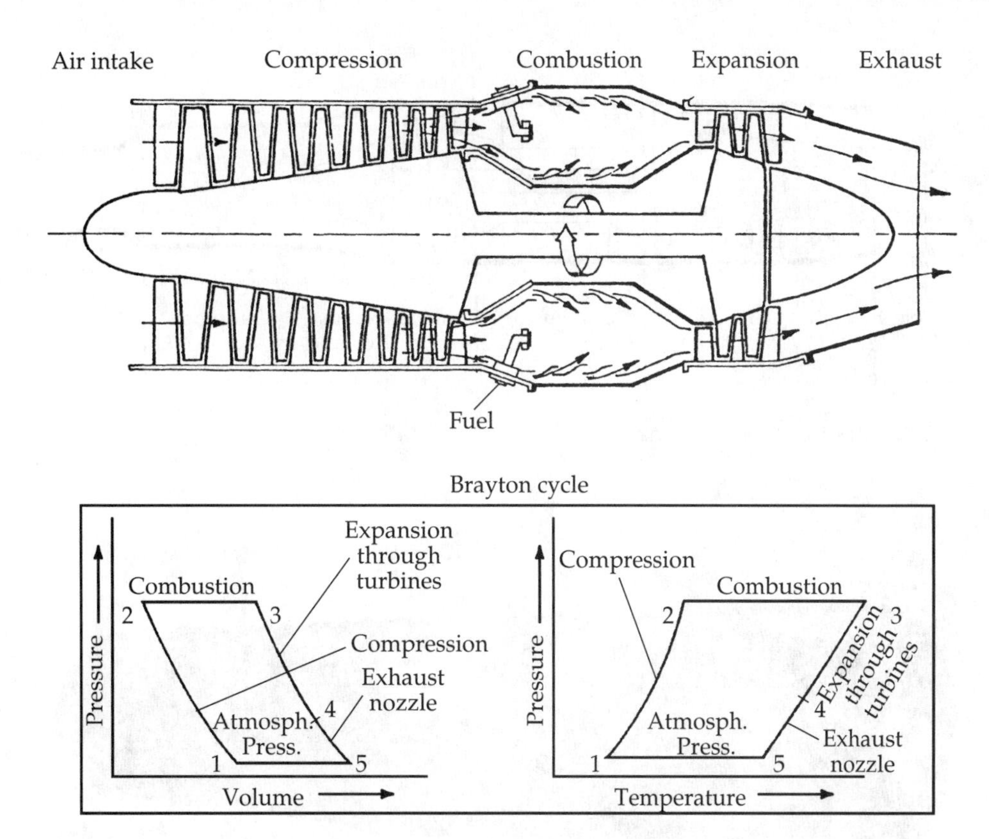

Fig. 6-2. Cycle events of the gas turbine engine.

The turbojet

The turbojet is a gas turbine engine in which the turbine extracts only the power required to drive the compressor and accessories necessary for continuous operation. Figure 6-3 shows a basic turbojet engine. The high velocity imparted to the exhaust gases by the exhaust nozzle provides the thrust for propulsion. Approximately 65 percent of the developed energy is used to drive the compressor, while only the remaining 35 percent is available for propulsion. Figure 6-4 shows a General Electric J-79 turbojet engine. This afterburning turbojet powers the McDonnell F-4 Phantom.

When additional turbine stages are added to extract more energy from the exhaust gas, the engine is classified as one of the following variations of the basic gas turbine engine: turboshaft, turboprop (propjet), or turbofan (bypass) engines.

The turboshaft (shaft turbine)

In a turboshaft engine, the turbine provides power in excess of that required to drive the compressor. The excess power is applied as driving torque available at an output shaft. The power to drive the output shaft can be extracted either

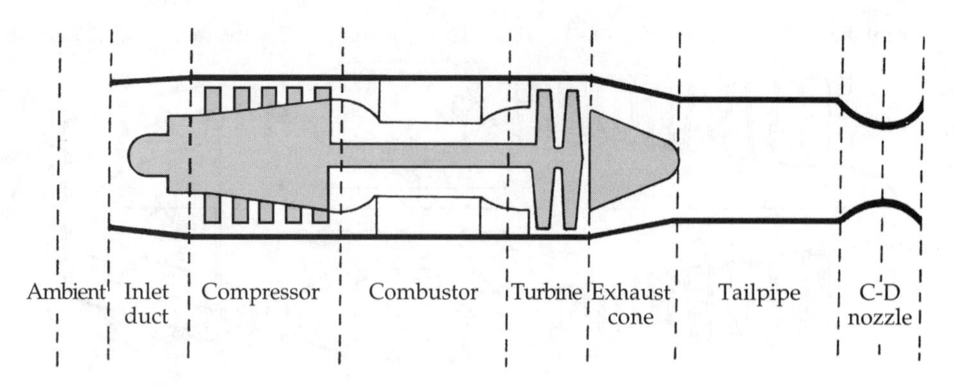

Fig. 6-3. The basic turbojet engine.

Fig. 6-4. The General Electric J-79 afterburning turbojet engine that powered the Mach 2.0 Lockheed F-104 and McDonnell F4 Phantom fighters of the 1950s and 1960s.

from the turbine that drives the compressor or from a separate, free-wheeling turbine. The turbine is usually connected, forward or aft, through a gearbox to the propeller or rotor blades (Fig. 6-5). Turboshaft engines are used primarily in helicopter aircraft.

The turboprop (propjet)

The turboshaft engine in which the output shaft drives a propeller is referred to as a turboprop (Fig. 6-6). The propeller produces approximately 90 percent of the thrust; approximately 10 percent of the thrust is developed by ejecting the gases through the exhaust nozzle. Many of today's short-haul and small transport aircraft are powered by turboprop engines.

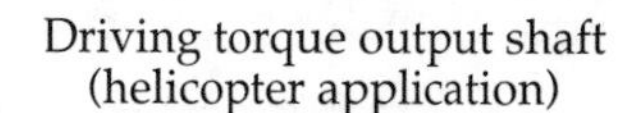

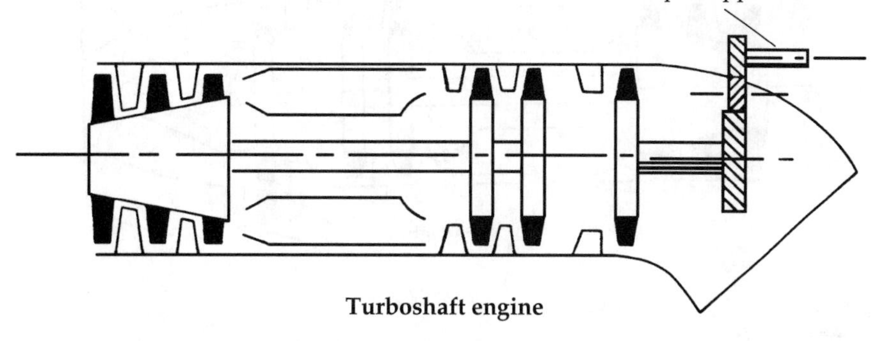

Fig. 6-5. The turboshaft gas turbine is used primarily in helicopter aircraft.

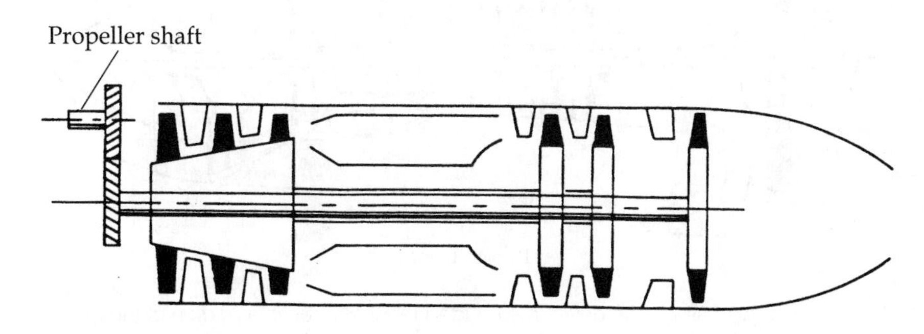

Turboprop engine

Fig. 6-6. A turboprop engine drives a propeller that produces approximately 90 percent of the thrust. Only about 10 percent of the thrust is developed by the turbine engine's exhaust.

The turbofan (bypass)

A gas turbine engine that drives a shrouded fan and causes a portion of the fan airflow to bypass the gas turbine is known as a turbofan engine. The bypass fan might be either of the front or aft configuration (Figs. 6-7 and 6-8). This engine powers most of today's commercial and military airplanes.

BASIC ENGINE COMPONENTS AND THEIR OPERATION

Compressor functions

Air enters the gas turbine engine at the engine inlet and flows from there into the engine compressor (Fig. 6-9). The function of the compressor is to raise the pressure and reduce the volume of the air as it pumps the air through the engine.

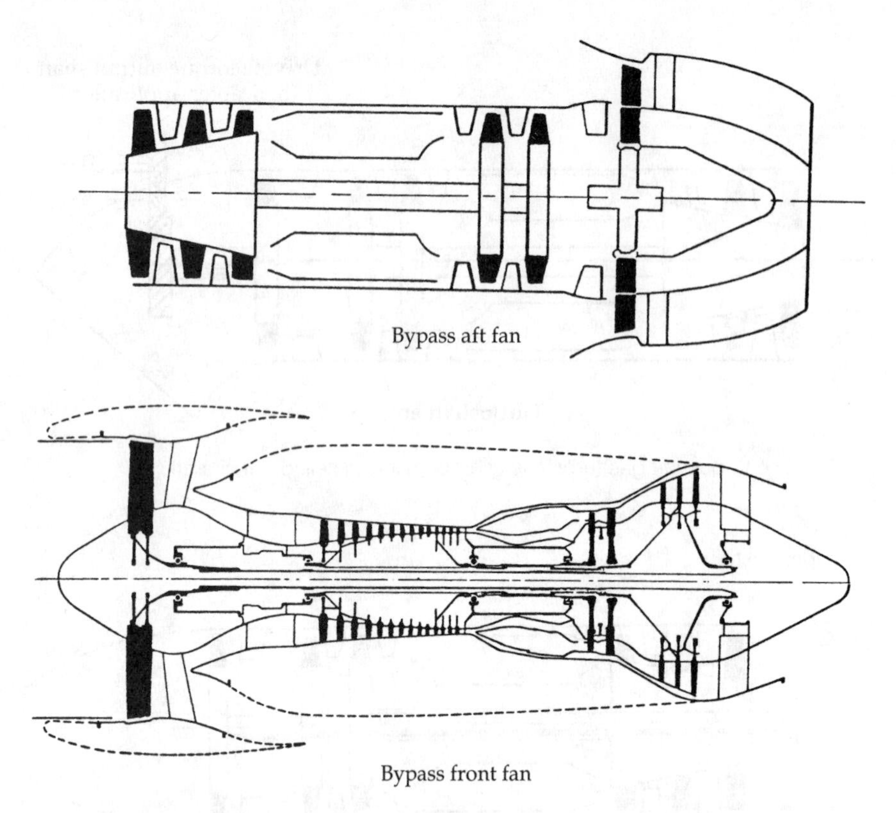

Fig. 6-7. The bypass engine drives a shrouded fan that causes a portion of the fan airflow to bypass the gas turbine.

Fig. 6-8. The 37,000-pound-thrust Pratt & Whitney PW2037 high-bypass turbofan powers the Boeing 757 airliner.

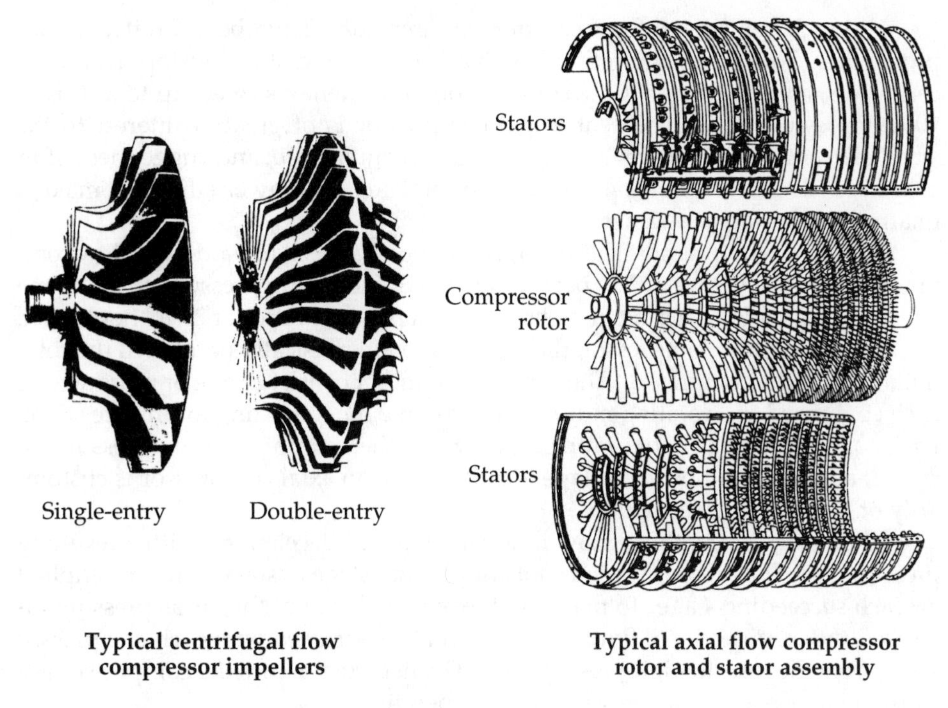

Fig. 6-9. Typical centrifugal and axial flow compressors.

One of the key factors affecting compressor and engine efficiency is *compressor pressure ratio*, which is the ratio of the air pressure at the compressor discharge to the compressor inlet air pressure (discharge pressure-to-inlet pressure ratio). The higher the pressure ratio, the more efficient the engine.

For a specific engine, the mass flow is determined primarily by the frontal area and the speed of the compressor. Mass flow (M) is one of the major factors in the generation of thrust (F = Ma). An ideal compressor would have a small frontal area for low air resistance, a high-pressure ratio for high-cycle efficiency, a light weight, and a resistance to stall.

Compressor types

The two basic compressor types are the centrifugal-flow and the axial-flow compressor (Fig. 6-9).

Centrifugal-flow compressor. Centrifugal-flow compressors, with either single- or double-entry impeller, were widely used in the early engine designs. Centrifugal compressor airflow starts out as an axial flow near the compressor hub. The high rotational velocity of the compressor rotor accelerates the air radially, imparting high velocity or kinetic energy to the air. The air flows into a diffuser section, where the air velocity is reduced, converting the kinetic energy into pressure energy. From there the airflow might go to the combustor or to the inlet of another compressor stage.

Although centrifugal-flow compressors generally have been limited to approximately 5-to-1 pressure ratio, there has been considerable development in recent years, and centrifugal-flow compressors have demonstrated up to at least a 10-to-1 pressure ratio. The centrifugal compressor is of greatest interest to the small-size engines, where its low cost, ease of manufacture, and ruggedness offer attractive advantages; however, its large frontal area and lower efficiency make it unattractive for larger engines.

Axial-flow compressor. The general evolution has been toward axial-flow compressors because the trend has been toward ever-increasing pressure ratios. Some advanced-technology engines now have overall pressure ratios of 25-to-1 or higher.

In an axial-flow compressor, the airflow remains basically parallel to the rotational axis of the compressor. The compressor might consist of one or more stages, each comprised of a rotating multiblade rotor and a nonrotating multivane stator. Each vane and blade are of airfoil section. Since the amount of work (pressure increase) accomplished by each stage is quite small, an axial compressor is customarily of multistage construction.

Within each stage, the airflow is accelerated and decelerated with a resulting pressure rise. The pressure ratio is not added from stage to stage, but is multiplied by each succeeding stage. To maintain the axial velocity of the air as pressure increases, the cross-sectional flow area is gradually decreased with each compressor stage, from the low- to high-pressure end. The net effect across the compressor is a substantial increase in air pressure and temperature.

Although compressors have been designed that have a discharge velocity the same as the inlet velocity, usually the discharge velocity is lower than the inlet velocity, so that excessive diffusion is not required to reduce the velocity to the low level required for efficient combustion.

A compressor is rated according to its pressure ratio and mass airflow. Because the compressor is a *volume flow device*, the ratings are achieved only at a specified RPM under standard inlet conditions.

As with any airfoil device, effectiveness gradually increases up to an optimum point, followed by a rapid decay known as *stall* or *surge*. With high-performance engines, the compressor operation and design are somewhat like walking a tightrope: to achieve the highest possible pressure ratio and still avoid compressor stall.

As RPM is increased from zero to rated speed, pressure ratio and mass airflow increase proportionally until rated performance has been reached. Top performance is not necessarily at top speed, but at the aerodynamic design point.

COMPRESSOR STALL

Generally speaking, if the air within the compressor, for any reason, reaches an unstable condition, stall will result. In various degrees, stall is a characteristic shared by all types of gas turbine compressors under certain operating conditions. It is most noticeable in high-pressure-ratio axial-flow compressors.

Whenever the relationship among air pressure, velocity, and compressor rotational speed is altered, the effective angle of attack of the compressor blades changes. If the relationship becomes incompatible, the effective angle of attack of the compressor blades becomes excessive, causing the blades to stall in very much the manner as the wing of an airplane will stall when its angle of attack is too high. Airflow over stalling compressor blades tends to become very turbulent, destroying the smooth airflow back through the compressor.

Different types of compressor stall, ranging from mild to severe cases, can be encountered when operating gas turbine engines.

As compared to a fixed-geometry single-rotor axial compressor, a dual-rotor compressor (Fig. 6-10) provides greater operating flexibility over a wider speed range and at higher compression ratios. Because of inherent airflow regulating features, a dual-rotor compressor also has a greatly improved stall margin throughout the operating range of the engine.

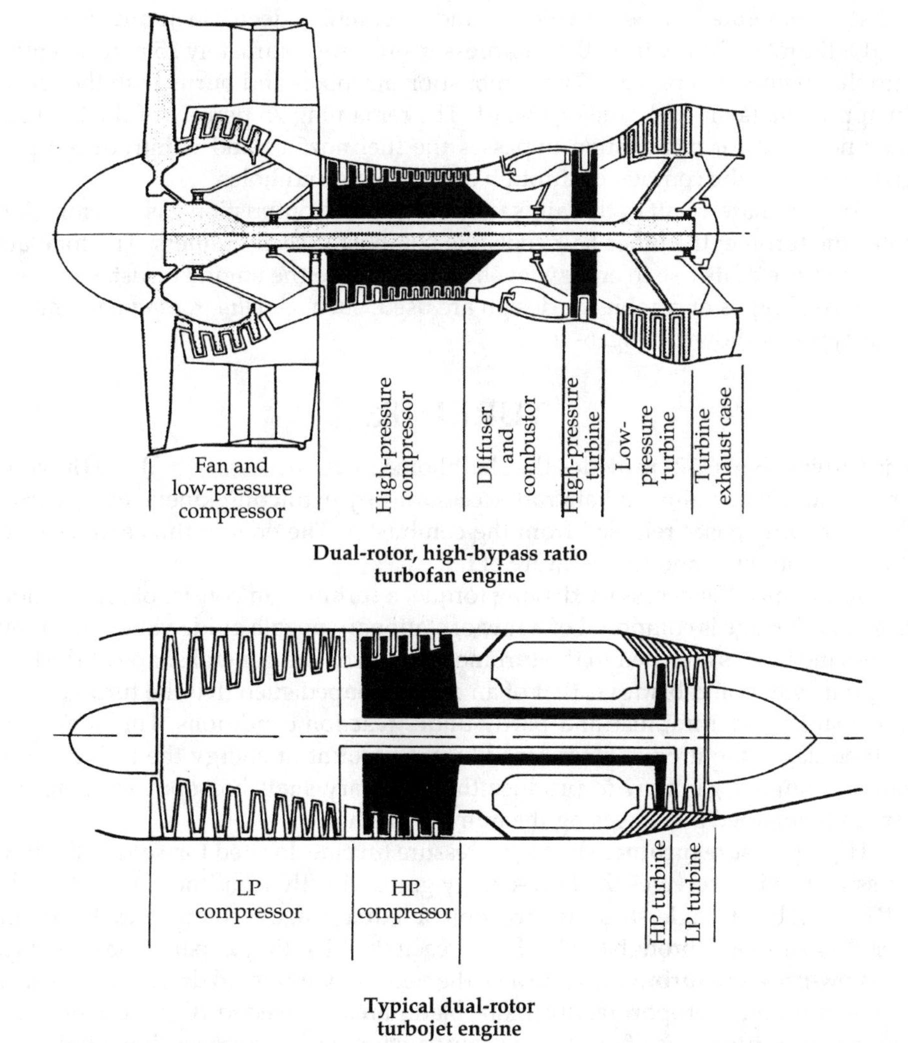

Fig. 6-10. Dual-rotor compressor types.

A dual-rotor compressor consists of two independent rotor systems, each driven by a separate turbine through coaxial shafts and each free to rotate at its own optimum speed. The rear rotor, the *high-pressure compressor* (HPC), is driven by the *high-pressure turbine* (HPT) and its speed is governed by the *main fuel control* (MFC) of the engine. The front rotor, or *low-pressure compressor* (LPC), is driven by the *low-pressure turbine* (LPT).

COMBUSTORS

From the compressor, the air flows to the combustor, into which fuel is injected in spray form, mixed with the airstream, and ignited. The resultant combustion causes an increase in gas temperature proportionate to the amount of fuel being injected, a moderate increase in velocity, and a negligible decrease in pressure.

Of the total airflow from the compressor, only approximately 25 percent enters into the combustion process. The combustion air mixes and burns with the fuel at an approximate air-fuel ratio of 15-to-1. The remaining 75 percent of the total airflow not used for combustion bypasses the fuel nozzles and is introduced progressively into the combustor through cooling slots and holes.

Approximately half of this air is used to cool the combustion gases before they enter the turbine; the other half cools the combustor section liners. The total airflow, because of its rise in energy level, contributes to the engine thrust.

Several types of combustor design are used, such as *cellular*, *cannular*, and *annular* types as shown in Fig. 6-11.

TURBINES

The turbine (Fig. 6-12) provides the shaft horsepower necessary to drive the compressor and the engine and aircraft accessories by extracting kinetic energy from the expanding gases released from the combustor. The energy thus extracted reduces the pressure and temperature of the gases.

To produce the necessary driving torque, a turbine can consist of one or more *stages*. Each stage is composed of a row of stationary nozzle guide vanes and a row of moving blades attached to the turbine rotor disc. The general shape of the nozzle guide vanes and blades is that of an airfoil, shaped such that the turbine functions partly under impulse and partly under reaction conditions. The number of turbine stages required is determined by the amount of energy the turbine must extract from the gas flow to produce the necessary shaft horsepower. Increased torque is obtained by increasing the number of stages.

High-pressure turbine. The high-pressure turbine, located forward of the low-pressure turbine, receives the high-energy gases directly from the combustor. The HPT might be of single-stage construction or a dual-stage turbine of smaller diameter. As gases pass through the HPT and reach the LPT, they expand considerably.

Low-pressure turbine. To produce the necessary power to drive the low-pressure compressor, proportionally more blade area is needed. Consequently, the low-pressure turbine is of multistage construction with a cross-sectional flow area increasing with each stage.

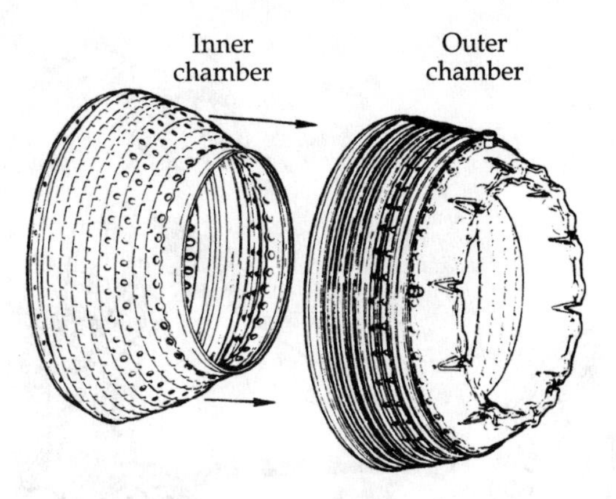

Annular combustion chamber

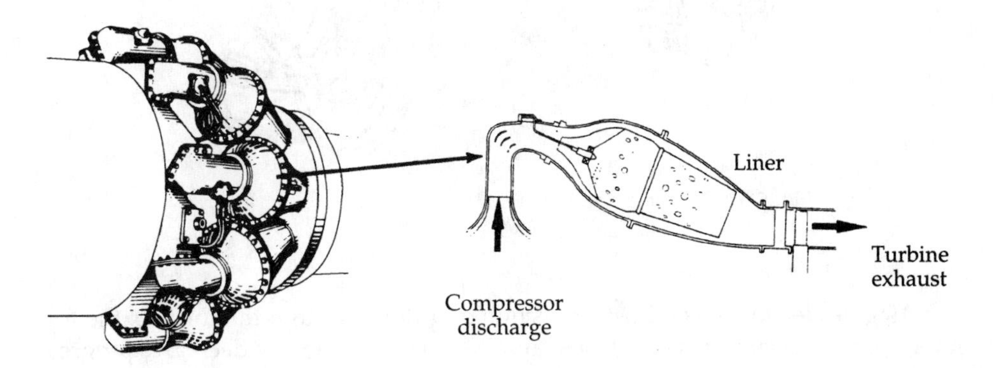

Typical multiple-can combustion chamber assembly

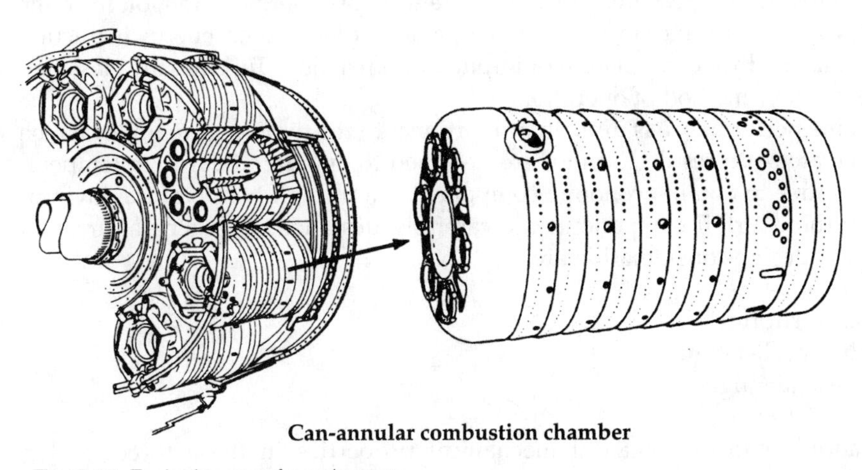

Can-annular combustion chamber

Fig. 6-11. Typical types of combustors.

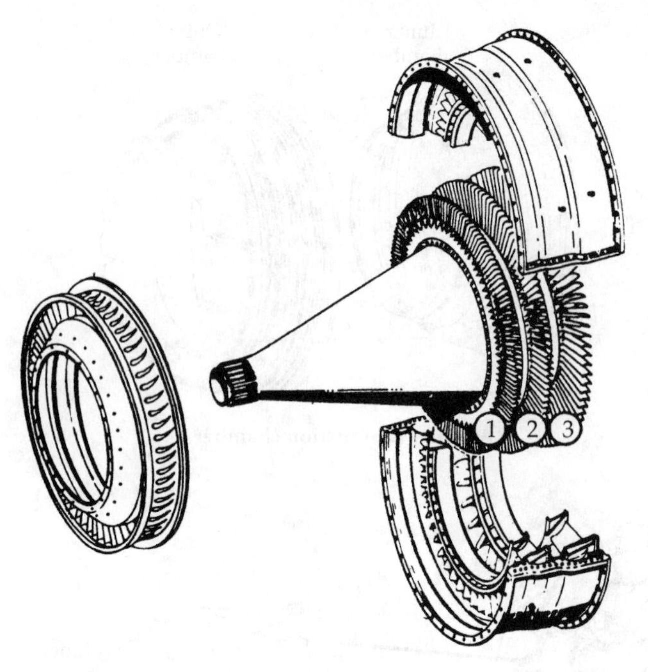

Fig. 6-12. A three-stage turbine.

As gases leave the combustion section and flow through the turbine, the following events occur: Pressure drops rapidly while temperature decreases progressively. Velocity initially increases significantly, followed by progressive decreases and increases across each turbine stage. The decreases in velocity, temperature, and pressure occur in the turbine as the cross-sectional area increases. The reduction in temperature, pressure, and velocity reflects the energy extraction from the hot gases by each turbine stage. In actual operation of a turbine, energy extraction is accomplished by a combination of impulse and reaction. Turbines are classified by their primary method of operation.

Turbine performance is directly proportional to the temperature under which it can operate. As a result, turbines are designed to operate under high temperatures. In order to obtain economic component life in high-temperature environments, turbine airfoil components are carefully designed. The general areas of mechanical design consideration are:

- Stress rupture
- High cycle fatigue
- Thermal fatigue

In addition to the material mechanical properties in these three modes, the oxidation, erosion, and corrosion characteristics of the airfoil material are considered.

Turbine airfoil cooling

The turbine nozzle guide vanes and rotor blades are exposed to the most severe temperatures. To permit the turbines to operate at gas-stream temperatures that are higher than the materials normally can tolerate, some form of cooling is required. Three forms of air-cooling of the vanes and blades are used, either singly or in combination, depending on the level of gas temperatures to which they are subjected:

- Convection
- Impingement
- Film cooling

Exhaust system

The extreme aft portion of a gas turbine aircraft engine contains the exhaust system, consisting basically of the exhaust duct (tailpipe), exhaust cone, and nozzle. The basic turbojet exhaust system is shown in Fig. 6-13.

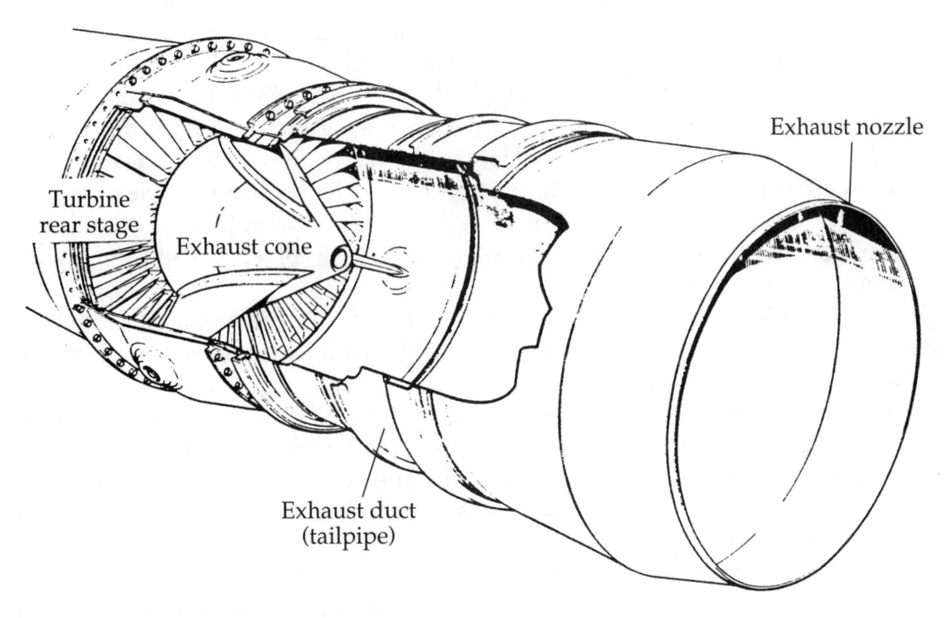

Fig. 6-13. Basic turbojet exhaust system.

The function of the exhaust system is to direct the turbine discharge gases aft to the atmosphere at a velocity and density necessary to produce the required thrust. The velocity of turbine discharge gases is relatively low, but as the gases pass through the exhaust duct, velocity is increased before the gases are discharged from the exhaust nozzle. This process converts energy available in the flowing gases, in the form of heat and pressure, to production of thrust.

The area of the exhaust nozzle (and fan discharge nozzle on the turbofan engines) is of extreme importance because this is the determining factor of the efficiency with which thrust is produced.

The exit velocity of gases at the exhaust nozzle is subsonic only at low-thrust conditions. At most operating thrust levels, the exit velocity of gases at the nozzle reaches Mach 1 in relation to the gas temperature. When the gas exit velocity has reached the speed of sound, the nozzle has become *choked* and no further velocity increase is possible unless the gas temperature is increased.

As upstream total pressure increases above the value at which the nozzle becomes choked, static pressure at the nozzle exit will exceed atmospheric pressure. This pressure differential will provide thrust in addition to that created by the velocity change of the gases. This additional thrust is referred to as *pressure thrust*.

Variable-area nozzle. Some engines are equipped with variable-area nozzles capable either of modulating in step function, or of being infinitely variable within the design operating range of the engine. Variable-area nozzle mechanisms can be controlled either by crew command or automatically in accordance with engine performance requirements.

For engines with narrow operating ranges, little performance is gained by use of variable-area nozzles because of their weight and complexity. In high-performance engines with broad operating ranges, noise, thrust, and fuel economy benefits can be achieved by use of variable-area nozzles.

Variable-area exhaust nozzles have, in the past, been used mostly on engines having some sort of thrust augmentation, such as afterburner or preturbine injection. The additional fuel used by the augmentation system would increase exhaust gas temperatures beyond allowable limits; however, by using a variable-area nozzle system, the exhaust area can be increased and thereby maintain the exhaust gas temperatures within allowable limits. Afterburning is a subject in itself and is subsequently discussed in this chapter.

The variable-area nozzle is opened during low-altitude takeoffs. At the appropriate altitude after takeoff, depending on the airframe, the nozzle is closed to achieve the necessary cruise thrust.

Convergent nozzle. The convergent nozzle (Fig. 6-14) is commonly used for subsonic flight. From the turbine discharge to the nozzle exit, the flow area decreases. The gas velocity might approach the speed of sound relative to its temperature, but generally remains subsonic because a loss in efficiency would occur if the velocity became sonic upstream of the nozzle exit.

Convergent-divergent nozzle. For supersonic flight, a convergent-divergent (C-D) exhaust nozzle is required (Fig. 6-15). The entire nozzle assembly consists of a conventional convergent nozzle extended by attaching a divergent nozzle, allowing controlled expansion of the gases, which causes the velocity to become supersonic.

At high supersonic flight Mach numbers, a well-designed C-D nozzle might have a pressure ratio in excess of 2-to-1, causing the exhaust velocity at the divergent nozzle exit to become greater than Mach 2.0. At low supersonic Mach numbers, the amount of thrust gained by use of a C-D nozzle might not be sufficient to compensate for the extra weight of the equipment.

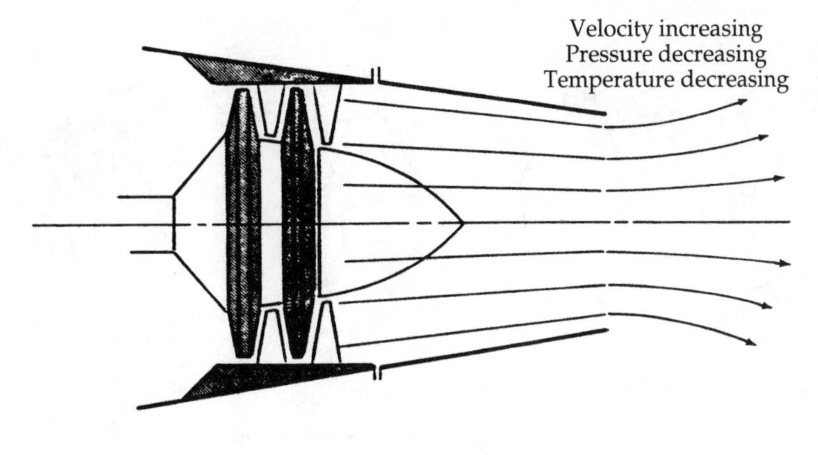

Fig. 6-14. A convergent exhaust nozzle is commonly used for subsonic flight.

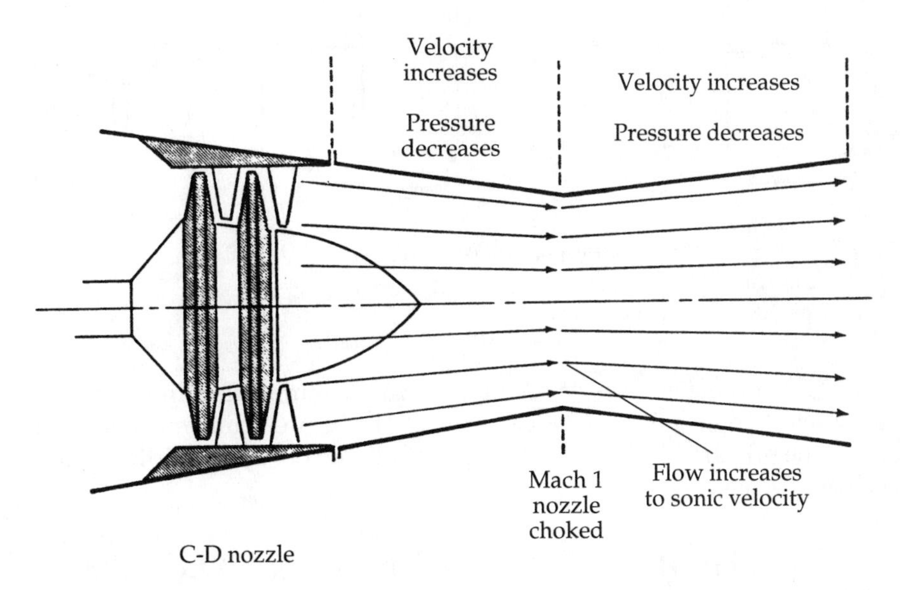

Fig. 6-15. For supersonic flight, a convergent-divergent exhaust nozzle is required.

Fixed C-D nozzles have a narrow optimum operating range and lose efficiency very rapidly if operated outside the design envelope. To compensate for this characteristic, C-D nozzles are frequently designed as variable area nozzles. They are usually automatically controlled per a predetermined schedule by the engine control as required by the flight environment. Figure 6-16 shows typical variable area nozzles.

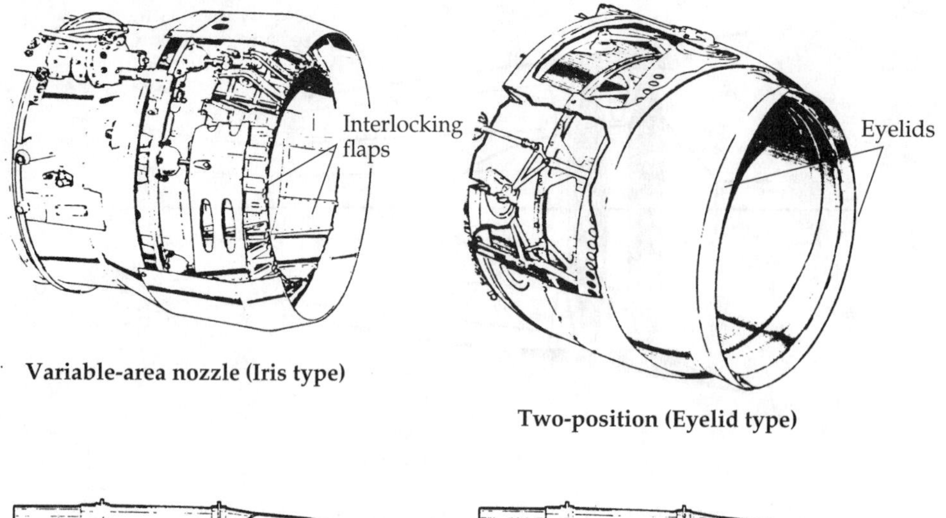

Variable-area nozzle (Iris type)

Two-position (Eyelid type)

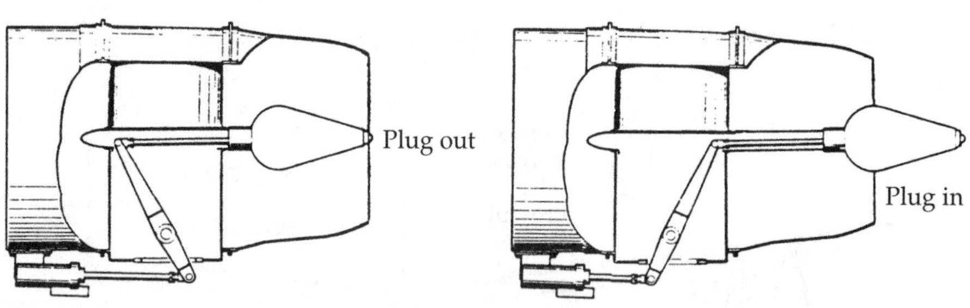

Two-position nozzle (Plug type)

Fig. 6-16. Typical variable-area nozzles.

Basic gas turbine engines have only one gas stream to eject through the exhaust system to atmosphere, whereas a bypass turbofan engine must accommodate both the primary turbine discharge gases and the cool bypass airflow through the exhaust system.

THE HIGH-BYPASS TURBOFAN ENGINE

The performance of a jet engine is based not only on its thrust-producing capability, but also on its efficiency in converting the heat energy of the fuel to kinetic energy. The energy conversion capabilities of an engine are limited by the mechanical and thermal stresses the engine components are designed to tolerate.

Propulsion efficiency

Propulsion efficiency, the process of converting kinetic energy to thrust or driving torque, depends on the amount of energy wasted by the propulsion method used. Kinetic energy wasted is proportional to the mass airflow and its relative velocity to the surrounding atmosphere. At slow aircraft speeds, a turbojet engine with its high-velocity jet exhaust is less efficient (a large percentage of the energy is being wasted in relation to the amount of fuel being converted) than a turboprop engine, which displaces a large air mass at relatively low velocity.

As aircraft speed increases, the high velocity of the turbojet exhaust becomes relatively less than the surrounding atmosphere; therefore, less energy is wasted and the turbojet becomes more efficient.

By the same token, at higher aircraft speeds, a turboprop engine quickly loses its propulsion efficiency and begins to create drag. The drag is caused by turbulence around the propeller, and drag increases as the propeller blades' tip speed approaches the speed of sound.

By using the ducted-fan and bypass principle, some of the better propulsion characteristics of both the turbojet and turboprop engine have been combined in the development of the improved propulsive efficiency engine referred to as the *high-bypass turbofan* (Fig. 6-17).

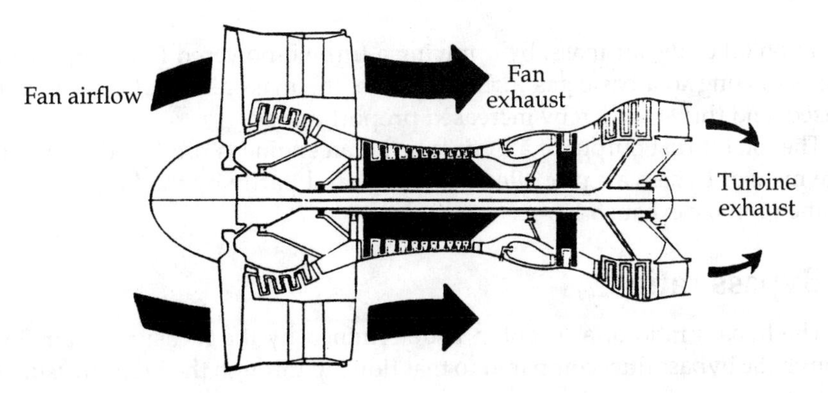

Fig. 6-17. The high-bypass-ratio turbofan is the standard engine type used in subsonic transport airplanes because it is more fuel-efficient than a turbojet engine.

Thrust calculations of the turbojet and the bypass turbofan engine for high-speed flight show that at speeds below Mach 1, the bypass turbofan is superior to the turbojet. For the commercial aircraft of today, operating at subsonic speeds, the high-bypass turbofan has definite advantages. As flight speeds increase and exceed Mach 1, however, the turbojet engine gains in thrust and becomes the more advantageous means of propulsion (Fig. 6-18).

Bypass and fan principle

By referring to the basic thrust formula F = Ma, based on Newton's second law of motion, where F = Force (thrust), M = Mass, and a = Acceleration, it is established that the thrust of an engine can be increased either by increasing the mass airflow (M) that is being accelerated through the engine, or by increasing the ac-

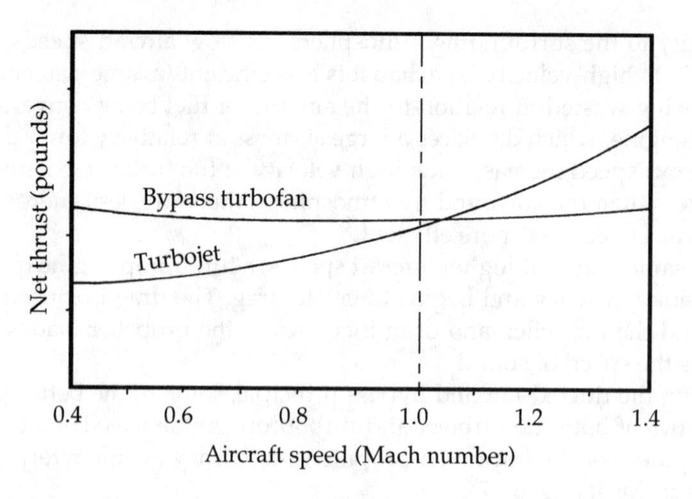

Fig. 6-18. Net thrust comparison of the bypass turbofan and turbojet engine, relative to flight speed.

celeration (a) of the air mass. By applying a turbine-powered fan component and bypass ducting to a basic gas turbine engine, the mass airflow (M) is greatly increased, and thrust is thereby increased proportionally.

The total thrust output of a bypass turbofan engine is produced partly by the large mass of bypass air propelled by the fan and partly by the high-velocity gas turbine primary exhaust.

Bypass ratio

The bypass ratio of a turbofan is determined by the measure of air flowing through the bypass duct compared to that flowing through the high-pressure compressor.

The less thrust that an engine must develop per pound of air, the better its *specific fuel consumption* (SFC) becomes. Consequently, the propulsive efficiency of a turbofan engine increases directly in proportion to its bypass ratio.

As indicated by Fig. 6-18 and previously discussed, the turbofan and turbojet have advantages based on aircraft Mach number. For commercial aircraft operating below Mach 1, a high bypass ratio provides more efficient operations. Many military aircraft are powered by pure turbojet engines; however, some later designs utilize a low-bypass engine as a compromise with efficiency (Fig. 6-19).

A typical high-bypass ratio turbofan for commercial aircraft has a bypass ratio of 5-to-1, which means a bypass airflow that is 5 times more than the primary gas turbine exhaust flow. The typical military low-bypass turbofan has a bypass ratio of 0.7-to-1.0, which is a bypass airflow equal to 0.7 of the primary gas turbine exhaust flow.

The propfan

A recent development that employs some features of both the turboprop and the turbofan is the propfan. It is also known as the *ultrahigh-bypass* engine, or *un-*

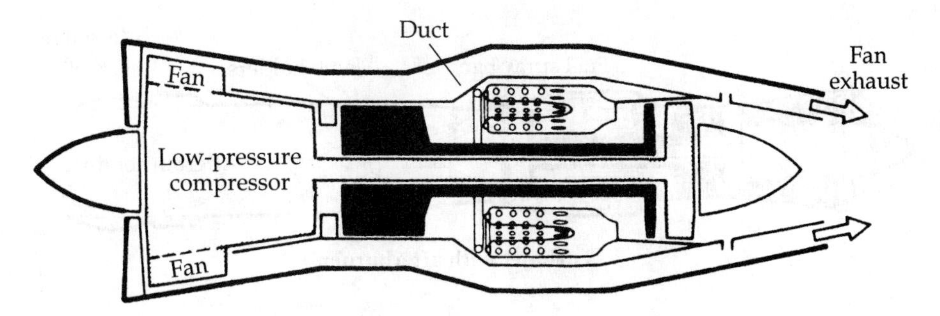

Fig. 6-19. A dual-rotor axial-flow low-bypass turbofan with long ducts and separate exhaust.

ducted fan. This concept utilizes a multibladed propeller, or fan, with swept blades of relatively short radius. Most designs employ two counterrotating fans. The propfan is capable of operating at cruise speeds equivalent to those of the turbofan, while retaining much of the efficiency of the turboprop at low speeds.

Although various propfan or ultrahigh-bypass engines have been developed and flight-tested, airline and military customers have shown little interest. No production aircraft employ the propfan engine. The propfan is further discussed in chapter 11.

Jet engines with afterburners

When a particular type of aircraft, such as a military fighter, needs extra bursts of speed during takeoff and climb, or for an intercept mission, each of its powerplants is usually provided with an afterburner. An engine equipped with an afterburner can develop 50 percent or more additional thrust when the afterburner is operating. Turbojet and low bypass-ratio turbofan engines can be equipped with an afterburner, no matter what type of compressor they use. The afterburner, when added to a turbofan engine, is sometimes called an *augmentor*.

Essentially, an afterburner (Fig. 6-20) is a pipe attached to the rear of an engine instead of a tail pipe and jet nozzle. Afterburning is possible because only about 25 percent of the air (oxygen) entering the basic engine at the compressor is used to support combustion in the basic engine. The remaining air through the engine is used only for cooling.

In afterburning engines, fuel is injected through a fuel nozzle arrangement, called *spray bars*, into the forward section of the afterburner and is ignited. Combustion takes place because 75 percent of the air (oxygen) entering the afterburner that is no longer needed for cooling is still available for burning with fuel. As the excess air and the fuel burn together in the afterburner duct, the engine exhaust gases are again expanded by heat and accelerated. Their increased velocity provides the additional thrust.

The afterburner is provided with flame holders downstream of the spray bars to prevent the flame from being blown out of the afterburner. The afterburner has an adjustable nozzle that opens automatically whenever the afterburner is on. When the afterburner is not operating, the nozzle remains in the closed (minimum opening) position on some models and continues to vary on others.

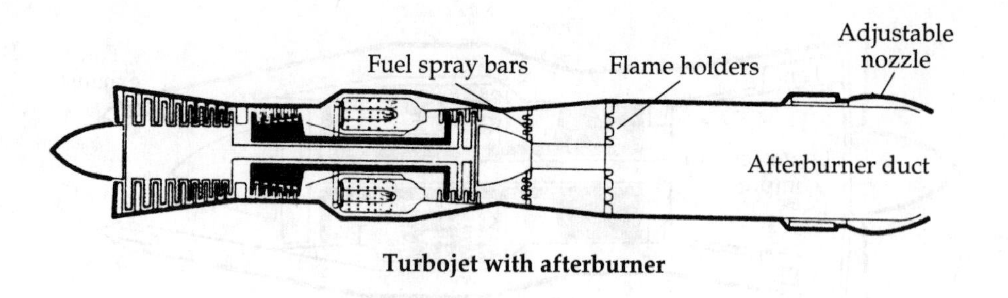

Turbojet with afterburner

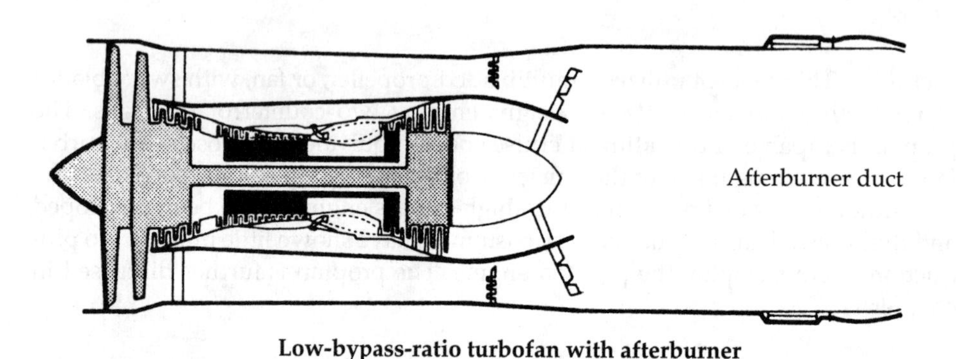

Low-bypass-ratio turbofan with afterburner

Fig. 6-20. An afterburner or augmenter can provide 50 percent or more additional thrust for military airplanes.

Some types of turbofan engines also use afterburning to increase thrust. In general, these engines are divided into two groups:

- Engines that use a conventional afterburning process for the mixed primary (hot) and secondary (fan) exhaust gases and airstream (Fig. 6-20).
- Engines that burn the extra fuel only in the fan air exhaust stream, which is called a *duct heater*.

Although afterburning more than doubles the engine fuel consumption, the use of an afterburner is profitable in those aircraft that might urgently require increased speed for short periods of time or extra power for short takeoffs with heavy loads. Figure 6-21 shows a cross-sectional view of a low-bypass turbofan that powers the F-16 fighter.

AIRCRAFT ENGINES FOR SHORT TAKEOFF AND VERTICAL TAKEOFF AND LANDING

Some mention should be made of engine configurations that allow fixed-wing aircraft to take off from short runways, make tighter maneuvers while in flight, or to take off and land vertically. The engines are gas turbines and operate according to

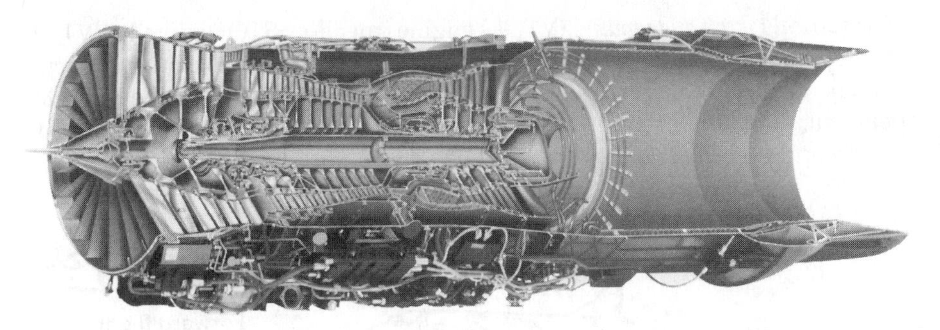

Fig. 6-21. The Pratt & Whitney F-100-PW-229 low-bypass, afterburning turbofan engine produces 29,100 pounds of thrust (afterburning) and 17,800 pounds of thrust (nonafterburning). This engine powers the General Dynamics F-16 fighter; General Dynamics' Fort Worth plant and F-16 production was taken over by Lockheed in 1992.

the fundamentals already discussed, but their placement in the aircraft, their exhaust duct and nozzle designs, and at times their configurations, differ from the conventional jet engine.

Short takeoff and landing (STOL) capability is partly dependent on the design of an airplane's wing, which can be optimized for high lift at slower runway speeds. After the wing and fuselage design, the engine is the deciding factor. Engines can provide STOL ability with a variable-angle exhaust nozzle. The exhaust nozzle is flexible and can deflect the exhaust gases to help "roll" the nose of the aircraft upward and then push the plane upward at a steep angle of climb.

The U.S. Air Force's new C-17A airlifter uses an alternative method. The exhaust gases from the turbofans mounted on the wings are deflected through special *slats*, which are essentially small wings that extend from the trailing edges of the wings. The slats turn the exhaust over the back edge of the wings, which helps to accelerate the airflow over the wings, creating greater lift than would be possible without the additional airflow (Fig. 6-22). Note that some STOL devices can be used while the airplane is flying to allow sharp, rapid maneuvers.

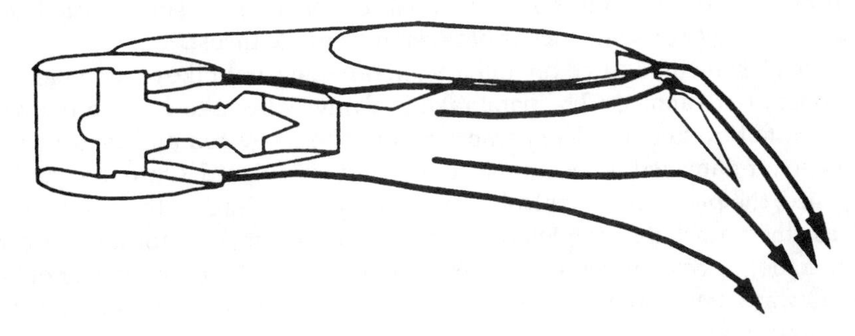

Fig. 6-22. Thrust that is diverted across wings and flaps enhances lift for short takeoff and landing. This concept is used on the U.S. Air Force McDonnell Douglas C-17 airlifter.

The *vertical takeoff and landing* (VTOL) engine can allow STOL as well as VTOL. Depending on the need, the exhaust of the jet nozzle or turbofans can be deflected at an angle to push the aircraft at a steep climb. Or the exhaust can be deflected perpendicularly to the ground to allow the airplane to take off straight upward like a helicopter (Fig. 6-23).

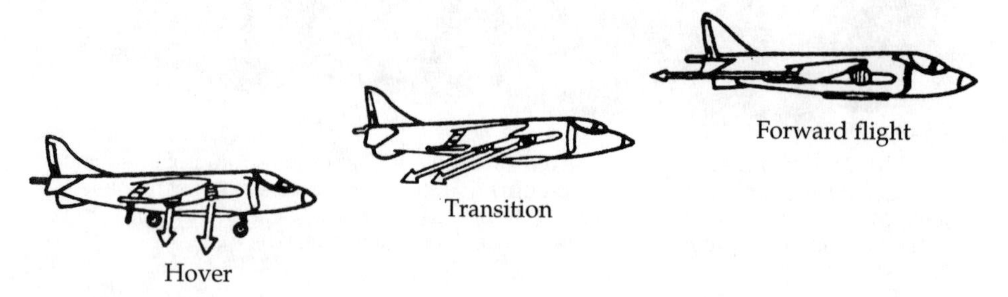

Fig. 6-23. Vertical to horizontal thrust is used for VSTOL of the AV-8, a version of the British Harrier. The AV-8 by McDonnell Douglas is powered by the Rolls-Royce Pegasus, vectored-thrust turbofan engine with 21,500 pounds of thrust.

The Rolls Royce Pegasus-powered McDonnell Douglas AV-8, based on the British Harrier, is a successful example of the V/STOL concept. For vertical takeoff and hovering, the vectored thrust of the engines must be greater than the aircraft weight; more than 25,000 pounds of thrust would be necessary to allow a 25,000-pound aircraft to take off vertically. By contrast, a conventional aircraft taking off from a conventional airfield requires about one-quarter to one-third of its weight in engine thrust.

THRUST REVERSERS

The problem of stopping an aircraft after landing increases as the gross weight, wing loading, and landing speed increases. Wheel brakes alone are not sufficient to stop large aircraft immediately after touchdown. The reversible-pitch propeller solved the problem for piston-engine and turboprop-powered airplanes. Aircraft with turbojets and turbofans, however, must rely on a device such as *parabrake* or *runway arrester* gear or some means of reversing engine thrust.

Although sometimes used on military aircraft, the parabrake or drag parachute has distinct disadvantages. The parabrake is always subject to either a premature opening or failure to open. The parabrake must be recovered and repacked after each use, or, if damaged or lost, must be repaired or replaced. Once the parabrake has opened, the pilot has no control over the amount of drag on the aircraft except to release the parachute completely. Arrester gear is primarily for aircraft carrier deck operation, although it is sometimes used at military bases as overshoot barriers on runways. It would hardly be suitable for commercial airline operation at municipal airports.

An engine thrust reverser, on the other hand, provides an effective braking force on the ground, and some reversers are suitable for use in flight. The thrust reverser directs the primary thrust (jet nozzle), secondary thrust (fan nozzle), or both at a forward angle to slow the aircraft. Some reversers can be used in the air to slow the aircraft's rate of descent, allowing it to land at steeper angles. More commonly, reversers are used after the wheels of the plane have touched the runway. The turbine exhaust gases and/or fan exit air is diverted in the reverse direction by an inverted cone, half sphere, turning vanes, or other devices introduced into the exhaust flow.

A thrust reverser must not have a significant effect on engine operation when the reverser is in use or when it is stowed. It must be able to withstand high temperatures if it is used in the turbine exhaust, and it must be mechanically strong, relatively light in weight, reliable, and fail-safe. When not in use, it should not add appreciably to the frontal area of the engine and must be streamlined into the engine nacelle. In order to satisfy the minimum braking requirements after a landing, a thrust reverser should be able to produce in reverse at least 40 percent of the full forward thrust.

Clamshell and *cascade* reversers are two common types that satisfy the general requirements of a reverser. The clamshell (Fig. 6-24) is used on nonafterburning engines. At an aircraft throttle position called *reverser idle*, the reverser automatically opens to form a clamshell shape to the rear of the engine exhaust nozzle. The reverser is a turning barrier in the path of the exhaust gases that reverses the forward thrust in controllable amounts from partial to full reverse to reduce the length of the landing roll. When the reverser is not in use, the clamshell doors retract and stow around the engine exhaust duct, sometimes forming the rear section of the engine nacelle.

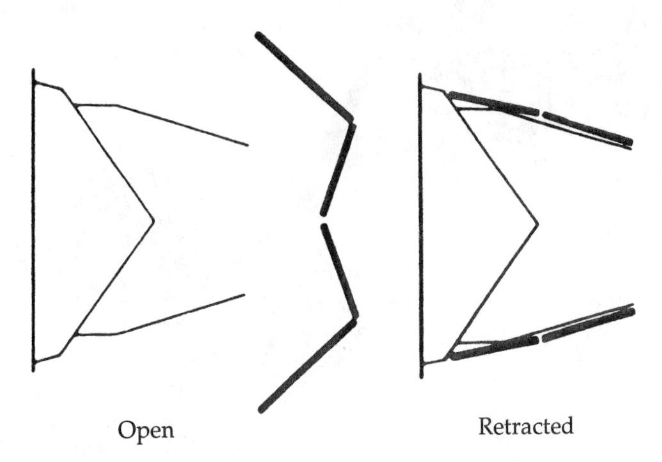

Fig. 6-24. Deployment of a clamshell thrust reverser reduces the landing roll by diverting the turbine exhaust.

The cascade reverser (Fig. 6-25) uses numerous turning vanes in the gas path to direct the gas flow outward and forward during operation. Fan air reversers are generally of this type. One type of fan reverser utilizes a sleeve to cover the fan cascade during forward-thrust operation. In the reverse-thrust mode, blocker doors seal off

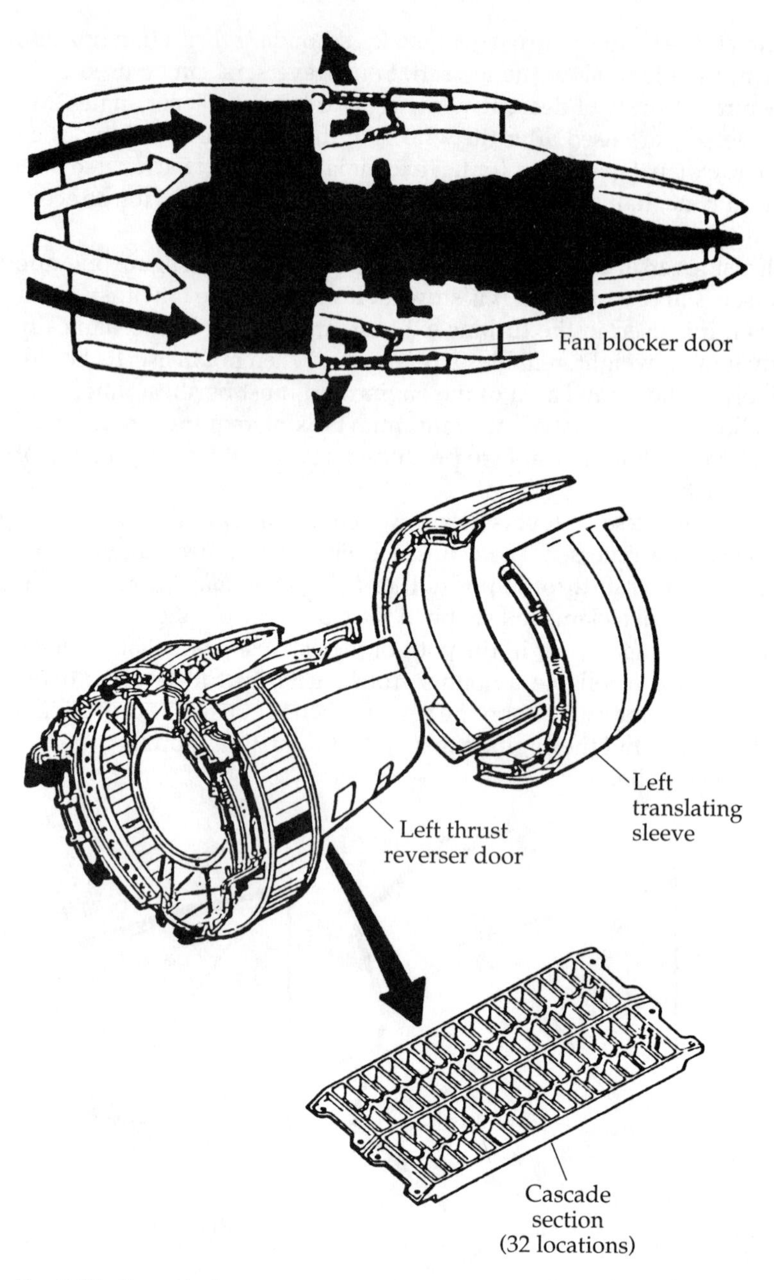

Fig. 6-25. Cascade thrust reverser as applied to fan air on a turbofan engine.

the fan exit, and a sleeve moves to expose the cascade vanes that direct the fan air through the cascades. The cascades are rows of airfoils that turn the air forward.

In some installations, the cascade turning vanes are used in conjunction with a clamshell to reverse the turbine exhaust gases (Fig. 6-26). The cascade and the

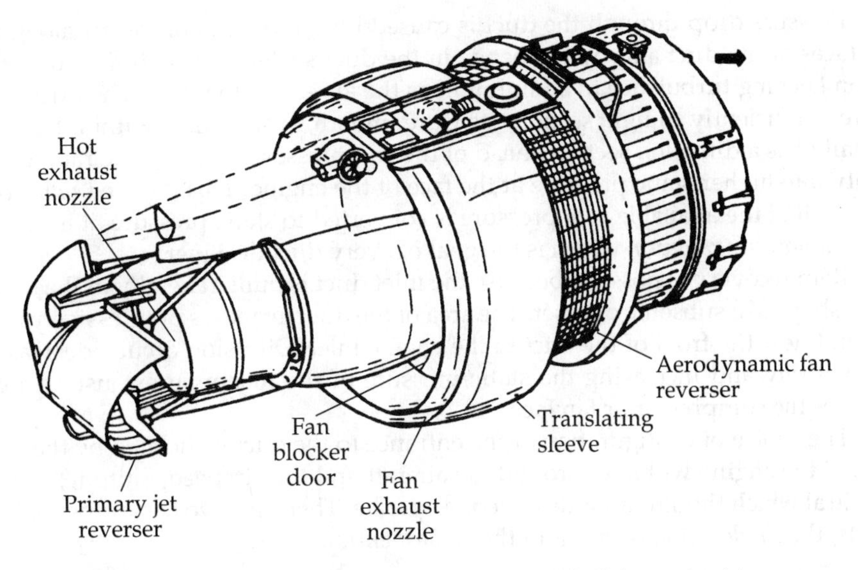

Fig. 6-26. A turbofan engine in reverse-thrust mode with both hot and cold reversers deployed.

clamshell in this configuration are located forward of the turbine exhaust nozzle. The clamshell covers the turbine cascade during forward-thrust operation. For reverse-thrust, the clamshell moves rearward, blocking the flow of exhaust gases and exposing the cascade vanes, causing the exhaust flow to be directed forward. In this case, the cascade acts as an exhaust nozzle as well as a reverser when the engine is operating in the reverse mode.

Thrust reversers of all types are most effective at high forward speed; high forward speed also prevents the ingestion of recirculating hot exhaust gases and the ingestion of foreign objects by the engine; therefore, reversers are normally used only during the first part of the landing roll. As a general rule, the use of thrust reversers below a forward speed of approximately 60 knots is not recommended.

AIR INLET DUCT

An engine air inlet duct is normally considered an airframe part, not a part of the engine; however, the duct is so important to engine performance that it must be taken into consideration in any discussion of the complete engine. The engine inlet and the inlet ducting direct the outside air to the face of the compressor. Any inefficiencies in the duct result in successively magnified losses through other components of the engine.

The inlet duct has two engine functions and one aircraft function. First, it must be able to recover as much of the total pressure of the free airstream as possible and deliver this pressure to the front of the engine with minimum loss. Second, the duct must deliver air to the compressor inlet under all flight conditions with as little turbulence and pressure variation as possible. As far as the aircraft is concerned, the inlet must create minimal drag.

Pressure drop through the duct is caused by the friction of the air along the surfaces of the duct and by the bends in the duct system. Smooth flow depends upon keeping turbulence to a minimum as the air enters the duct. The duct must have a sufficiently straight section to ensure smooth, even airflow within. The duct usually has a diffusion section ahead of the compressor to change the ram air velocity into higher static pressure at the face of the engine. This is called *ram recovery*. If all of the available ram pressure is converted to static pressure, it is known as *total pressure recovery*, which is the goal of every duct designer.

Ram recovery is possible because the inlet duct is built generally in the divergent shape of a subsonic diffuser. The area of the duct increases progressively from a point near the front of the duct to the engine inlet. Diffusion occurs, decreasing the velocity and increasing the static pressure of the incoming air just before it reaches the compressor or fan face.

The choice of configuration of the entrance to the duct is dictated by the location of the engine within or around the aircraft and the airspeed, altitude, and attitude at which the aircraft is designed to operate. There are two basic types of inlet ducts, the *single-entrance duct* and the *divided-entrance duct*.

Single-entrance duct

The single-entrance duct is the most common and the simplest type of duct. It can also be the most effective if the duct inlet is located directly ahead of the engine in such a position that it scoops undisturbed air from the front of the airplane. The single-entrance duct is standard for all multiengine, subsonic aircraft. Such ducts are relatively short (Fig. 6-27), which results in a minimum pressure loss. Nevertheless, short ducts have to be carefully designed to avoid air distortion at the face of the compressor when the aircraft is flying at slow speed and steep angles of attack.

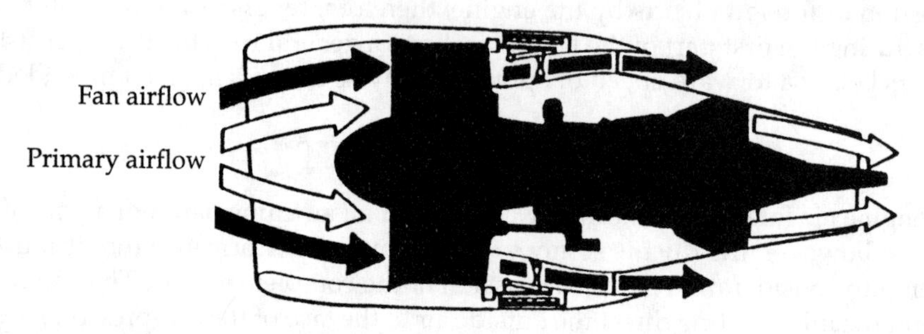

Fig. 6-27. A single entrance inlet duct for multiengine turbofan-powered aircraft.

In a single-engine fighter where the engine is mounted amidships, the duct must be long of necessity (Fig. 6-28). While some pressure drop is caused by friction through the length of the duct, the duct still provides a reasonably smooth flow of air to the engine. Fighter aircraft with two engines have a separate duct for

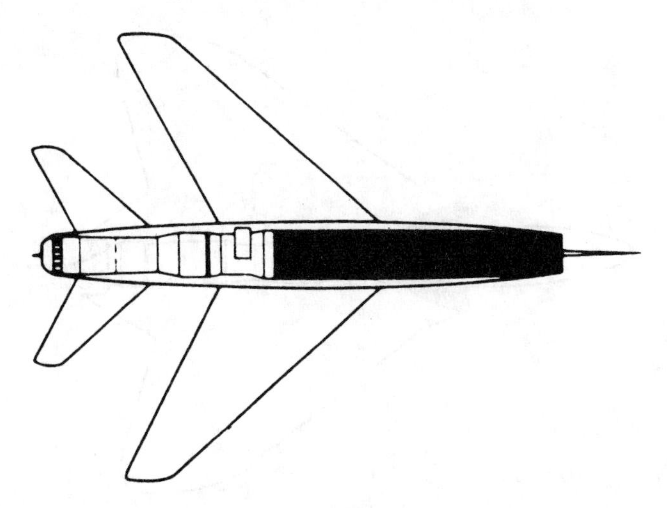

Fig. 6-28. Single-entrance duct in single-engine fighter aircraft.

each engine. These are usually scoops on each side of the fuselage or mounted in the root of the wings. The duct curvature problems explained in the following paragraphs for a divided entrance duct also apply to side-scoop and wing-root ducts and to the center engine duct on an aircraft powered by three engines.

Blow-in doors

The single-entrance ducts for some engines have blow-in doors. The perimeter of the inlet duct has a number of ports to deliver more air to the face of the engine during high thrust operation at slow airspeed, such as takeoff. The ports have hinged, spring-loaded doors. The doors open automatically at slow airspeed to permit more air to enter the inside of the duct. As airspeed increases, ram-air pressure closes the doors, returning the duct to its normal, flight operating configuration. Blow-in doors are an aircraft part, not an engine part.

Divided-entrance duct

The requirements of high-speed, single-engine aircraft, in which the pilot sits low in the fuselage and close to the nose, render it difficult to use the single-entrance duct. Some form of a divided duct, which takes air from either side of the fuselage, might be required. This divided duct can be either a wing-root inlet (Fig. 6-29) or a scoop at each side of the fuselage (Fig. 6-30).

Either type of duct presents more problems to the aircraft designer than a single-entrance duct because of boundary layer problems and the difficulty of obtaining sufficient air-scoop area without imposing prohibitive amounts of drag. Internally, the problem is the same as that encountered with the single-entrance duct; that is, to construct a duct of reasonable length, yet with as few bends as possible.

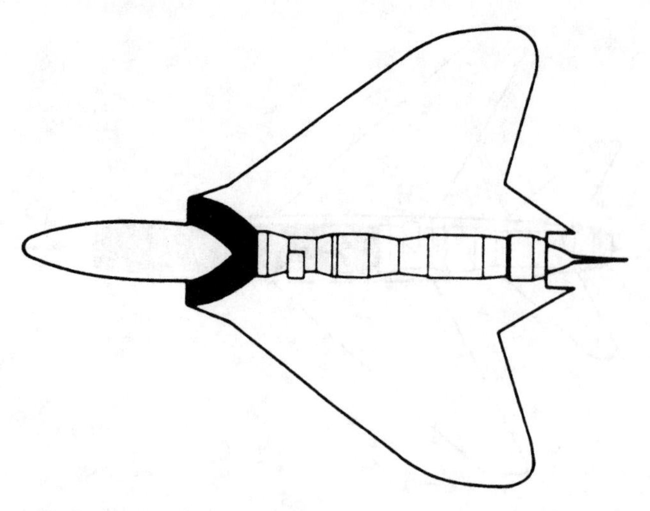

Fig. 6-29. Divided entrance duct in single-engine fighter aircraft.

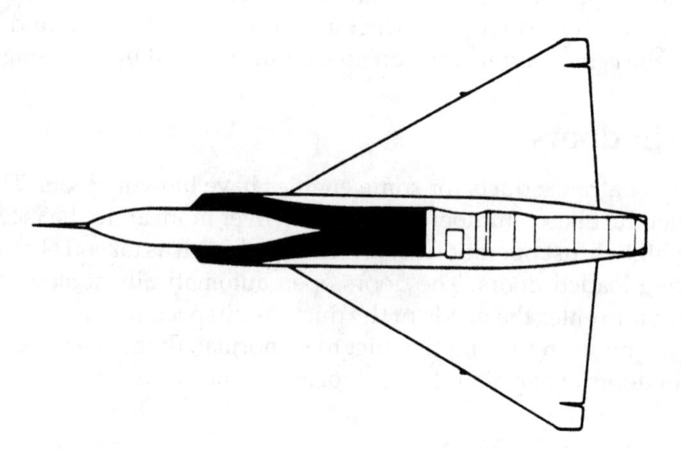

Fig. 6-30. Another version of the divided-entrance duct in single-engine fighter aircraft.

The wing-root inlet on aircraft on which the wing is located fairly far aft presents a design problem because, although short, the duct must have considerable curvature to deliver air properly to the compressor inlet. Scoops at the sides of the fuselage are often used. These side scoops are placed as far apart as possible to permit a gradual bend toward the compressor inlet, making the airflow characteristics approach those of a single-entrance duct. A series of small rods is sometimes placed in the side-scoop inlet to assist straightening the incoming airflow and prevent turbulence.

Variable-geometry duct for supersonic aircraft

In a supersonic aircraft flying at speeds above Mach 1, the inlet duct is designed to slow incoming air to subsonic velocity before it reaches the compressor. Sometimes this is accomplished, in part, by giving the inlet duct the shape of a combined supersonic and subsonic diffuser (Fig. 6-31). The forward, supersonic part of the duct slows the velocity of the incoming air to Mach 1, after which the subsonic part further decreases the velocity and increases the pressure of the air before it enters the engine. For very high-speed aircraft, the inside area or configuration of the duct is often changed by a mechanical device as the speed of the aircraft increases or decreases. A duct of this type is known as a *variable-geometry duct.*

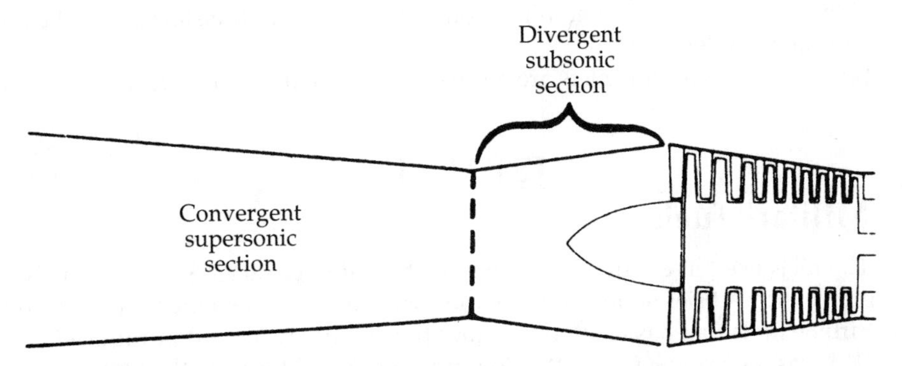

Fig. 6-31. Supersonic aircraft require a supersonic inlet duct to slow down compressor inlet air to subsonic speed.

Two different methods are commonly used to diffuse the inlet air and slow the inlet airflow at supersonic flight speed. One is to vary the area, or geometry, of the inlet duct either by using a movable spike within the duct itself, or by using some form of movable restriction, such as a ramp or wedge, inside the duct; variable-geometry systems might also include scoops to increase airflow and spill valves to prevent turbulence at the face of an engine. These techniques ensure that the demand for engine airflow is met precisely to optimize efficiency at all flight conditions. Variable-geometry ducts usually operate automatically as the Mach number varies.

The other method to diffuse and slow inlet airflow during supersonic flight is the use of a shock wave in the airstream. A shock wave is a thin region of discontinuity in a flow of air or gas, during which the speed, pressure, density, and temperature of the air or gas undergo a sudden change. The stronger the shock wave, the larger will be the change produced in the properties of the air or gas. A shock wave is intentionally set up in the supersonic flow of the air entering the duct by means of some restriction or small obstruction that automatically protrudes into the duct at high Mach numbers. The shock wave results in diffusion of the airflow that, in turn, slows down the velocity of the airflow.

In at least one aircraft installation, both the shock method and the variable-geometry method of causing diffusion are used in combination. The same device that changes the area of the duct also sets up a shock wave that further reduces the speed of the incoming air within the duct. The amount of change in duct area and the magnitude of the shock are varied automatically with the airspeed of the aircraft.

With a variable-geometry inlet, the so-called inlet *buzz*, which sometimes occurs during flight at high Mach number, can often be prevented by changing the amount of inlet area variation that takes place when the variable-geometry inlet system is in operation. Buzz is an airflow instability that occurs when a shock wave is alternately swallowed and regurgitated by the inlet. At its worst, the condition can cause violent fluctuations in pressure through the inlet, which might result in damage to the inlet structure or possibly to the engine. A suitable variable-geometry duct will eliminate buzz by increasing the stability of the airflow within the inlet duct.

Supersonic engine air inlets are discussed in more detail in chapter 8.

JET FUELS

Military fuels

The fuels used in gas turbine engines by the military services have been given the prefix "JP." A brief review of these fuels will show the development of aircraft gas turbine fuels that has taken place since the engines were first introduced.

JP-1 was a kerosene fuel with a low freezing point. Because the low-freezing-point requirement made procurement difficult, production of this fuel was rather limited. JP-2 was somewhat more volatile than JP-1, but was never used extensively. The JP-2 specification existed late in World War II to relieve the potential shortage of JP-1. JP-1 and JP-2 fuels are no longer available.

JP-3 was developed to provide maximum availability and was more volatile than JP-2. Because of its volatility, great losses of fuel occurred in flight due to evaporation at high altitude and during high rates of climb.

The specification for JP-4 began in 1951, featuring a lower volatility than JP-3. JP-4 was at first a step backward toward JP-2 because the JP-4 specification was more restrictive as far as availability was concerned. Later, some of the properties of the fuel, other than volatility, were changed to increase the supply.

JP-5 is a heavy kerosene blended with gasoline to produce a fuel similar to JP-4. This procedure was developed for aircraft carriers in which limited storage space was available for large quantities of JP-4. The JP-5 fuel could be blended, as needed, with the supply of aviation gasoline carried aboard ship for aircraft piston engines. The JP-5 could be stored in any available tanks normally used for the ship's diesel engine fuel. In emergencies, the JP-5 could be used in the ship's own engines; however, as aircraft gas turbine development progressed, even JP-4 became too volatile for some missions, so engines were designed to use JP-5 directly. The JP-5 has a high flash point (140°F) and very low volatility.

JP-6 was developed by the U.S. Air Force for land-based, high-speed, supersonic aircraft. The fuel is slightly more volatile than JP-5 and has a low freezing point (–65°F) for flight in cold climates and high altitudes.

Commercial fuels

When it was assured in 1956 that gas turbine engines would be used in commercial aircraft, a definition was needed for a fuel suitable for commercial airline operation. The American Society for Testing Materials (A.S.T.M.) has published two specifications defining grades of fuel suitable for commercial gas turbine engine use. The first is known as the *A Specification* and is for a kerosene-type fuel, not unlike JP-5. The A.S.T.M. *B Specification* is for a JP-4 type of fuel. Each operator will use a slightly different purchase specification to suit the needs of a particular operation. Commercial gas turbine engine fuel will therefore embody a range of volatility characteristics similar to that encompassed by the JP-4 and JP-5 military fuels.

ENGINE INSTRUMENTATION AND CONTROLS

When a turbojet or turbofan engine is installed in an airframe, various instruments and controls, plus indicating and warning devices, become necessary for normal control and operation of the engine. These requirements will differ somewhat from one engine or aircraft type to another. The airframe manufacturer will determine and be responsible for many of the items required in the final installation configuration.

Figure 6-32 shows the basic instrumentation and control requirements for an installed commercial high-bypass turbofan, such as the General Electric CF6 front fan engine.

TYPICAL GAS TURBINE ENGINES

Figures 6-33, 6-34, and 6-35 show examples of typical gas turbine (jet) engines as installed in military and civil airplanes.

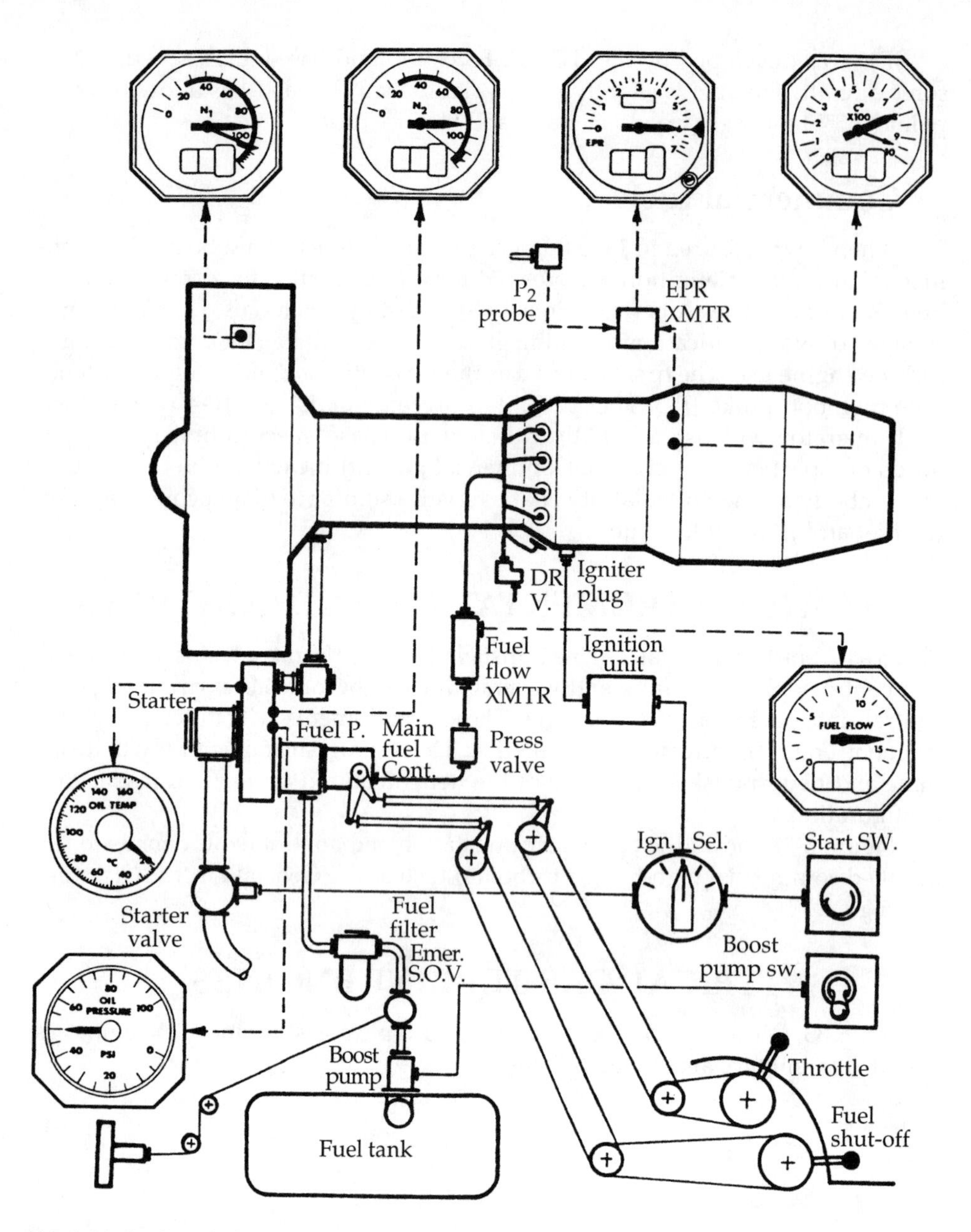

Fig. 6-32. Basic turbofan engine instrumentation and controls.

Fig. 6-33. Typical General Electric high-bypass (top) and low-bypass (bottom) turbofan engines. The CF-6 series high-bypass engines produce 50,000–60,000 pounds takeoff thrust with a bypass ratio of 5.15. These engines power Airbus 300 series airliners and various models of the Boeing 767 and 747 airliners. The lower photo shows the F-404 series, low-bypass, augmented (afterburning) turbofan engine of 16,000 pounds thrust. Two of these engines power the U.S. Navy McDonnell Douglas F/A-18 airplanes.

Fig. 6-34. Examples of Pratt & Whitney turbofan engines are the PW 4000 series high-bypass (top) and F100 low-bypass (bottom) augmented (afterburning) engines. The PW 4000 turbofan produces 56,700 pounds thrust with a bypass ratio of 4.8. Growth versions produce up to 84,600 pounds thrust with bypass ratios up to 6.4. These engines power versions of the Airbus 300 series airliners, Boeing 767, 747, and 777 as well as McDonnell Douglas MD-11 airliners. The F100, which produces 27,000–30,000 pounds of thrust, powers F-15 and F-16 U.S. Air Force fighter airplanes.

Fig. 6-35. The CFE738 turbofan engine produces 5,990 pounds of thrust. It was developed jointly by General Electric and Garrett. A typical installation is the Dassault Falcon 2000 business jet as well as other 0.85 Mach number business jets and regional airliners.

7

Stability and control

AN AIRCRAFT MUST HAVE satisfactory handling qualities in addition to adequate performance. It must have adequate stability to maintain a uniform flight condition and recover from the various disturbing influences. It is necessary to provide sufficient stability to minimize the workload of the pilot. Also, the aircraft must have proper response to the controls so that it can achieve the required performance. There are certain conditions of flight that provide the most critical requirements of stability and control. These conditions must be understood and respected to accomplish safe and efficient operation of the aircraft.

Simply defined, *stability* is the tendency, or lack of it, of an airplane to fly a prescribed flight condition. *Control* is the response of an aircraft to the directions of the pilot. For an aircraft to respond to the controls, its stability must be overcome. Stability and control are often at odds. The more stability an aircraft has, the less controllability it has, and vice versa.

Complex high-performance aircraft have stability problems that are very complicated and beyond the scope of this book. The discussion here is very basic using simplified assumptions.

To begin with, we need to define the basic axes of an airplane and the motions about them. To visualize the forces and moments on the aircraft, it is necessary to establish a set of mutually perpendicular reference axes originating at the center of gravity. Figure 7-1 illustrates a conventional axes system. The longitudinal X axis is located in a plane of symmetry and is given a positive direction pointing into the wind; a moment about the longitudinal X axis is a *rolling moment*. The vertical Z axis also is in a plane of symmetry and is established positive downward; a moment about the vertical Z axis is a *yawing moment*. The lateral Y axis is perpendicular to the plane of symmetry and is given a positive direction out the right side of the aircraft; a moment about the lateral Y axis is a *pitching moment*.

For an airplane to be in equilibrium for a particular flight condition, the sum of all the forces and moments on it must be zero. There is no pitching, yawing, or rolling, nor any change in velocity. For example, consider an airplane flying straight and level, as in Fig. 7-2(A) (also refer to Fig. 7-3). Then the lift equals the

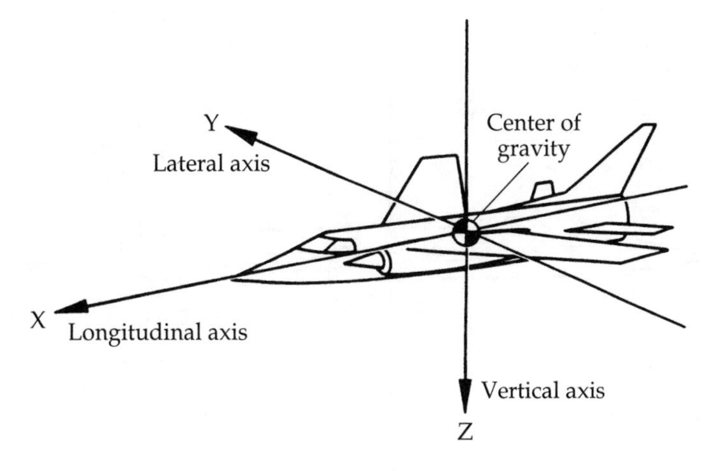

Fig. 7-1. Airplane reference axis system.

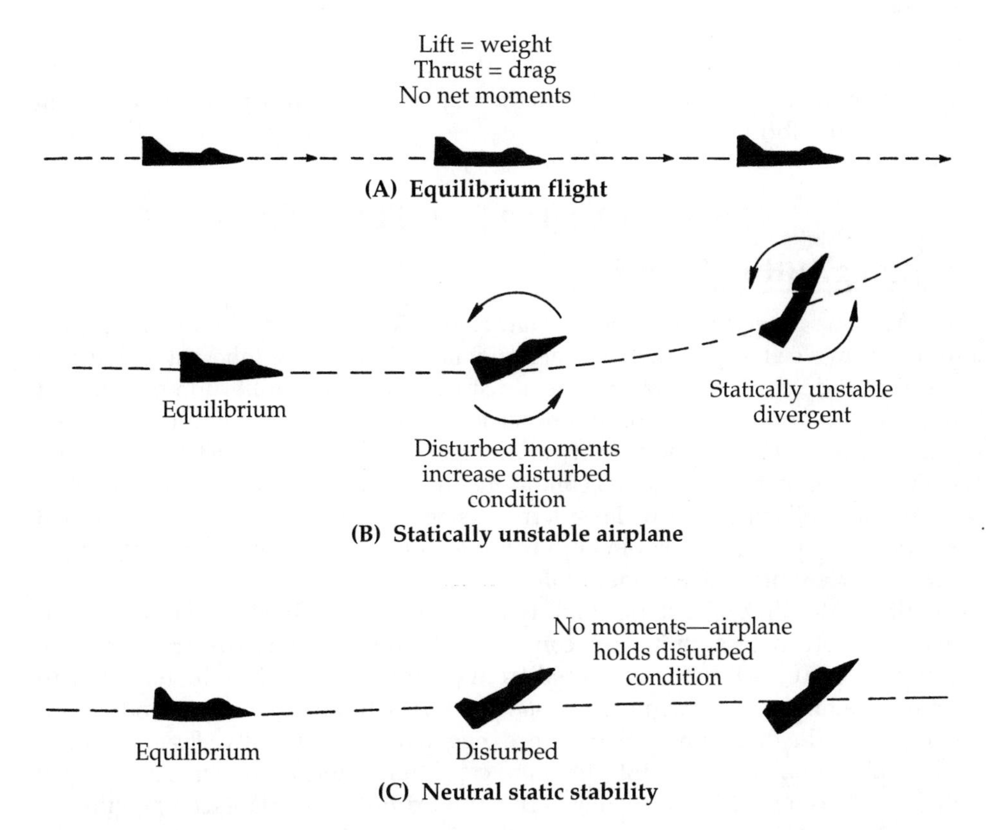

Fig. 7-2. Static stability. Typical motions of a statically unstable airplane when disturbed and typical motions of an airplane with neutral static stability.

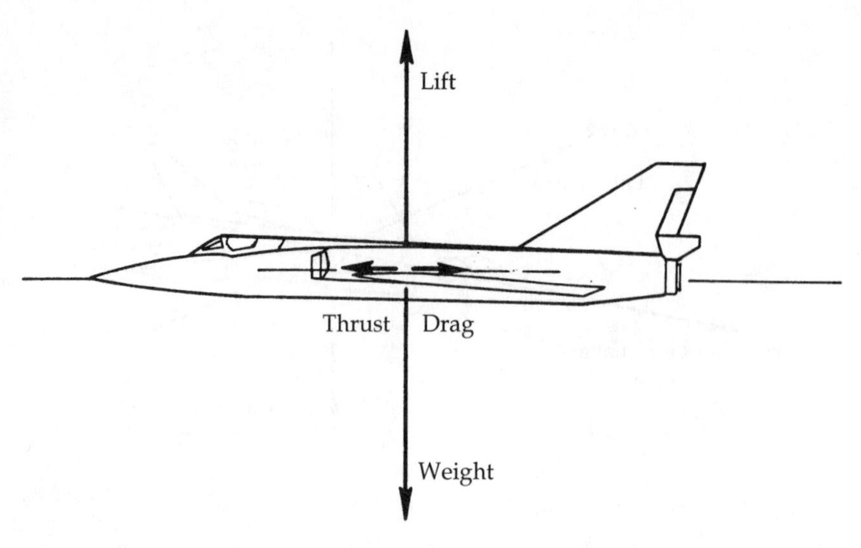

Fig. 7-3. Straight-and-level equilibrium flight. Lift equals weight and thrust equals drag.

weight, the thrust equals the drag, and there are no net rotating moments acting on it. It is in equilibrium.

STABILITY CHARACTERISTICS

Longitudinal stability

Airplane longitudinal stability characteristics result from the inherent static and dynamic stability of the basic aerodynamic design. The inherent stability of the aircraft determines its reaction to a disturbance from equilibrium or trimmed flight, either from external conditions such as gusts, or from internally generated pilot inputs. The higher the stability of the airplane, the greater the tendency to remain in or return to the trimmed condition. There is an optimum balance of stability and controllability that will result in satisfactory flying qualities. This also will minimize the pilot workload in both trimmed and maneuvering flight, as well as provide for a more efficient operation.

Static stability. The static stability of a system is defined by the initial tendency to return to equilibrium conditions following some disturbance from equilibrium. If an object is disturbed from equilibrium and has the tendency to return to equilibrium, positive *static stability* exists. If the object has a tendency to continue in the direction of disturbance, negative static stability (Fig. 7-2(B)), or *static instability*, exists. An intermediate condition could occur where an object displaced from equilibrium remains in equilibrium in the displaced position. If the object subject to the disturbance has neither the tendency to return nor the tendency to continue in the displacement direction, *neutral static stability* exists (Fig. 7-2(C)).

The term static is applied to this form of stability since only the tendency to return to equilibrium conditions is considered, not the resulting motion. The static longitudinal stability of an aircraft is appreciated by displacing the aircraft from some trimmed angle of attack. If the aerodynamic pitching moments created by this displacement tend to return the aircraft to the equilibrium angle of attack, the aircraft has positive static longitudinal stability.

Dynamic stability. While static stability is concerned with the tendency of a displaced body to return to equilibrium, dynamic stability is defined by the resulting motion with time. If an object is disturbed from equilibrium, the time history of the resulting motion indicates the dynamic stability of the system (Fig. 7-4).

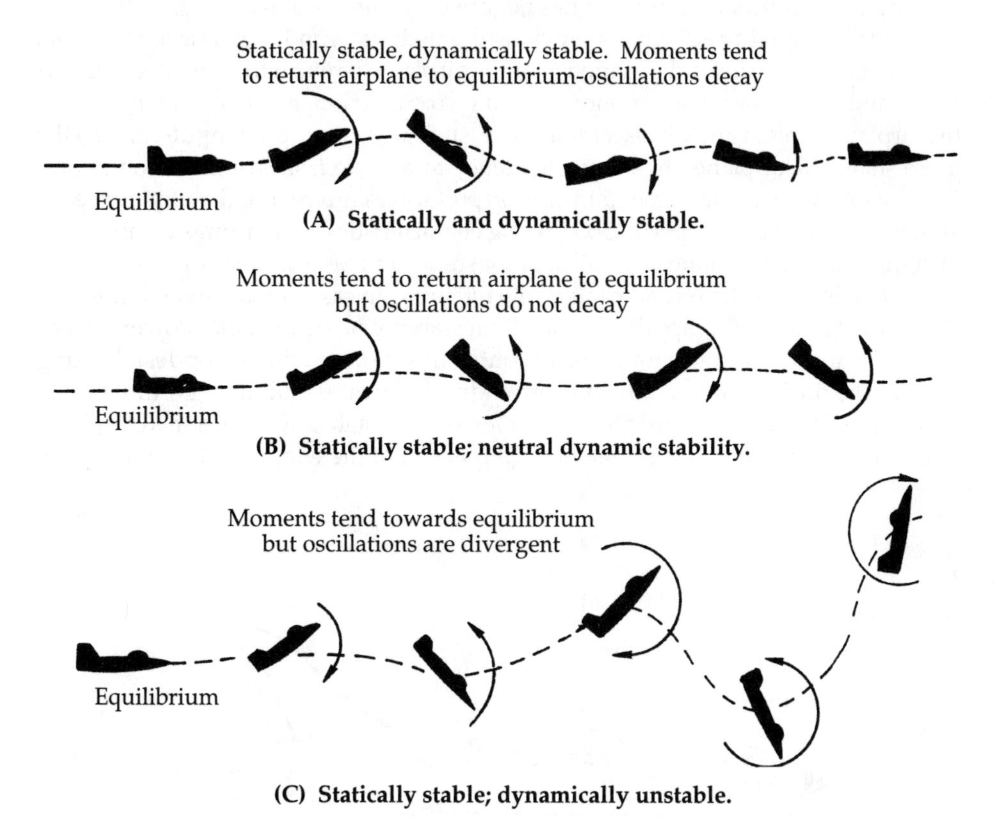

Fig. 7-4. Interrelation of static and dynamic stability.

The first three dynamic stability modes described are referred to as *nonoscillatory modes.* If the airplane has negative static stability, it will continue to diverge from the trimmed condition and will also have *negative dynamic stability* (be dynamically unstable). This behavior is termed *divergence.* If the airplane has neutral static stability, the disinclination for motion also indicates *neutral dynamic stability.*

If an airplane has positive static stability, any of four dynamic stability modes can occur. If the airplane returns to and maintains the trimmed condition after the initial disturbance without overshooting, it has *positive dynamic stability*. This specific behavior is usually called *subsidence* or *dead beat return*.

If the airplane passes through the original trim condition after its initially stable static reaction, any of the three oscillatory dynamic modes might ensue. First, if the airplane regains the trim condition after one or more overshoots of decreasing amplitude, positive dynamic stability is being exhibited (Fig. 7-4(A)). Second, if the airplane continues undamped oscillation around the trim point without any amplitude change, the aircraft has neutral dynamic stability (Fig. 7-4(B)). Third, if the airplane continues oscillating about the trim point while increasing its displacement from trim each time, it has negative dynamic stability (Fig. 7-4(C)).

Handling quality effects. As previously stated, static and dynamic longitudinal stability are a measure of the tendency of the airplane to return to trim after a disturbance and the characteristic motion in doing so. Static longitudinal stability causes the airplane to resist any displacement force, such as gusts or pilot inputs, and so the more stable the airplane, the less its tendency to deviate from trimmed conditions. For a stable airplane, incremental push or pull forces are required to fly at speeds above or below the trim speed. Good flying characteristics result from a combination of static and dynamic longitudinal stability sufficient to ensure that the airplane remains at the desired trim conditions while retaining an ease of maneuverability.

Many factors influence the longitudinal stability of an airplane. Power effects, engine location, fuselage, and wing all contribute to the stabilizing or destabilizing effects. The horizontal tail, however, provides the greatest stabilizing influence.

Figure 7-5 shows the basic balance of forces for a stable, well-designed airplane. The complete aircraft *center of gravity* (c.g.) is ahead of the wing aerodynamic center,

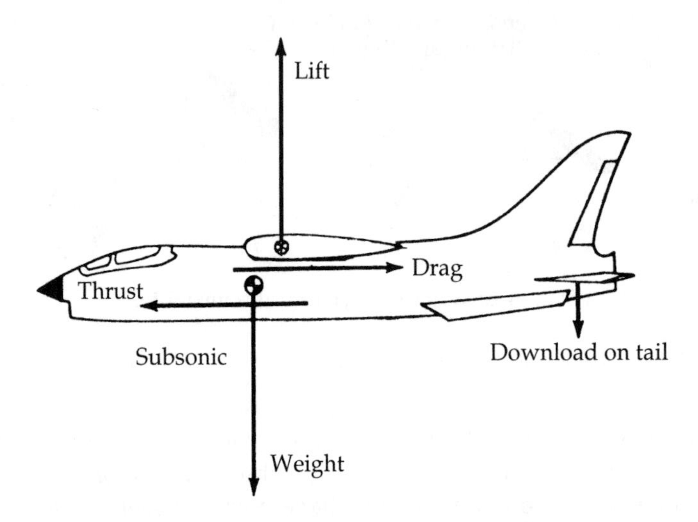

Fig. 7-5. Typical balance of forces for a stable, well-designed airplane in subsonic flight. The center of gravity is forward of the wing's aerodynamic center, and a tail download provides the final balance.

which produces a nose-down pitch tendency. Thrust and drag forces can produce either pitch-up or pitch-down moments depending on their lines of action. The horizontal stabilizer (tail), however, provides the final balancing as well as control forces.

Transonic trim change. At speeds below critical Mach number, the pressure distribution over the wings is such that the center of lift is at approximately the 25-percent chord position. As flight speed increases above the critical Mach number into the transonic range and supersonic flow occurs over the wing, with its attendant shock waves, the pressure distribution changes and the center of lift starts to move rearward. This change tends to introduce a gradually increasing nose-down pitching moment, the magnitude of which will be dependent on airplane stability, as well as other factors. Additional effects also can contribute to this condition, although the pressure change is the most prominent cause.

The overall effect of these pressure changes is generally destabilizing in terms of required counterforces, and the phenomenon has come to be known as *transonic tuck* (or *Mach tuck*). In some airplanes, as Mach number is further increased, stability might tend to become more positive. If the conditions were such that the airplane continued to diverge (stability continued to be negative) and the pilot took no corrective action, the airplane would continue to nose over and, with increasing speed, become continuously more unstable.

Figure 7-6 shows the rearward shift in aerodynamic center and the subsequent increase in elevator power required to prevent an uncontrollable dive. This particular control requirement can be most critical for an airplane in supersonic flight. Supersonic flight is usually accompanied by a reduction in the effectiveness of control surfaces. In order to cope with these trends, powerful all-movable surfaces

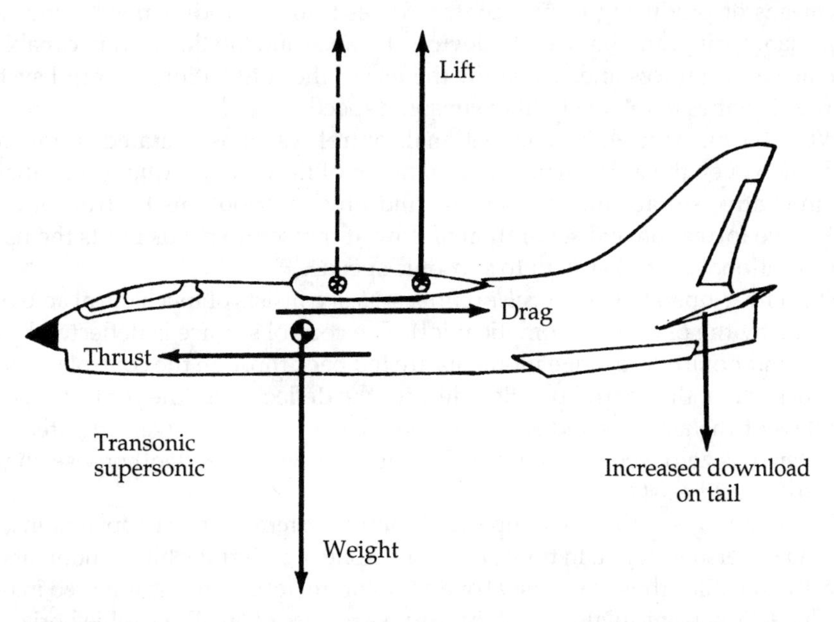

Fig. 7-6. At transonic and supersonic speeds, the rearward shift of the wing's aerodynamic center requires an increase in the tail download to counteract the nose-down tendency.

must be used. Figure 7-7 shows the all-movable *stabilators* on the supersonic McDonnell Douglas F-4 Phantom and Grumman F-14 Tomcat jet fighters. A stabilator is the stabilizer and elevator in one unit.

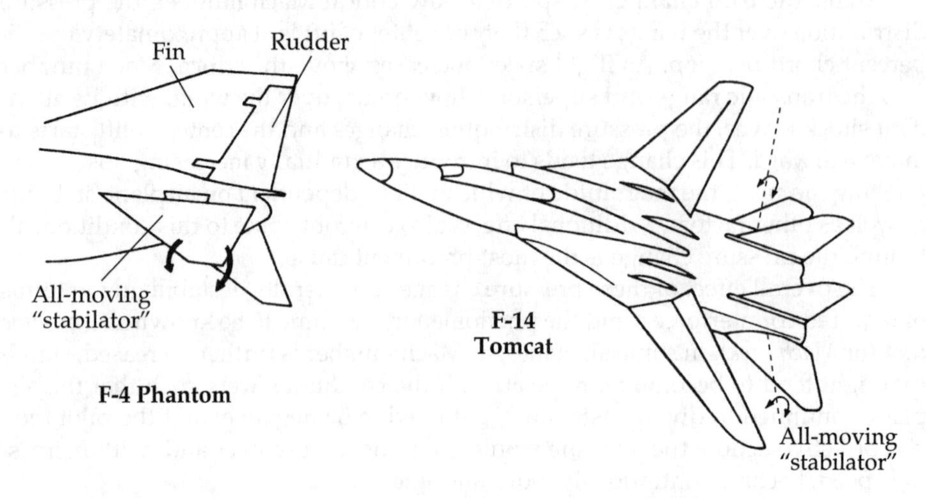

Fig. 7-7. The powerful one-piece stabilators on the supersonic F-4 Phantom and F-14 Tomcat are powered by an irreversible hydraulic actuator system.

The conventional control system (Fig. 7-8) consists of direct mechanical linkages from the controls to the control surfaces. For the subsonic airplane, the principal means of producing proper control forces utilize aerodynamic balance and various tab, spring, and bobweight devices. Balance and tab devices are capable of reducing control forces and will allow the use of the conventional control system on large airplanes to relatively high subsonic speeds.

When the airplane with a conventional control system is operated at transonic speeds, however, the great changes in character of flow can produce great aberrations in control surface hinge moments and any contributions by trim devices. Shock wave formation and separation of flow at transonic speeds limits the use of the conventional control system to subsonic speeds.

The power-operated, *irreversible control system* consists of mechanical actuators controlled by the pilot (or automatic pilot). The control surface is deflected by the actuator and none of the hinge moments are fed back through the controls. In such a control system, the control position decides the deflection of the control surfaces regardless of the airloads and hinge moments. Because the power-operated control system has zero feedback, control feel must be synthesized; otherwise an infinite boost would exist.

The advantages of the power-operated control system are most apparent in transonic and supersonic flight. In transonic flight, none of the erratic hinge moments are fed back to the pilot; thus, no unusual or erratic control forces are encountered in transonic flight. Supersonic flight generally requires the use of an all-movable horizontal surface to achieve the necessary control effectiveness. Such control surfaces must then be actuated and positively positioned by an irreversible device.

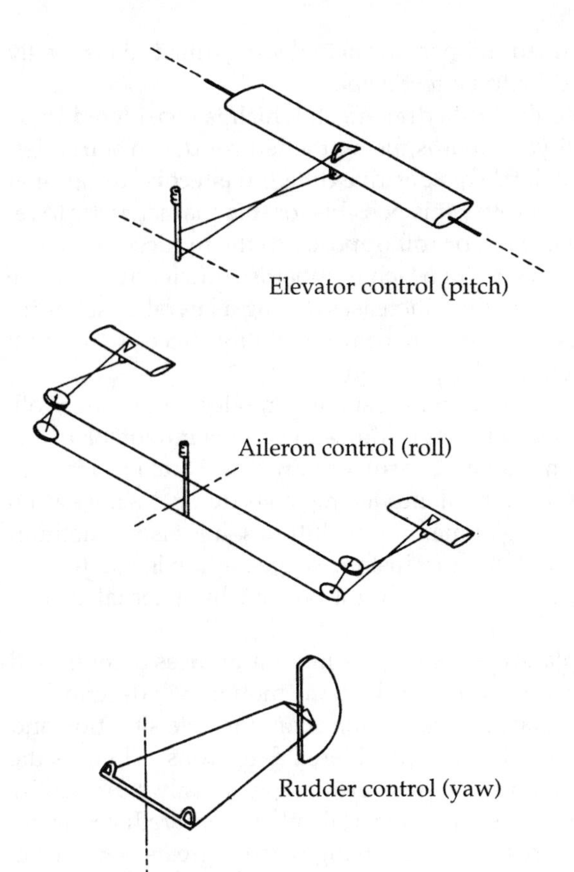

Fig. 7-8. A typical manual control system for low-speed airplanes. At transonic and supersonic speeds, control forces are too high for direct pilot-operated controls.

Lateral/directional stability

Some of the basic discussion on longitudinal stability applies as well to lateral-directional stability. Lateral and directional stability are usually considered together as a result of their influence upon each other. There is also some effect on longitudinal motion, but this discussion pertains to only the more obvious and familiar lateral-directional effects.

Dynamic lateral-direction stability is particularly important for a high-speed jet airplane. The two modes of interest in connection with dynamic stability are the Dutch roll and spiral modes.

Dihedral effects. The rolling tendency of an airplane due to yaw, referred to as *dihedral effect,* is desirable because of the tendency for the wings to return to a level condition after a lateral upset. The ability to pick up a wing with rudder as a result of yawing is also desirable. The swept-wing airplane usually has *positive dihedral effect* because during yaw the leading wing has less effective sweep than the trailing wing. This condition provides higher lift on the leading wing, thus rolling the airplane to a motion that will tend to subsequently return the airplane to a wings-level, balanced flight condition. In addition, some actual wing dihedral is usually provided to ensure an adequate dihedral effect level.

The previous discussion described dihedral effect, which is considered to be positive by definition of induced yaw; that is, using the left rudder to start a left yaw will induce a tendency to roll left. This positive dihedral effect is strongest at slower speeds. At higher speeds, however, it is possible for this characteristic to reverse, resulting in *negative dihedral effect*, or roll opposite to the induced yaw. The reason for this reversal is that when aircraft Mach number is sufficiently high, the effective Mach number on the leading wing increases during a lateral upset or induced yaw. Attendant with this is an increase in compressibility effects, which will result in loss of lift due to shock-induced separation.

Dutch roll. One of the most familiar dynamic stability modes is the Dutch roll, or lateral/directional oscillatory motion. This rolling and yawing motion of an aircraft generally follows a disturbance about the roll or yaw axis. Dutch roll is usually encountered on high-performance airplanes having considerable wing sweep and high wing loadings and operating at maximum altitudes. It is also sometimes referred to as a *nuisance mode* because, in most instances, the motion is not deliberately initiated by the pilot, but is inadvertently introduced by external disturbances.

Spiral mode. For many airplanes, a slight yaw motion induces a continued banked turn of ever-decreasing radius. If unchecked, the motion will describe the shape of a spiral, as viewed from above. This motion is an unstable situation and is often referred to as *spiral divergence*. It is not considered dangerous as long as the rate of divergence with time is not too great. In such a case, it is easily corrected by the pilot. Increased dihedral as well as smaller vertical tail area will reduce the degree of spiral instability; however, both of these changes cause greater Dutch roll tendency.

Pitch-up

Pitch-up is an unstable flight condition resulting in an uncontrollable nose-up moment. It can occur at high angle of attack either at high subsonic speeds at high altitude or at low speeds and is peculiar to the high-speed airplane. This sort of instability implies that an increase in angle of attack produces nose-up moments that tend to bring about further increase in angle of attack; hence, the term pitch-up is applied.

Several items might contribute to a pitch-up tendency. Sweepback of the wing planform can contribute to unstable moments when separation or stall occurs at the tips first. The combination of sweepback and taper alters the lift distribution to produce high local lift coefficients and a low-energy boundary layer near the tip; thus, the tip stall is an inherent tendency of such a planform. In addition, if high local lift coefficients exist near the tip, the tendency will be to incur the shock-induced separation first in these areas. Generally, the wing will contribute to pitch-up only when there is large sweepback (Fig. 7-9).

Of course, the wing is not the only item contributing to the longitudinal stability of the airplane. Another item important as a source of pitch-up is the down-

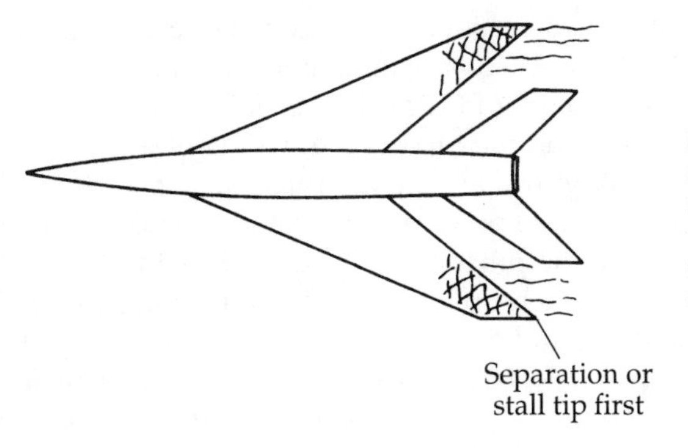

Fig. 7-9. A contributing factor resulting in an uncontrollable nose-up moment (pitch-up) is the forward travel of the center of lift due to tip stall of a sweepback wing.

wash at the horizontal tail. The contribution of the tail to stability depends on the change in tail lift when the airplane is given a change in angle of attack.

Since the downwash at the tail reduces the change in angle of attack at the tail, any increase in downwash at the tail is destabilizing. For certain low-aspect-ratio airplane configurations, an increase in airplane angle of attack can physically locate the horizontal tail in the wing flow field where higher relative downwash exists (Fig. 7-10); thus, a decrease in stability would take place.

Certain changes in the flow field behind the wing at high angles of attack can produce large changes in the tail contribution to stability. If the wing tips stall first, the vortices shift inboard and increase the local downwash at the tail. Also, the fuselage at high angle of attack can produce strong cross-flow separation vortices

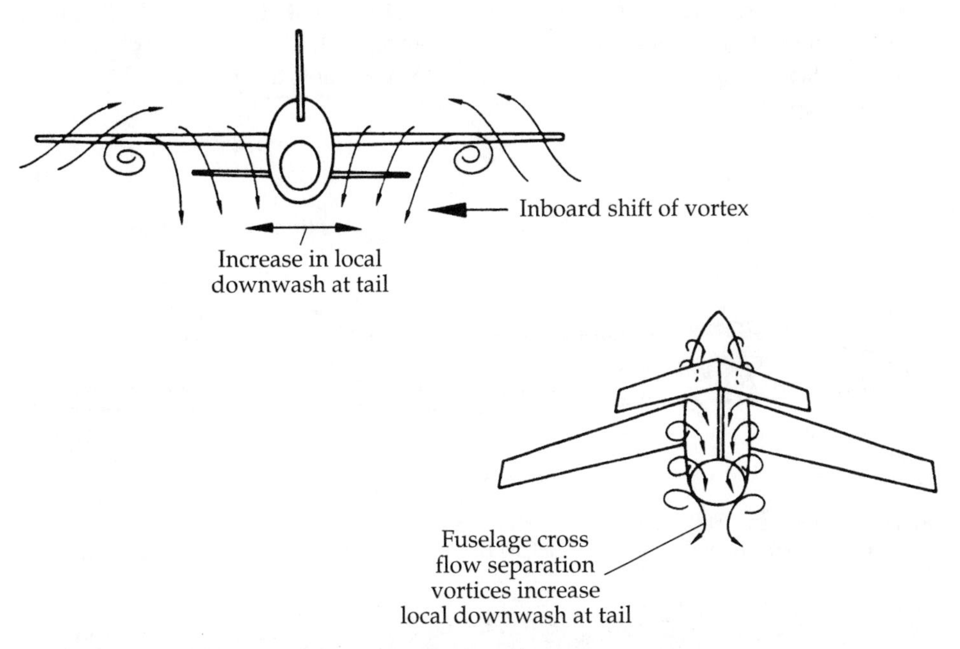

Fig. 7-10. For certain low-aspect-ratio airplanes, an increase in airplane angle of attack can physically locate the horizontal stabilizer in the wing flow field, resulting in pitch-up.

that increase the local downwash for a horizontal tail placed above the fuselage. Either one of these downwash influences, or a combination of the influences, might provide a large unstable contribution of the horizontal tail.

The pitch-up instability is usually confined to the high angle-of-attack range and might be a consequence of a configuration that otherwise has very desirable flying qualities. In such a case, it would be necessary to provide some automatic control function to prevent entry into the pitch-up range or to provide synthetic stability for the condition. Because the pitch-up is usually a strong instability with a high rate of divergence, most pilots would not be capable of contending with the condition. At high speed, pitch-up would be of great danger because structural failure could easily result. At slow speed, failing flight loads might not result, but the strong instability might preclude a successful recovery from the ensuing motion of the airplane, since this can occur at low altitudes during landing approach.

As discussed in chapter 5, low-aspect ratio requires high angle of attack to achieve high lift coefficients; thus, the high-speed airplane with its combination of swept, low-aspect-ratio wings is very susceptible to pitch-up.

During the 1950s, the Century Series fighters were developed (F-100, F-101, F-102, etc.). These were the first operational supersonic airplanes. At that time, high-speed aerodynamics was in its infancy and some of these airplanes experienced pitch-up, which threatened their operational use. This was mainly the result of the horizontal stabilizer position.

On the F-101 and F-104, for example, the horizontal stabilizer was placed high, on top of the vertical fin, a very desirable location above the normal wing downwash (Fig. 7-11); however, at high angles of attack, these swept-wing, low-aspect-ratio airplanes experienced pitch-up due to the turbulent flow over the T-tail. A complex angle-of-attack sensing system, combined with a stick pusher to limit angle of attack, eventually solved the problem. Subsequent designs placed the horizontal tail as low as possible to minimize pitch-up.

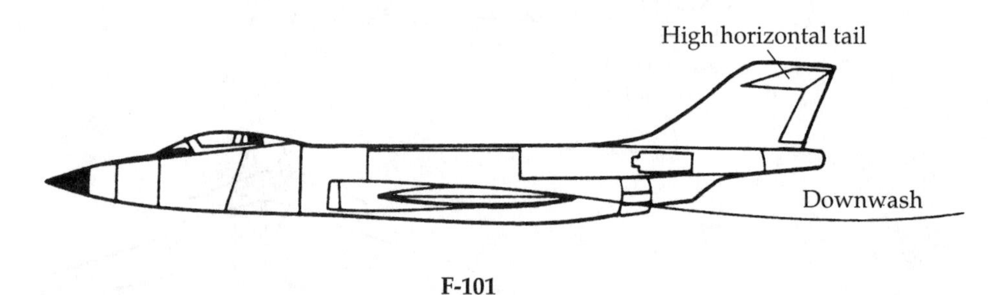

Fig. 7-11. A high horizontal stabilizer is normally desirable; however, at high angles of attack, the high stabilizer might experience turbulent flow due to wing downwash resulting in pitch-up.

8

Trade-offs for design

EVERY AIRPLANE IS A COMPROMISE. Airplane designers usually conduct trade-off studies to determine the best compromise between high speed, slow-speed capability, and stability and control, all with regard to an airplane's mission. Chapter 4 examines high-speed flight in a highly simplified form, mainly from the point of view of drag and drag reduction; chapter 5 examines the high-speed airplane at slow speed; chapter 7 examines stability and control.

This chapter will attempt to bring it all together by reviewing the previous themes and discussing their relationships to actual flying qualities and aircraft configuration.

COMPRESSIBILITY

Because compressibility of air is the determining factor in the design and performance of the high-speed airplane, a detailed discussion to supplement chapter 4 is in order.

Recall that at slow flight speeds the study of aerodynamics is greatly simplified by the fact that air might experience relatively small changes in pressure with only negligible changes in density. This airflow is termed *incompressible* because the air can undergo changes in pressure without apparent changes in density. Such a condition of airflow is analogous to the flow of water, hydraulic fluid, or any other incompressible fluid; however, at fast flight speeds, the pressure changes that take place are quite large, and significant changes in air density occur. The study of airflow at fast speeds must account for these changes in air density and must consider that the air is compressible and that there will be "compressibility effects."

A factor of great importance in the study of high-speed airflow is the *speed of sound*. The speed of sound is the rate at which small pressure disturbances will be propagated through the air, and this propagation speed is solely a function of air temperature. Figure 4-6 (chapter 4) illustrates the variation of the speed of sound in the standard atmosphere.

As an object moves through the air mass, velocity and pressure changes occur that create pressure disturbances in the airflow surrounding the object. Of course, these pressure disturbances are propagated through the air at the speed of sound in "waves." These waves are similar to sound waves and they travel at the speed of sound (about 661 knots at sea level conditions). The relationship between pressure waves and sound waves is simple because sound is a pressure wave set up by some local compression of the air, and the speed of sound is simply the speed of propagation of refractions and compressions of small amplitude in the air.

Thus, if a body travels through the speed of sound, there is no time for the wave to get ahead and the air will come up against the body with a "shock." If the object is traveling at slow speed, the pressure disturbances are propagated ahead of the object, and the airflow immediately ahead of the object is influenced by the pressure field on the object. Actually, these pressure disturbances are transmitted in all directions and extend indefinitely in all directions. Evidence of this "pressure warning" is seen in the typical subsonic flow pattern of Fig. 8-1 where there is upwash and flow direction change well ahead of the leading edge.

If the object is traveling at some speed faster than speed of sound, the airflow ahead of the object will not be influenced by the pressure field on the object because pressure disturbances cannot be propagated ahead of the object; thus, as the flight speed nears the speed of sound, a compression wave will form at the leading edge, and all changes in velocity and pressure will take place quite sharply and suddenly. The airflow ahead of the object is not influenced until the air particles are suddenly forced out of the way by the concentrated pressure wave set up by the object. Evidence of this phenomenon is seen in the typical supersonic flow pattern of Fig. 8-1.

At this point it should become apparent that all compressibility effects depend upon the relationship of airspeed to the speed of sound. The term used to describe this relationship is the Mach number, M, and this term is the ratio of the true airspeed to the speed of sound.

$$M = \frac{V}{a}$$

where

M = Mach number
V = true airspeed, knots
a = speed of sound, knots (shown in Fig. 4-6)

It is important to note that compressibility effects are not limited to flight speeds at and above the speed of sound. Because any aircraft will have some aerodynamic shape and will be developing lift, there will be local flow velocities on the surfaces that are greater than the flight speed; thus, an aircraft can experience compressibility effects at flight speeds well below the speed of sound. Because there is the possibility of subsonic and supersonic flows existing on the aircraft, it is convenient to define certain regimes of flight. These regimes are defined approximately as follows:

Subsonic Mach numbers below 0.75
Transonic Mach numbers from 0.75–1.20
Supersonic Mach numbers from 1.20–5.00
Hypersonic Mach numbers above 5.00

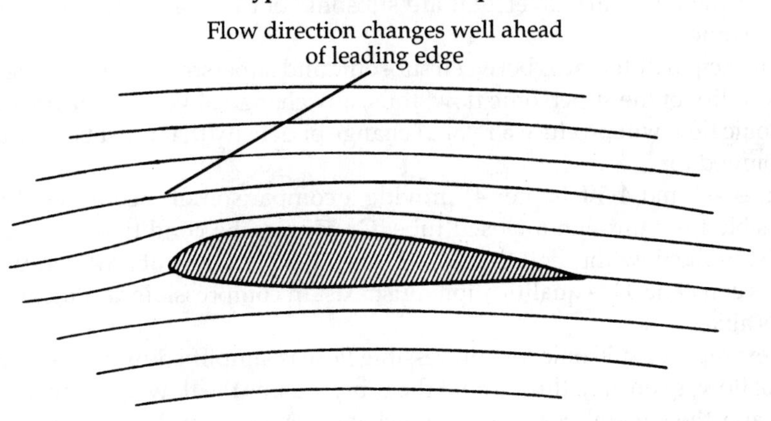

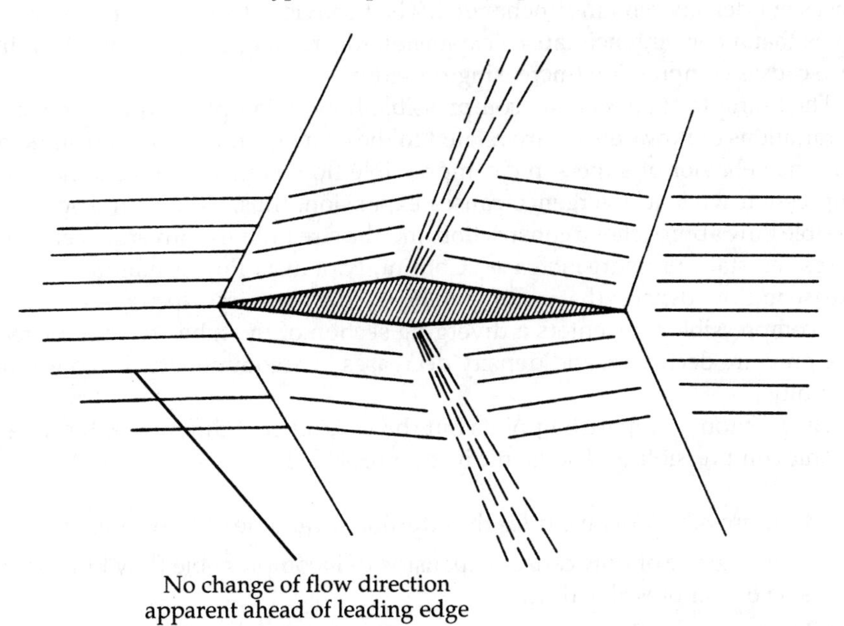

Fig. 8-1. Comparison of subsonic and supersonic flow patterns.

While the flight Mach numbers used to define these regimes of flight are quite approximate, it is important to appreciate the types of flow existing in each area. In the subsonic regime, it is most likely that pure subsonic airflow exists on all parts of the aircraft. In the transonic regime it is very probable that flow on the aircraft components can be partly subsonic and partly supersonic. The supersonic and hypersonic flight regimes will provide definite supersonic flow velocities on all parts of the aircraft. Of course, in supersonic flight there will be some portions of the boundary layer that are subsonic but the predominating flow is still supersonic.

The principal differences between subsonic and supersonic flow are due to the compressibility of the supersonic flow; thus, any change of velocity or pressure of a supersonic flow will produce a related change of density that must be considered and accounted for.

Figures 4-1 and 4-5 (chapter 4) provide a comparison of incompressible and compressible flow through a closed tube. Of course, the condition of continuity must exist in the flow through the closed tube; the mass flow at any station along the tube is constant. This qualification must exist in compressible and incompressible cases alike.

The example of subsonic incompressible flow is simplified by the fact that the density of flow is constant throughout the tube; thus, as the flow approaches a constriction and the streamlines converge, velocity increases and static pressure decreases. In other words, a convergence of the tube requires an increasing velocity to accommodate the continuity of flow. Also, as the subsonic incompressible flow enters a diverging section of the tube, velocity decreases and static pressure increases but density remains unchanged. The behavior of subsonic incompressible flow is that a convergence causes expansion (decreasing pressure) while a divergence causes compression (increasing pressure).

The example of supersonic compressible flow is complicated by the fact that the variations of flow density are related to the changes in velocity and static pressure. The behavior of supersonic compressible flow is that a convergence causes compression while a divergence causes expansion; thus, as the supersonic compressible flow approaches a constriction and the streamlines converge, velocity decreases and static pressure increases. Continuity of mass flow is maintained by the increase in flow density that accompanies the decrease in velocity. As the supersonic compressible flow enters a diverging section of the tube, velocity increases, static pressure decreases, and density decreases to accommodate the condition of continuity.

The previous comparison points out three significant differences between supersonic compressible and subsonic incompressible flow:

- Compressible flow includes the additional variable of flow density.
- Convergence of flow causes expansion of incompressible flow but compression of compressible flow.
- Divergence of flow causes compression of incompressible flow but expansion of compressible flow.

TYPICAL SUPERSONIC FLOW PATTERNS

When supersonic flow is clearly established, all changes in velocity, pressure, density, flow direction, etc., take place quite suddenly and in relatively confined areas. The areas of flow change are generally distinct and the phenomena are referred to as *wave formations*. All compression waves occur suddenly and are wasteful of energy; hence, the compression waves are distinguished by the sudden "shock" type of behavior. All expansion waves are not so sudden in their occurrence and are not wasteful of energy, like the compression shock waves. Various types of waves can occur in supersonic flow and the nature of the wave formed depends upon the airstream and the shape of the object causing the flow change. Essentially, there are three fundamental types of waves formed in supersonic flow:

- Oblique shock wave (compression)
- Normal shock wave (compression)
- Expansion wave (no shock)

Oblique shock wave

Consider the case where a supersonic airstream is turned into the preceding airflow. Such would be the case of a supersonic flow "into a corner" as shown in Fig. 8-2. A supersonic airstream passing through the oblique shock wave will experience five changes:

- The airstream is slowed down; the velocity and Mach number behind the wave are reduced but the flow is still supersonic.
- The flow direction is changed to flow along the surface.
- The static pressure of the airstream behind the wave is increased.
- The density of the airstream behind the wave is increased.
- Some of the available energy of the airstream (indicated by the sum of dynamic and static pressure) is dissipated and turned into unavailable heat energy; hence, the shock wave is wasteful of energy.

A typical case of oblique shock wave formation is that of a wedge pointed into a supersonic airstream. The oblique shock wave will form on each surface of the wedge, and the inclination of the shock wave will be a function of the free-stream Mach number and the wedge angle. As the free-stream Mach number increases, the shock wave angle decreases; as the wedge angle increases the shock wave angle increases and, if the wedge angle is increased to some critical amount, the shock wave will detach from the leading edge of the wedge. It is important to note that detachment of the shock wave will produce subsonic flow immediately after the central portion of the shock wave. Figure 8-3 illustrates these typical flow patterns and the effect of Mach number and wedge angle.

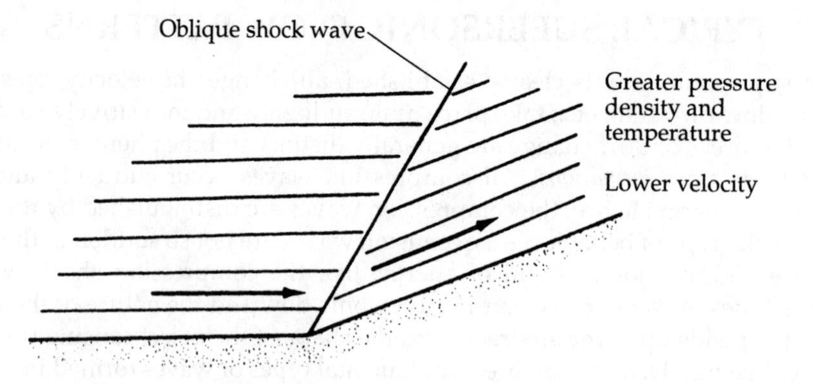

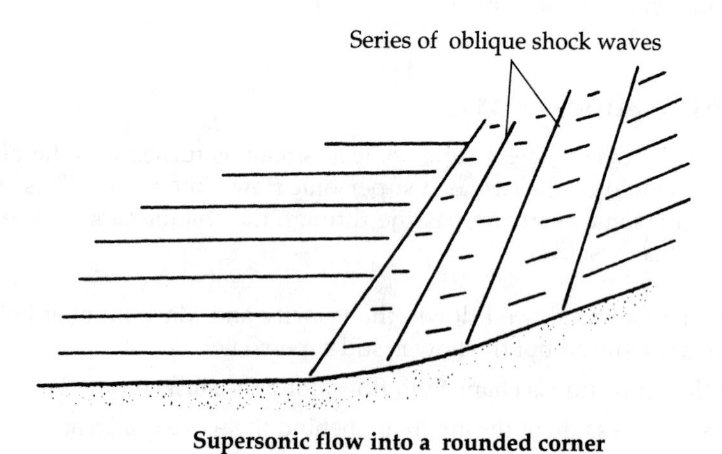

Fig. 8-2. Oblique shock wave formation.

Flow across a wedge in a supersonic airstream allows a flow in two dimensions. If a cone were placed in a supersonic airstream, the airflow would occur in three dimensions and there would be some noticeable differences in flow characteristics. Three-dimensional flow for the same Mach number and a flow direction change would produce a weaker shock wave with less change in pressure and density. Also, this conical wave formation allows changes in airflow that continue to occur past the wave front and the wave strength varies with distance away from the surface. Figure 8-4 depicts the typical three-dimensional flow past a cone.

Oblique shock waves can be reflected like any pressure wave and this effect is shown in Fig. 8-4, a model in a wind tunnel. This reflection appears logical and

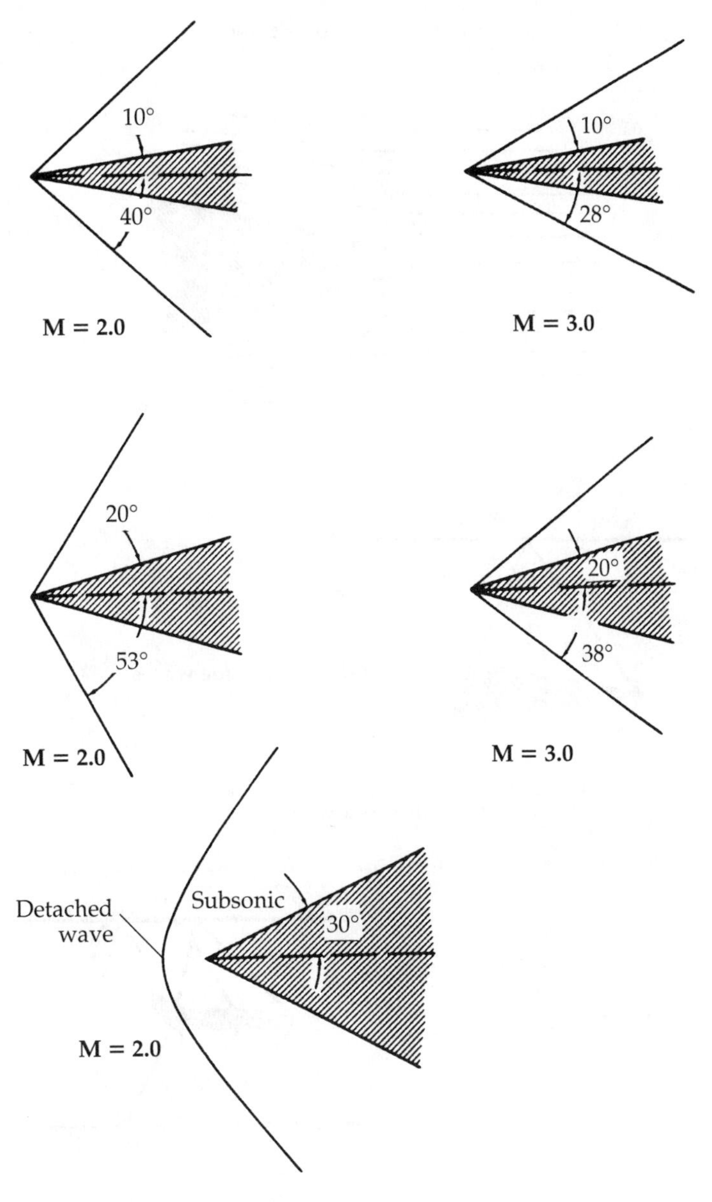

Fig. 8-3. Shock waves formed by various wedge shapes.

necessary because the original wave changes the flow direction toward the wall and the reflected wave creates the subsequent flow change to cause the flow to remain parallel to the wall surface. This reflection phenomenon places definite restrictions on the size of a model in a wind tunnel because a wave reflected back to the model would cause a pressure distribution not typical of free flight.

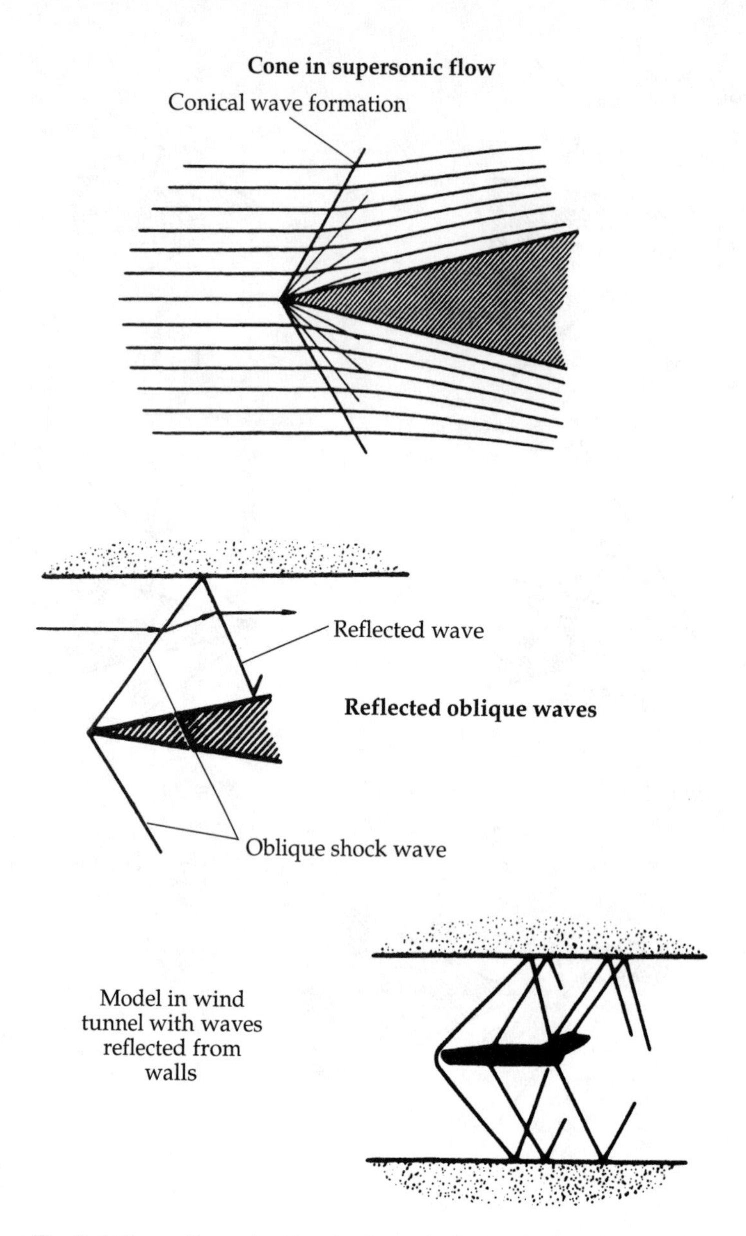

Fig. 8-4. Three-dimensional and reflected shock waves.

Normal shock wave

If a blunt-nosed object is placed in a supersonic airstream, the shock wave that is formed will be detached from the leading edge. This detached wave also occurs when a wedge or cone angle exceeds some critical value. Wherever the shock wave forms perpendicular to the upstream flow, the shock wave is termed a *normal shock wave* and the flow immediately behind the wave is subsonic. Any relatively blunt object in a supersonic airstream will form a normal shock wave immediately ahead of the leading edge, slowing the airstream to subsonic so the airstream can feel the presence of the blunt nose and flow around it. Past the blunt nose, the airstream might remain subsonic or accelerate back to supersonic depending on the shape of the nose and the Mach number of the free stream.

In addition to the formation of normal shock waves described above, this same type of wave might be formed in an entirely different manner when there is no object in the supersonic airstream. Whenever a supersonic airstream is slowed to subsonic without a change in direction, a normal shock wave will form as a boundary between the supersonic and subsonic regions. This is an important fact because aircraft usually encounter some *compressibility effects* before the flight speed is sonic.

Figure 8-5 illustrates the manner in which an airfoil at fast subsonic speeds has local flow velocities that are supersonic. As the local supersonic flow moves aft, a normal shock wave forms slowing the flow to subsonic. The transition of flow from subsonic to supersonic is smooth and is not accompanied by shock waves if the transition is made gradually with a smooth surface. The transition of flow from supersonic to subsonic without directional change always forms a normal shock wave.

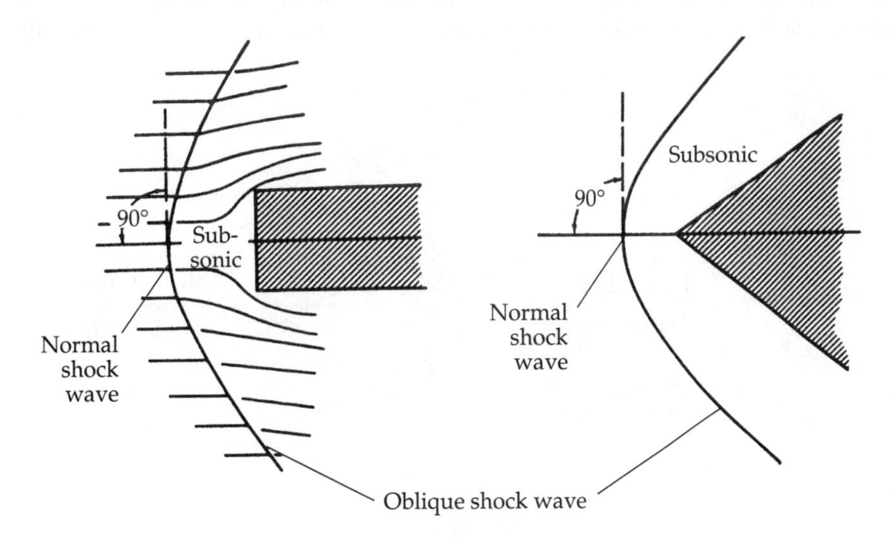

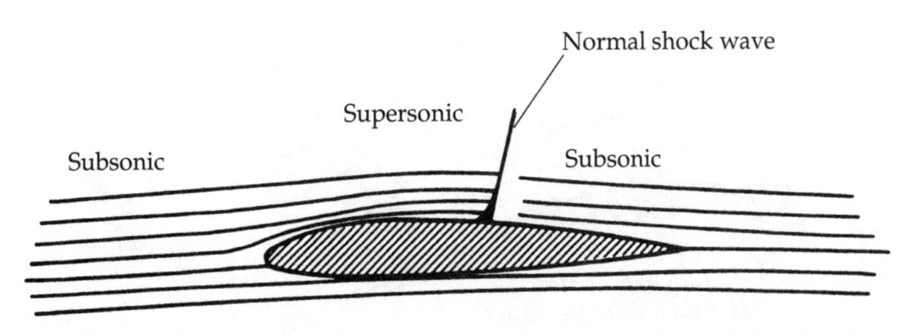

Fig. 8-5. Normal shock wave formation.

A supersonic airstream passing through a normal shock wave will experience these changes:

- The airstream is slowed to subsonic, and the local Mach number behind the wave is approximately equal to the reciprocal of the Mach number ahead of the wave; for instance, if the Mach number ahead of the wave is 1.25, the Mach number of the flow behind the wave is approximately 0.80.
- The airflow direction immediately behind the wave is unchanged.
- The static pressure of the airstream behind the wave is considerably increased.
- The density of the airstream behind the wave is considerably increased.
- The energy of the airstream (indicated by total pressure—dynamic plus static) is greatly reduced. The normal shock wave is very wasteful of energy.

Expansion wave

If a supersonic airstream were turned away from the preceding flow, an expansion wave would form. The flow "around a corner" shown in Fig. 8-6 will not cause sharp, sudden changes in the airflow except at the corner itself and thus is

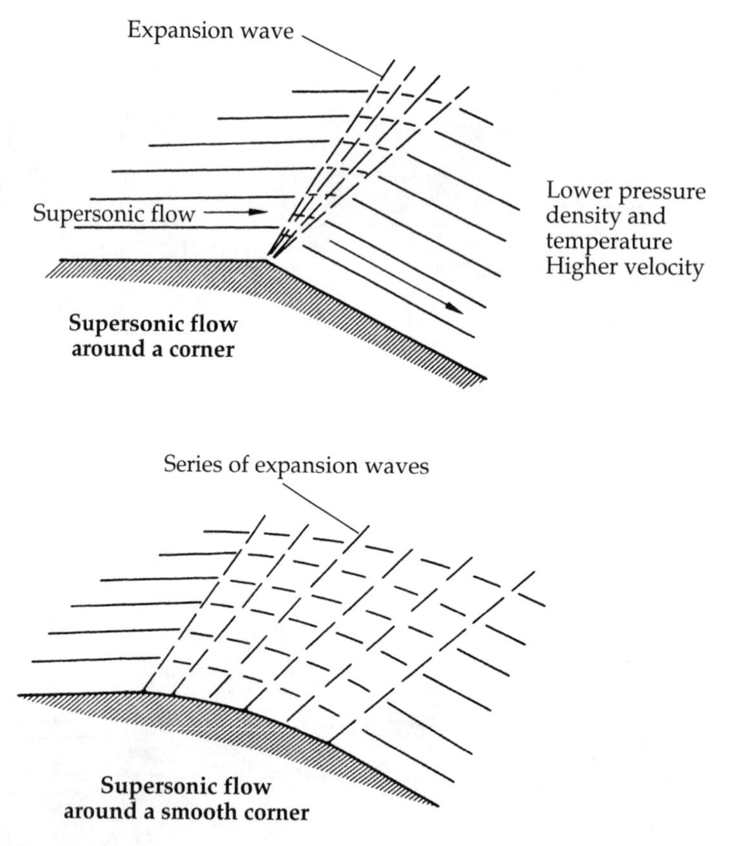

Fig. 8-6. Expansion wave formation.

not actually a "shock" wave. A supersonic airstream passing through an expansion wave will experience these changes:

- The airstream is accelerated; the velocity and Mach number behind the wave are greater.
- The flow direction is changed to flow along the surface, if separation does not occur.
- The static pressure of the airstream behind the wave is decreased.
- The density of the airstream behind the wave is decreased.
- Because the flow changes in a rather gradual manner, there is no "shock" and no loss of energy in the airstream. The expansion wave does not dissipate airstream energy.

The expansion wave in three dimensions is a slightly different case, and the principal difference is the tendency for the static pressure to continue to increase past the wave.

Table 8-1 summarizes the characteristics of the three principal wave forms encountered with supersonic flow.

TABLE 8-1. Supersonic wave characteristics

Type of wave formation	Oblique shock wave	Normal shock wave	Expansion wave
Flow direction change	"Flow into a corner," turned into preceding flow.	No change.	"Flow around a corner," turned away from preceding flow.
Effect on velocity and Mach number.	Decreased but still supersonic.	Decreased to subsonic.	Increased to higher supersonic.
Effect on static pressure and density.	Increase.	Great increase.	Decrease.
Effect on energy or total presure.	Decrease.	Great decrease.	No change (no shock).

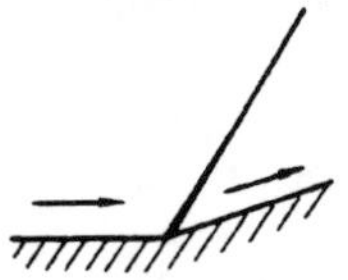

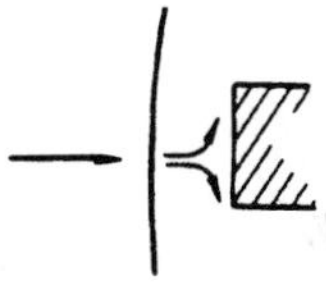

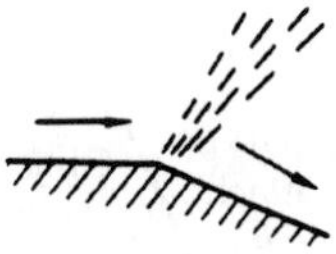

AIRFOIL SECTIONS IN SUPERSONIC FLOW

Since straight lines and sharp corners appear to be at least as good as curves, the simplest airfoil section for supersonic flight would be a flat plate at a small angle of attack. In order to appreciate the effect of these various wave forms on the aerodynamic characteristics in supersonic flow, inspect Fig. 8-7. Parts (A) and (B) show the wave pattern and resulting pressure distribution for a thin flat plate at a positive angle of attack.

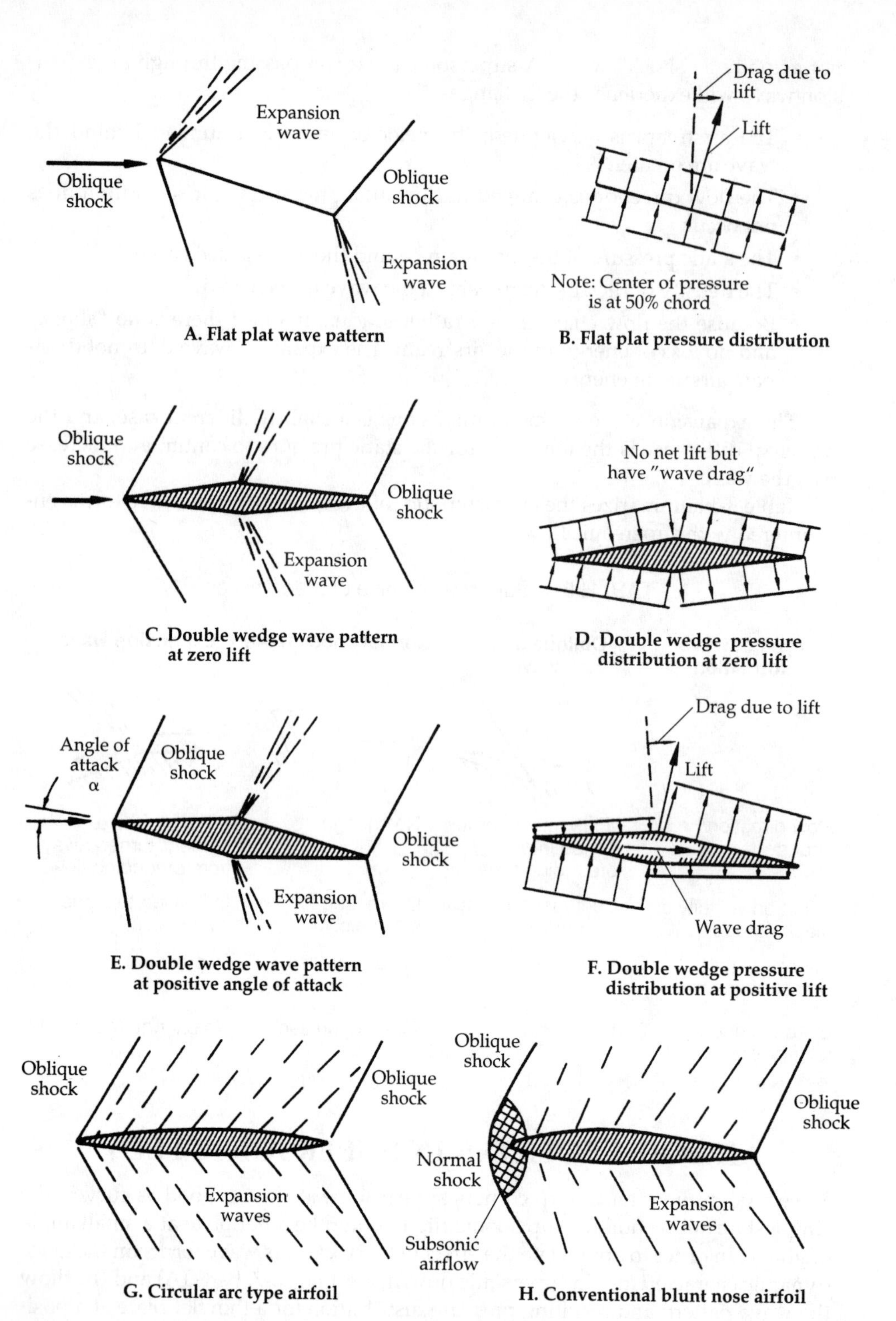

Fig. 8-7. Typical supersonic flow patterns and distribution of pressure.

The airstream moving over the upper surface passes through an expansion wave at the leading edge and then an oblique shock wave at the trailing edge; thus, a uniform suction pressure exists over the upper surface. The airstream moving underneath the flat plate passes through an oblique shock wave at the leading edge and then an expansion wave at the trailing edge. This produces a uniform positive pressure on the underside of the section. This distribution of pressure on the surface will produce a net lift and incur a subsequent drag due to lift from the inclination of the resultant lift from a perpendicular to the free stream.

A flat plate is definitely not practical for an airfoil, but the double wedge is. Parts (C) and (D) of Fig. 8-7 show the wave pattern and resulting pressure distribution for a double-wedge airfoil at zero lift. The airstream moving over the surface passes through an oblique shock, an expansion wave, and another oblique shock.

The resulting pressure distribution on the surfaces produces no net lift, but the increased pressure on the forward half of the chord along with the decreased pressure on the aft half of the chord produces a *wave drag*. This wave drag is caused by the components of pressure forces that are parallel to the free-stream direction. The wave drag is in addition to the drag due to friction, separation, lift, and the like, and can be a very considerable part of the total drag at the fast supersonic speeds.

Parts (E) and (F) of Fig. 8-7 illustrate the wave pattern and resulting pressure distribution for the double-wedge airfoil at a small positive angle of attack. The net pressure distribution produces an inclined lift with drag due to lift that is in addition to the wave drag at zero lift.

Part (G) of Fig. 8-7 shows the wave pattern for a circular arc airfoil. After the airflow traverses the oblique shock wave at the leading edge, the airflow undergoes a gradual but continual expansion until the trailing edge shock wave is encountered.

Part (H) of Fig. 8-7 illustrates the wave pattern on a conventional blunt-nose airfoil in supersonic flow. When the nose is blunt, the wave must detach and become a normal shock wave immediately ahead of the leading edge. Of course, this wave form produces an area of subsonic airflow at the leading edge with very high pressure and density behind the detached wave.

The drawings of Fig. 8-7 illustrate the typical patterns of supersonic flow and point out these facts concerning aerodynamic surfaces in two-dimensional supersonic flow:

- All changes in velocity, pressure, density, and flow direction will take place quite suddenly through the various wave forms. The shape of the object and the required flow direction change dictate the type and strength of the wave formed.
- As always, lift results from the distribution of pressure on a surface and is the net force perpendicular to the free-stream direction. Any component of the lift in a direction parallel to the windstream will be drag due to lift.

- In supersonic flight, the zero lift drag of an airfoil of some finite thickness will include a *wave drag*. The thickness of the airfoil will have an extremely powerful effect on this wave drag because the wave drag varies as the square of the thickness ratio; if the thickness is reduced 50 percent, the wave drag is reduced 75 percent. The leading edges of supersonic shapes must be sharp or the wave formed at the leading edge will be a strong detached shock wave.
- When the flow on the conventional airfoil is supersonic, the aerodynamic center of the surface will be located approximately at the 50-percent chord position. As this contrasts with the subsonic location for the aerodynamic center of the 25- percent chord position, significant changes in aerodynamic trim and stability might be encountered in transonic flight.

What should be the shape of the airfoil section? First, it must be thin, or more correctly, the airfoil section should have a low thickness-to-chord ratio. After that, shape doesn't matter much. Straight lines are just as good as curved surfaces. A flat plate would make an excellent airfoil section, but obviously it would not have sufficient strength or stiffness. It could therefore be thickened in the middle to provide the double wedge or circular section shape as previously discussed. These are as good as any other supersonic airfoil sections; however, subsonic flight is also required; therefore, variations of the wedge and circular sections must be considered.

CONFIGURATION EFFECTS

Transonic and supersonic flight

Any object in subsonic flight that has some finite thickness or is producing lift will have local velocities on the surface that are greater than the free-stream velocity; hence, compressibility effects can be expected to occur at flight speeds less than the speed of sound. The transonic regime of flight provides the opportunity for mixed subsonic and supersonic flow and accounts for the first significant effects of compressibility.

Consider a conventional airfoil shape as shown in Fig. 8-8. If this airfoil is at a flight Mach number of 0.50 and a slight positive angle of attack, the maximum local velocity on the surface will be greater than the flight speed but most likely less than sonic speed.

Assume that an increase in flight Mach number to 0.72 would produce first evidence of local sonic flow. This condition of flight would be the highest flight speed possible without supersonic flow and would be termed the *critical Mach number*; thus, critical Mach number is the boundary between subsonic and transonic flight and is an important point of reference for all compressibility effects encountered in transonic flight. By definition, critical Mach number is the "free-stream Mach number that produces first evidence of local sonic flow." Shock waves, buffet, airflow separation, etc., take place above critical Mach number.

As critical Mach number is exceeded, an area of supersonic airflow is created, and a normal shock wave forms as the boundary between the supersonic flow and

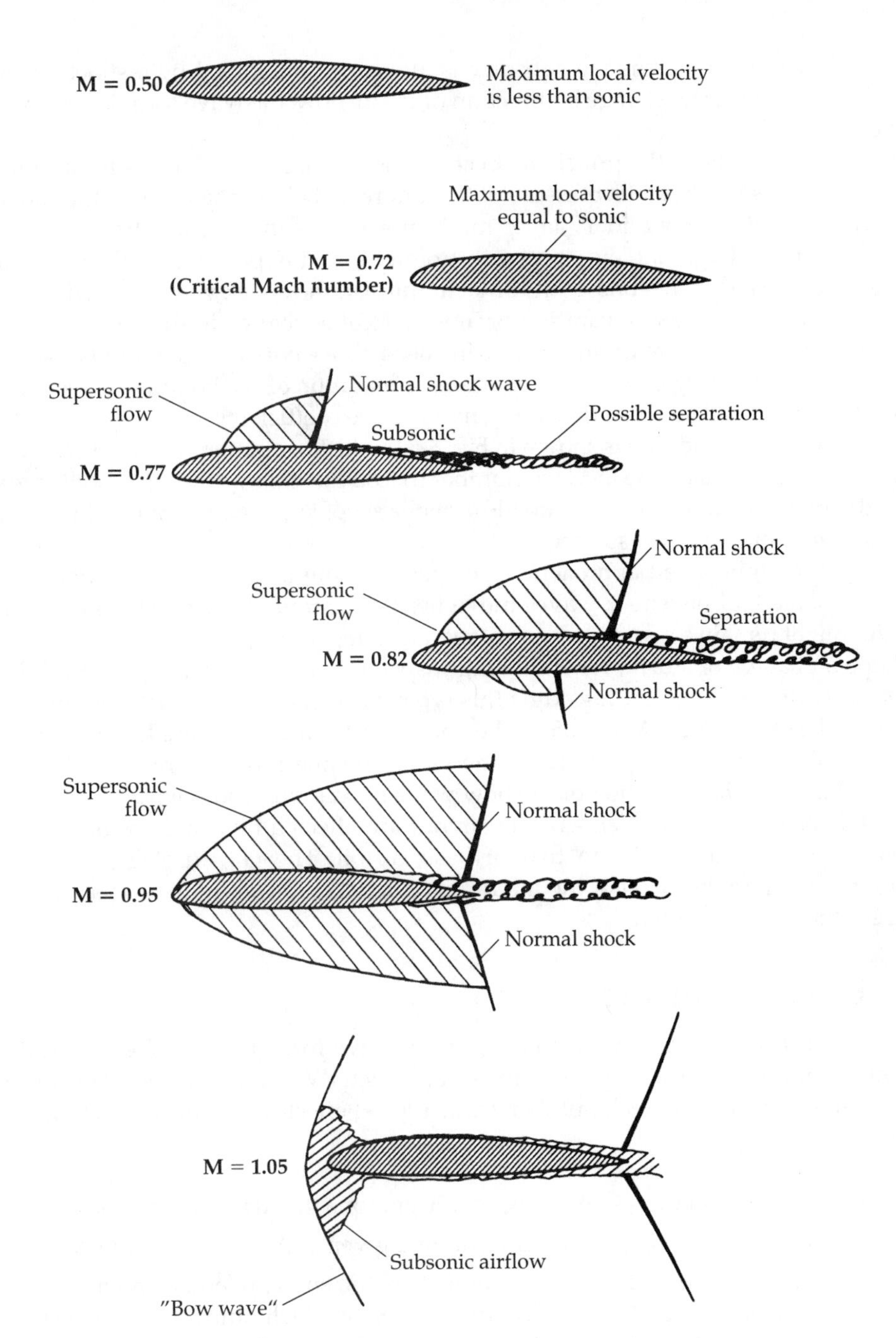

Fig. 8-8. Transonic flow patterns as velocity increases from M = 0.5 to M = 1.05.

the subsonic flow on the aft portion of the airfoil surface. The acceleration of the airflow from subsonic to supersonic is smooth and unaccompanied by shock waves if the surface is smooth and the transition gradual; however, the transition of airflow from supersonic to subsonic is always accompanied by a shock wave and, when there is no change in direction of the airflow, the wave form is a normal shock wave.

Recall that one of the principal effects of the normal shock wave is to produce a large increase in the static pressure of the airstream behind the wave. If the shock wave is strong, the boundary layer might not have sufficient kinetic energy to withstand the large, adverse pressure gradient, and separation will occur. At speeds only slightly beyond critical Mach number, the shock wave formed is not strong enough to cause separation or any noticeable change in the aerodynamic force coefficients; however, an increase in speed above critical Mach number sufficient to form a strong shock wave can cause separation of the boundary layer and produce sudden changes in the aerodynamic force coefficients.

Such a flow condition is shown in Fig. 8-8 by the flow pattern for $M = 0.77$. Notice that a further increase in Mach number to 0.82 can enlarge the supersonic area on the upper surface and form an additional area of supersonic flow and normal shock wave on the lower surface.

As the flight speed approaches the speed of sound, the areas of supersonic flow enlarge and the shock waves move nearer the trailing edge. The boundary layer might remain separated or might reattach, depending much upon the airfoil shape and angle of attack. When the flight speed exceeds the speed of sound the *bow wave* forms at the leading edge; this typical flow pattern is illustrated in Fig. 8-8 by the drawing for $M = 1.05$. If the speed is increased to some higher supersonic value, all oblique portions of the waves incline much more, and the detached normal shock portion of the bow wave moves closer to the leading edge.

Of course, all components of the aircraft are affected by compressibility in a manner somewhat similar to that of basic airfoil. The tail, fuselage, nacelles, canopy, etc., and the effect of the interference between the various surfaces of the aircraft must be considered.

Force divergence

The airflow separation induced by shock wave formation can create significant variations in the aerodynamic force coefficients. When the free-stream speed is greater than critical Mach number, some typical effects on an airfoil section are as follows:

- An increase in the section drag coefficient for a given section lift coefficient.
- A decrease in section lift coefficient for a given section angle of attack.
- A change on section pitching moment coefficient. A reference point is usually taken by a plot of drag coefficient versus Mach number for a constant lift coefficient. Such a graph is shown in Fig. 8-9. The Mach number that produces a sharp change in the drag coefficient is termed the *force divergence* Mach number and, for most airfoils, usually exceeds the critical Mach number at least 5–10 percent. This condition is also referred to as the *drag divergence* or *drag rise*.

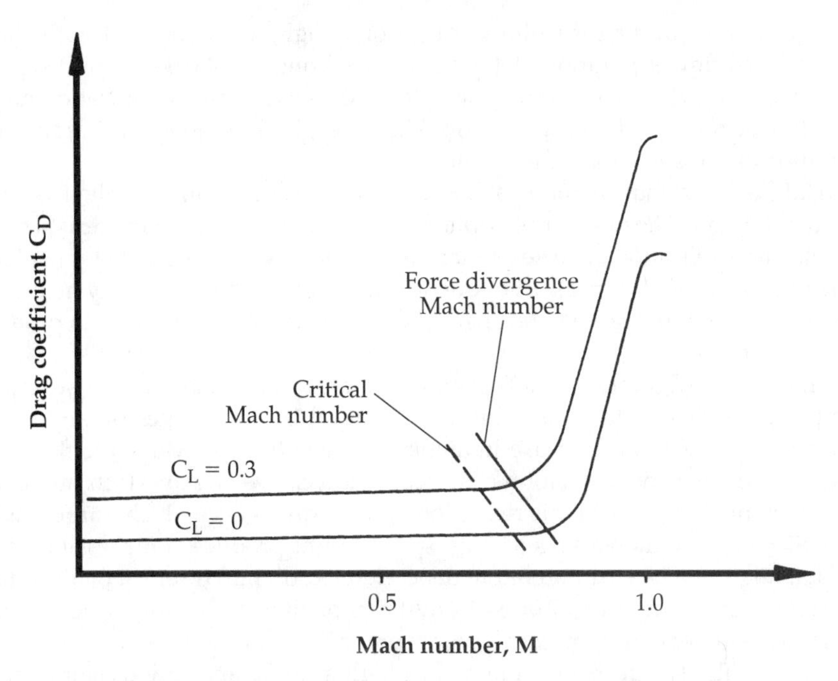

Fig. 8-9. Compressibility drag rise.

CHARACTERISTICS OF TRANSONIC FLIGHT

Buffet, trim, and stability changes, plus a decrease in control-surface effectiveness, are associated with drag rise. Conventional aileron, rudder, and elevator surfaces subjected to this high frequency buffet might "buzz," and changes in hinge moments might produce undesirable control forces. Of course, if the buffet is quite severe and prolonged, structural damage might occur if this operation is in violation of operating limitations.

When airflow separation occurs on the wing due to shock wave formation, there will be a loss of lift and subsequent loss of downwash aft of the affected area. If the wings shock unevenly due to physical shape differences or sideslip, a rolling moment will be created in the direction of the initial loss of lift and contribute to control difficulty; a *wing drop* occurs. If the shock-induced separation occurs symmetrically near the wing root, a decrease in downwash behind this area is a corollary of the loss of lift. A decrease in downwash on the horizontal tail will create a diving moment, and the aircraft will *tuck under*.

If these conditions occur on a swept-wing planform, the wing center-of-pressure shift contributes to the trim change. When the shock forms first at the wing root, the wing center-of-pressure moves aft and adds to the diving moment. When the shock forms first at the wingtips, the center of pressure moves forward, and the resulting climbing moment and tail downwash change can contribute to *pitch-up*.

Because most of the difficulties of transonic flight are associated with shock wave-induced flow separation, delaying or alleviating the shock-induced separation will improve the aerodynamic characteristics. An aircraft configuration might utilize thin surfaces of low aspect ratio with sweepback to delay and reduce the magnitude of transonic force divergence.

In addition, various methods of boundary layer control, high-lift devices, vortex generators, and the like can be applied to improve transonic characteristics. For example, the application of vortex generators to a surface can produce higher local surface velocities and increase the kinetic energy of the boundary layer; thus, a more severe pressure gradient (stronger shock wave) will be necessary to produce airflow separation.

When the configuration of a transonic aircraft is fixed, the pilot must respect the effect of angle of attack and altitude. The local flow velocities on any upper surface increase with an increase in angle of attack; hence, local sonic flow and subsequent shock wave formation can occur at lower free-stream Mach numbers.

A pilot must appreciate this reduction of force divergence Mach number with lift coefficient since maneuvers at high speed might produce compressibility effects that might not be encountered in unaccelerated flight. The effect of altitude is important because the magnitude of any force or moment change due to compressibility will depend upon the dynamic pressure of the airstream.

Compressibility effects encountered at high altitude and low dynamic pressure might be of little consequence in the operation of a transonic aircraft; however, the same compressibility effects encountered at low altitudes and high dynamic pressures will create greater trim changes, heavier buffet, and the like, and perhaps transonic flight restrictions that are of principal interest only at low altitude.

CHARACTERISTICS OF SUPERSONIC FLIGHT

Many of the particular effects of supersonic flight are subsequently detailed, but for the moment, many general effects can be anticipated. The airplane configuration must have aerodynamic shapes that will have low drag in compressible flow. Generally, this will require airfoil sections of low thickness ratio and sharp leading edges and body shapes of high fineness ratio to minimize the supersonic wave drag.

Due to the aft movement of the aerodynamic center with supersonic flow, the increase in static longitudinal stability will demand effective, powerful control surfaces to achieve adequate controllability for supersonic maneuvering. Figure 8-10 shows a Navy F/A-18 ready for a catapult launch from an aircraft carrier. Notice its all-movable stabilator in the nose-up position.

As a corollary of supersonic flight, the shock wave formation on the airplane might create special problems outside the immediate vicinity of the airplane surfaces. While the shock waves that are a great distance away from the airplane can be quite weak, the pressure waves can be of sufficient magnitude to create an audible disturbance; thus, sonic booms will be a simple consequence of supersonic flight.

Fig. 8-10. A McDonnell Douglas F/A-18 ready for a catapult launch from an aircraft carrier. Note the all-movable stabilator in the nose-up position.

The aircraft powerplants for supersonic flight must be of relatively high thrust output. Also, in many cases it might be necessary to provide the air-breathing powerplant with special inlet configurations that will slow the airflow to subsonic prior to reaching the compressor face or combustion chamber. Aerodynamic heating of supersonic flight can provide critical inlet temperatures for the gas turbine engine, as well as critical structural temperatures.

TRANSONIC AND SUPERSONIC CONFIGURATIONS

Aircraft configurations developed for high-speed flight will have significant differences in shape and planform when compared with aircraft designed for low-speed flight. One of the outstanding differences will be in the selection of airfoil profiles for transonic or supersonic flight.

Airfoil sections

It should be obvious that airfoils for high-speed subsonic flight should have high critical Mach numbers because critical Mach number defines the lower limit for shock wave formation and subsequent force divergence. An additional complication to airfoil selection in this speed range is that the airfoil should have a high maximum lift coefficient and sufficient thickness to allow application of high lift devices. Otherwise, an excessive wing area would be required to provide maneuverability and reasonable takeoff and landing speeds; however, if high-speed flight is the primary consideration, the airfoil must be chosen to have the highest practical critical Mach number.

Recall that critical Mach number is the flight Mach number that produces first evidence of local sonic flow; thus, the airfoil shape and lift coefficient, which determine the pressure and velocity distribution, will have a profound effect on critical Mach number. Conventional, low-speed airfoil shapes have relatively poor compressibility characteristics because of the high local velocities near the leading edge. These high local velocities are inevitable if both the maximum thickness and camber are well forward on the chord.

Compressibility characteristics can be improved by moving the points of maximum camber and thickness aft on the chord. This would distribute the pressure and velocity more evenly along the chord and produce a lower peak velocity for the same lift coefficient. Fortunately, the airfoil shape to provide extensive laminar flow and low-profile drag in low-speed subsonic flight will provide a pressure distribution that is favorable for high-speed flight.

In order to obtain a high critical Mach number from an airfoil at a certain low-lift coefficient, the section must have two properties:

- Low thickness ratio. The point of maximum thickness should be aft to smooth the pressure distribution.
- Low camber. The mean camber line should be shaped to help minimize the local velocity peaks.

As discussed in chapter 4, the super-critical airfoil developed by NASA meets these requirements. In addition, the higher the required lift coefficient, the lower the critical Mach number, and more camber is required of the airfoil. If supersonic flight is a possibility, the thickness ratio and leading edge radius must be small to decrease wave drag.

Figure 8-11 shows the flow patterns for two basic supersonic airfoil sections. The wave drag is the only factor of difference between the two airfoil sections. For the same thickness ratio, the circular-arc airfoil would have a larger wedge angle formed between the upper and lower surfaces at the leading edge. At the same flight Mach number, the larger angle at the leading edge would form the stronger shock wave at the nose and cause a greater pressure change on the circular-arc airfoil.

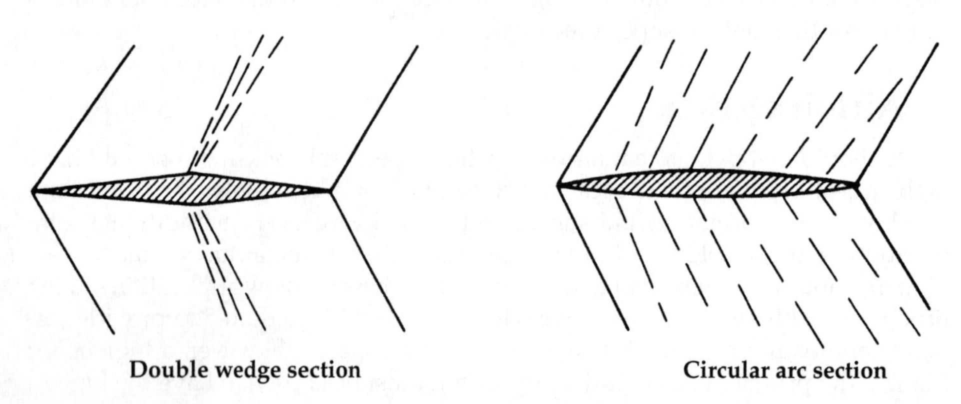

Fig. 8-11. Flow patterns for two basic supersonic airfoil sections.

This same principle applies when investigating the effect of airfoil thickness. The wave-drag coefficients for both airfoils vary as the *square* of the thickness ratio, which means that if the thickness ratio were doubled, the wave-drag coefficient would be four times as great. If the thickness were increased, the airflow at the leading edge will experience a greater change in direction, and a stronger shock wave will be formed.

This powerful variation of wave-drag-with-thickness ratio necessitates the use of very thin airfoils with sharp leading edges for supersonic flight. An additional consideration is that thin airfoil sections favor the use of low aspect ratios and high taper to obtain lightweight structures and preserve stiffness and rigidity.

Any aerodynamic surface becomes less sensitive to changes in angle of attack at higher Mach numbers. The decrease in lift curve slope with Mach number has tremendous implications on the stability and control of high-speed aircraft. The vertical tail becomes less sensitive to angles of sideslip, and the directional stability of the aircraft will deteriorate with Mach number.

The horizontal tail of the airplane experiences the same general effect and contributes less damping to longitudinal pitching oscillations. These effects can become so significant at high Mach numbers that the aircraft might require complete synthetic stabilization.

Planform effects

The development of surfaces for high speed involves consideration of many items in addition to the airfoil sections. Taper, aspect ratio, and sweepback can produce major effects on the aerodynamic characteristics of a surface in high-speed flight. Sweepback produces an unusual effect on the high-speed characteristics of a surface.

A grossly simplified method of visualizing the effect of sweepback is shown in Fig. 8-12, which is a variation of Fig. 4-11 in chapter 4. The swept-wing that is shown has the streamwise velocity broken down to a component of velocity perpendicular to the leading edge and a component parallel to the leading edge. The component of speed perpendicular to the leading edge is less than the free-stream speed (by the cosine of the sweep angle), and it is this velocity component that determines the magnitude of the pressure distribution.

The component of speed parallel to the leading edge could be visualized as moving across constant sections and, in doing so, does not contribute to the pressure distribution on the swept wing; hence, sweep of a surface produces a beneficial effect in high-speed flight because higher flight speeds can be obtained before components of speed perpendicular to the leading edge produce critical conditions on the wing.

This is one of the most important advantages of sweep because there is an increase in critical Mach number, force divergence Mach number, and the Mach number at which the drag rise will peak. In other words, sweep will delay the onset of compressibility effects.

Generally, the effect of wing sweep will apply to either sweepback or sweepforward. The swept-forward wing has been used in rare instances, but the aero-

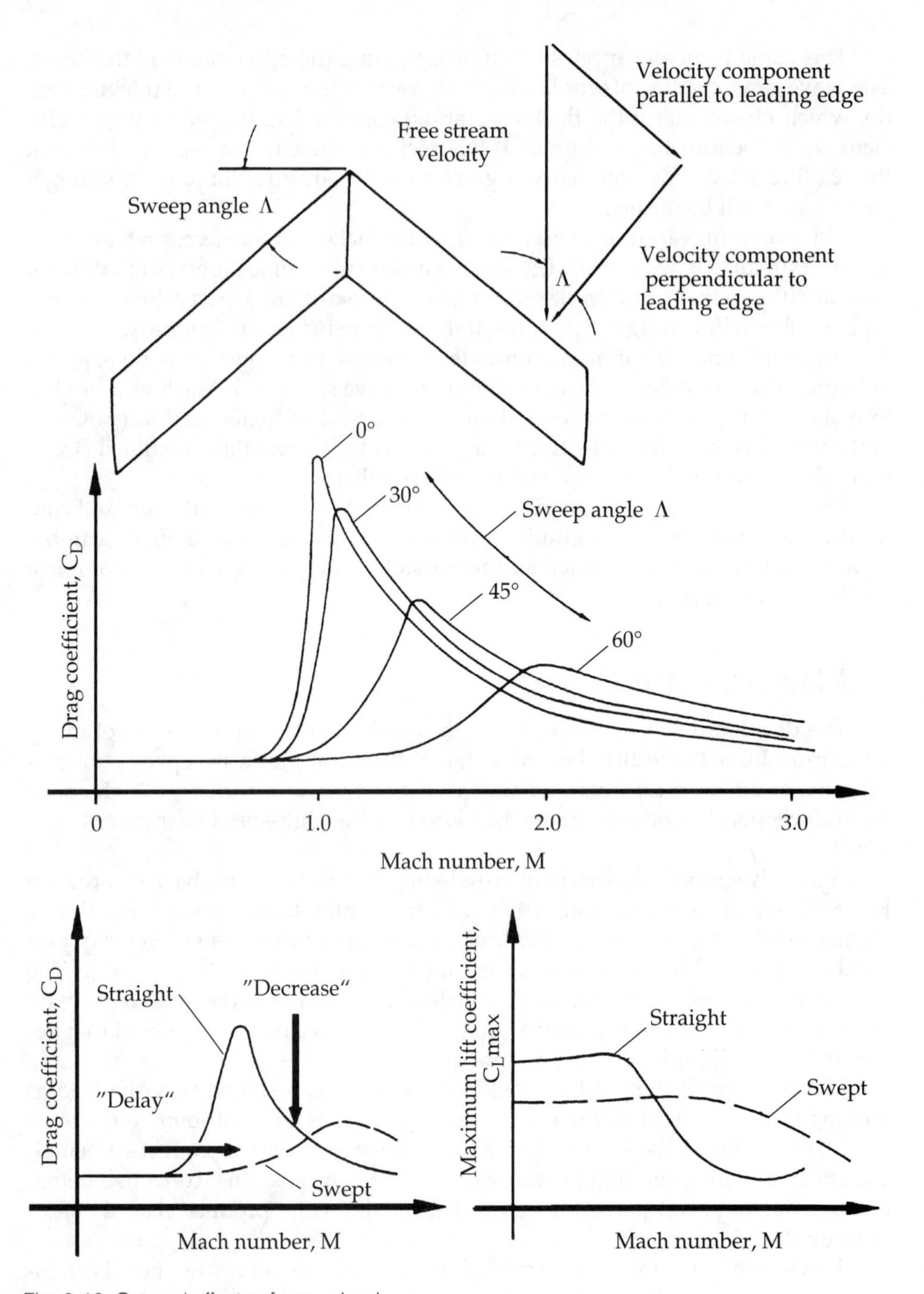

Fig. 8-12. General effects of sweepback.

elastic instability of such a wing creates such a problem that sweepback is more practical for ordinary applications. Chapter 5 details sweepforward wings.

In addition to the delay of the onset of compressibility effects, sweepback will reduce the magnitude of the changes in force coefficients due to compressibility. Since the component of velocity perpendicular to the leading edge is less than the free-stream velocity, the magnitude of all pressure forces on the wing will be reduced (approximately by the square of the cosine of the sweep angle).

Since compressibility force divergence occurs due to changes in pressure distribution, the use of sweepback will "soften" the force divergence. This effect is illustrated by the graph of Fig. 8-12, which shows the typical variation of drag coefficient with Mach number for various sweepback angles. The straight wing shown begins drag rise at M = 0.70, reaches a peak near M = 1.0, and begins a continual drop past M = 1.0. Note that the use of sweepback then delays the drag rise to some higher Mach number and reduces the magnitude of the drag rise.

In view of the preceding discussion, sweepback will have two principal advantages:

- First, sweepback will delay the onset of all compressibility effects. Critical Mach number and force divergence Mach number will increase because the velocity component affecting the pressure distribution is less than the free-stream velocity. Also, the peak of drag rise is delayed to some higher supersonic speed—approximately the speed that produces sonic flow perpendicular to the leading edge. Various sweeps applied to wings of moderate aspect ratio will produce these approximate effects in transonic flight:

Sweep angle (Λ)	Percent increase in critical Mach number	Percent increase in drag peak Mach number
0°	0	0
15°	2	4
30°	8	15
45°	20	41
60°	41	100

- Second, sweepback will reduce the magnitude of change in the aerodynamic force coefficients due to compressibility. Any change in drag, lift, or moment coefficients will be reduced by the use of sweepback. Various sweep angles applied to wings of moderate aspect ratio will produce these approximate effects in transonic flight:

Sweep angle (Λ)	Percent reduction in drag rise	Percent reduction in loss of $C_{L_{max}}$
0°	0	0
15°	5	3
30°	15	13
45°	35	30
60°	60	50

These advantages of drag reduction and preservation of the transonic maximum lift coefficient are illustrated in Fig. 8-12.

Thus, the use of sweepback on a transonic aircraft will reduce and delay the drag rise and preserve the maneuverability of the aircraft in transonic flight. It should be noted that a small amount of sweepback produces very little benefit. If sweepback is to be used at all, at least 30°–35° must be used to produce any significant benefit.

Also note from Fig. 8-12 that the amount of sweepback required to delay drag rise in supersonic flight is very large, that is, more than 60° necessary at M = 2.0. By comparison of the drag curves at high Mach numbers, it will be appreciated that extremely high (and possibly impractical) sweepback is necessary to delay drag rise and that the lowest drag is obtained with zero sweepback; therefore, the planform of a wing designed to operate continuously at high Mach numbers will tend to be very thin, low-aspect ratio, and unswept. An immediate conclusion is that sweepback is a device of greatest application in the regime of transonic flight.

Four less-significant advantages of sweepback are as follows:

- First, the wing lift-curve slope is reduced for a given aspect ratio. This is illustrated by the lift-curve comparison of Fig. 8-13 for the straight and swept wing. Any reduction of lift-curve slope implies that the wing is less sensitive to changes in angle of attack. This is a beneficial effect only when the effect of gusts and turbulence is considered.

  Since the swept wing has the lower lift-curve slope, it will be less sensitive to gusts and experience less "bump" due to gust for a given aspect ratio and wing loading. This is a consideration particular to the aircraft with a structural design that shows a predominating effect of the gust-load spectrum: transport, cargo, and patrol types.
- Second, divergence of a surface is an aeroelastic problem that can occur at high dynamic pressures. Combined bending and twisting deflections interact with aerodynamic forces to produce sudden failure of the surface at high speeds. Sweepforward will aggravate this situation by "leading" the wing into the windstream and tends to lower the divergence speed (chapter 5). On the other hand, sweepback tends to stabilize the surface by "trailing" and tends to raise the divergence speed. By this tendency, sweepback might be beneficial in preventing divergence within the anticipated speed range.
- Third, sweepback contributes slightly to the static directional, or *weathercock*, stability of an aircraft. This effect can be appreciated by inspection of Fig. 8-13, which shows the swept wing in a yaw or sideslip. The wing into the wind has less sweep and a slight increase in drag; the wing away from the wind has more sweep and less drag. The net effect of these force changes is to produce a yawing moment tending to return the nose into the relative wind. This directional stability contribution is usually small and of importance in tailless aircraft only.
- Fourth, sweepback contributes to lateral stability in the same sense as dihedral. When the swept-wing aircraft is placed in a sideslip, the wing into the wind experiences an increase in lift because the sweep is less, and the wing away from the wind produces less lift because the sweep is greater. As

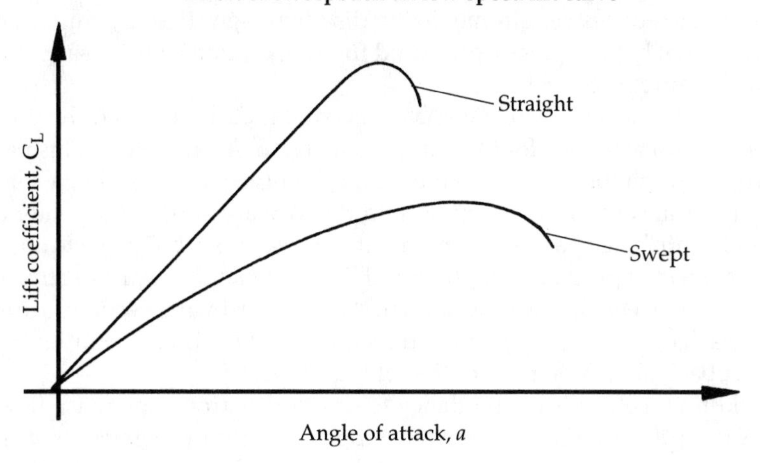

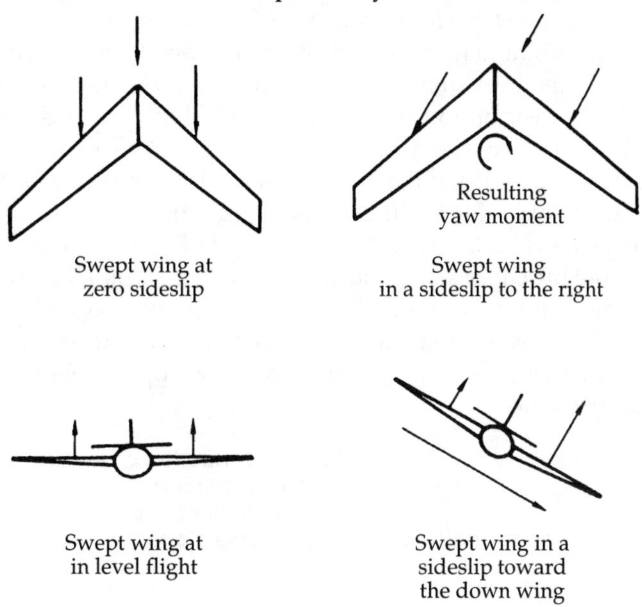

Fig. 8-13. Aerodynamic effects of sweepback, which tend to flatten the lift curve and provide an increase in lateral stability.

shown in Fig. 8-13, the swept-wing aircraft in a sideslip experiences lift changes and a subsequent rolling moment that tends to right the aircraft. This lateral stability contribution depends on the sweepback and the lift coefficient of the wing. A highly swept wing operating at a high-lift coefficient usually experiences such an excess of this lateral stability contribution that adequate controllability might be a significant problem.

As shown, the swept wing has certain important advantages; however, the use of sweepback produces certain inevitable disadvantages that are important from the standpoint of both airplane design and flight operations. Five disadvantages of sweepback should be considered.

First disadvantage: When sweepback is combined with taper, there is an extremely powerful tendency for the wing to stall tip-first (chapter 5). This pattern of stall is very undesirable because there would be little stall warning, a serious reduction in lateral control effectiveness, and the forward shift of the center of pressure would contribute to a nose-up moment (*pitch up* or *stick force lightening*). Taper has its own effect of producing higher local lift coefficients toward the tip, and one of the effects of sweepback is very similar. All outboard wing sections are affected by the upwash of the preceding inboard sections, and the lift distribution resulting from sweepback alone is similar to that of high taper.

An additional effect is the tendency to develop a strong spanwise flow of the boundary layer toward the tip when the wing is at high-lift coefficients. This spanwise flow produces a relatively low energy boundary layer near the tip that can be easily separated. The combined effect of taper and sweep present a considerable problem of tip stall, and this is illustrated by the flow patterns of Fig. 8-14.

Design for high-speed performance might dictate high sweepback, while structural efficiency might demand a highly tapered planform. When such is the case, the wing might require extensive aerodynamic tailoring to provide a suitable stall pattern and a lift distribution at cruise condition that reduces drag due to lift. Washout of the tip, variation of section camber throughout span, flow fences, slats, leading edge extension, and the like are typical devices used to modify the stall pattern and minimize drag due to lift at cruise condition.

Second sweepback disadvantage: The lift curve in Fig. 8-13 shows that sweepback will reduce the lift-curve slope and the subsonic maximum lift coefficient. It is important to note that this case is definitely subsonic because sweepback can be used to improve the transonic maneuvering capability. Various sweep angles applied to wings of moderate aspect ratio produce these approximate effects on the subsonic lift characteristics:

Sweep angle (Λ)	Percent reduction of subsonic maximum lift coefficient and lift-curve slope
0°	0
15°	4
30°	14
45°	30
60°	50

The reduction of the low-speed maximum lift coefficient (which is in addition to that lost due to tip stall) has very important implications in design. If wing loading is not reduced, stall speeds increase and subsonic maneuverability decreases. On the other hand, if wing loading is reduced, the increase in wing surface area

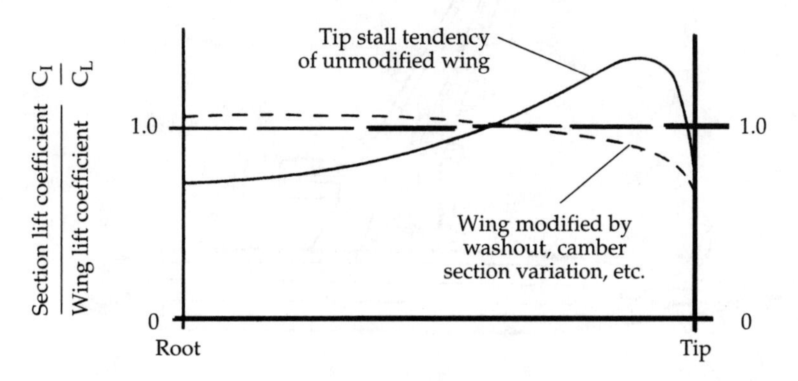

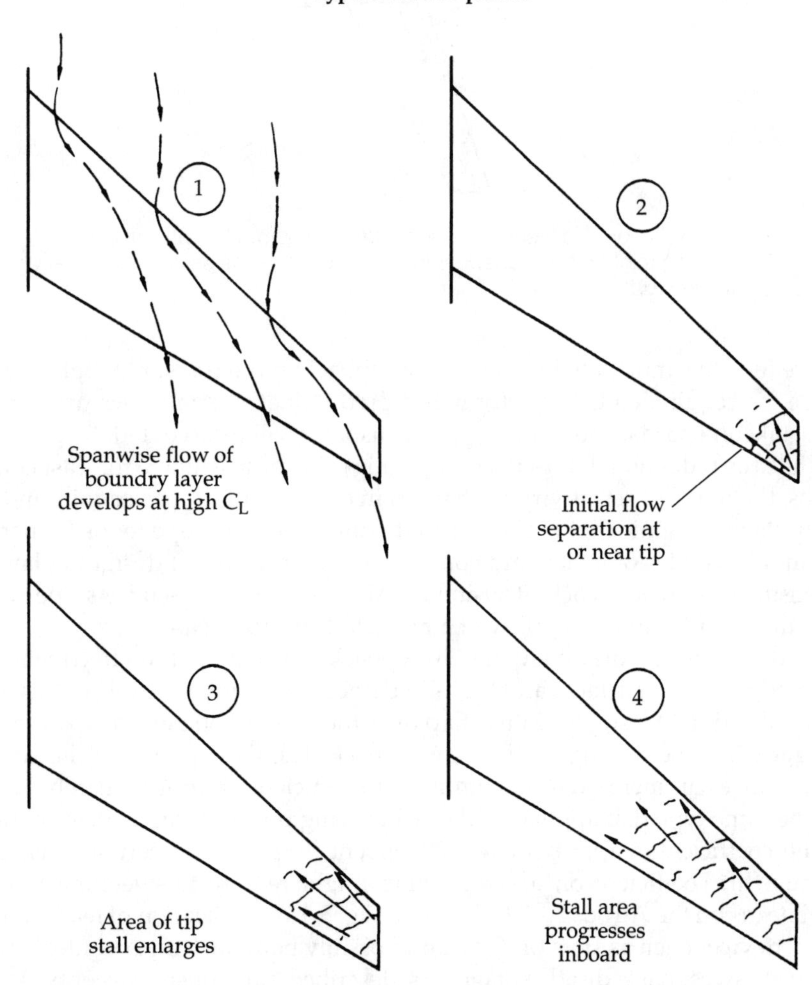

Fig. 8-14. Stall characteristics of a tapered swept wing.

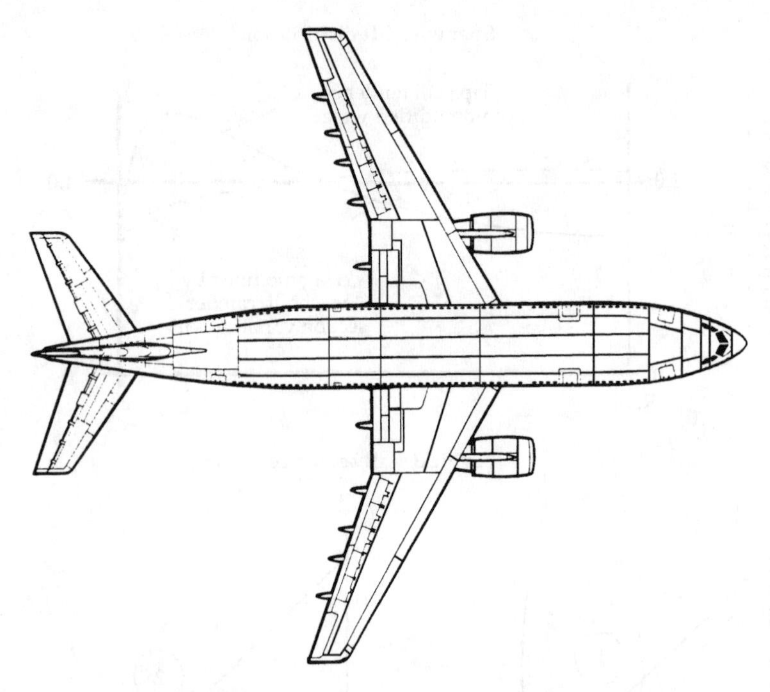

Fig. 8-15. To produce some reasonable maximum lift coefficient on a swept wing might require unsweeping the flap hinge line as shown on the Airbus A-310.

might reduce the anticipated benefit of sweepback in the transonic flight regime. Because the requirements of performance predominate, certain increases of stall speeds, takeoff speeds, and landing speeds usually will be accepted.

While the reduction of lift-curve slope might be an advantage for gust considerations, the reduced sensitivity to changes in angle of attack has certain undesirable effects in subsonic flight. The reduced wing lift-curve slope tends to increase maximum lift angles of attack and complicate the problems of designing landing gear, ensuring sufficient cockpit visibility. Also, the lower lift-curve slope would reduce the contribution to stability of a given tail surface area.

Third sweepback disadvantage: Sweepback will reduce the effectiveness of trailing-edge control surfaces and high-lift devices. A typical example of this effect is the application of a single slotted flap over the inboard 60 percent span to both a straight wing and a wing with 35° sweepback. The flap applied to the straight wing produces an increase in maximum lift coefficient of approximately 50 percent. The same type flap applied to the swept wing produces an increase in maximum lift coefficient of approximately 20 percent. To produce a certain reasonable maximum lift coefficient on a swept wing might require unsweeping the flap hinge line (as on the Airbus A-310 shown in Fig. 8-15), application of leading-edge high-lift devices such as slots or slats, and possibly boundary layer control.

Fourth sweepback disadvantage: As described previously, sweepback contributes to lateral stability by producing stable rolling moments with sideslip. The lateral stability contribution of sweepback varies with the amount of wing sweepback and wing lift coefficient; large sweepback and high-lift coefficients produce a large contribution to lateral stability. While stability is desirable, any excess of stability will reduce controllability. For the majority of airplane configurations, high lateral stability is neither necessary nor desirable, but adequate control in roll is absolutely necessary for good flying qualities.

An excess of lateral stability from sweepback can aggravate Dutch roll problems and produce marginal control during crosswind takeoff and landing where the aircraft must move in a controlled sideslip; therefore, it is not unusual to find swept-wing aircraft with negative dihedral and lateral control devices designed principally to meet crosswind takeoff and landing requirements. Most high-wing, swept-wing aircraft, such as the C-141 (Fig. 8-16), use negative dihedral to counteract excessive lateral stability.

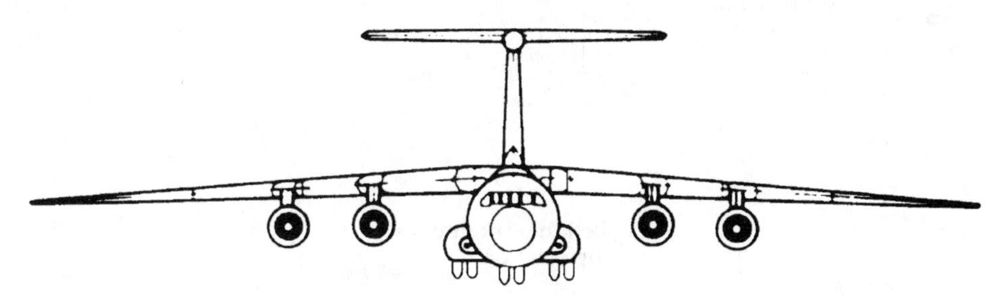

Fig. 8-16. Most high-wing swept-wing aircraft use negative dihedral, such as shown on the C-141, to counteract excessive lateral stability due to the swept wing.

Fifth sweepback disadvantage: The structural complexity and aeroelastic problems created by sweepback are of great importance. A swept wing has a greater structural span than a straight wing of the same area and aspect ratio (Fig. 8-17); this effect increases wing structural weight because greater bending and shear material must be distributed in the wing to produce the same design strength.

An additional problem is created near the wing root and carrythrough structure due to the large twisting loads and the tendency of the bending stress distribution to concentrate toward the trailing edge. Also shown in Fig. 8-17 is the influence of wing deflection on the spanwise lift distribution. Wing bending produces tip rotation that tends to unload the tip and move the center of pressure forward; thus, the same effect that tends to allay divergence can make an undesirable contribution to longitudinal stability.

Effect of aspect ratio and tip shape

In addition to wing sweep, planform properties such as aspect ratio and tip shape can produce significant effects on the aerodynamic characteristics at high speeds. There is no particular effect of aspect ratio on critical Mach number at high or medium aspect ratios. The aspect ratio must be less than 4 or 5 to produce any

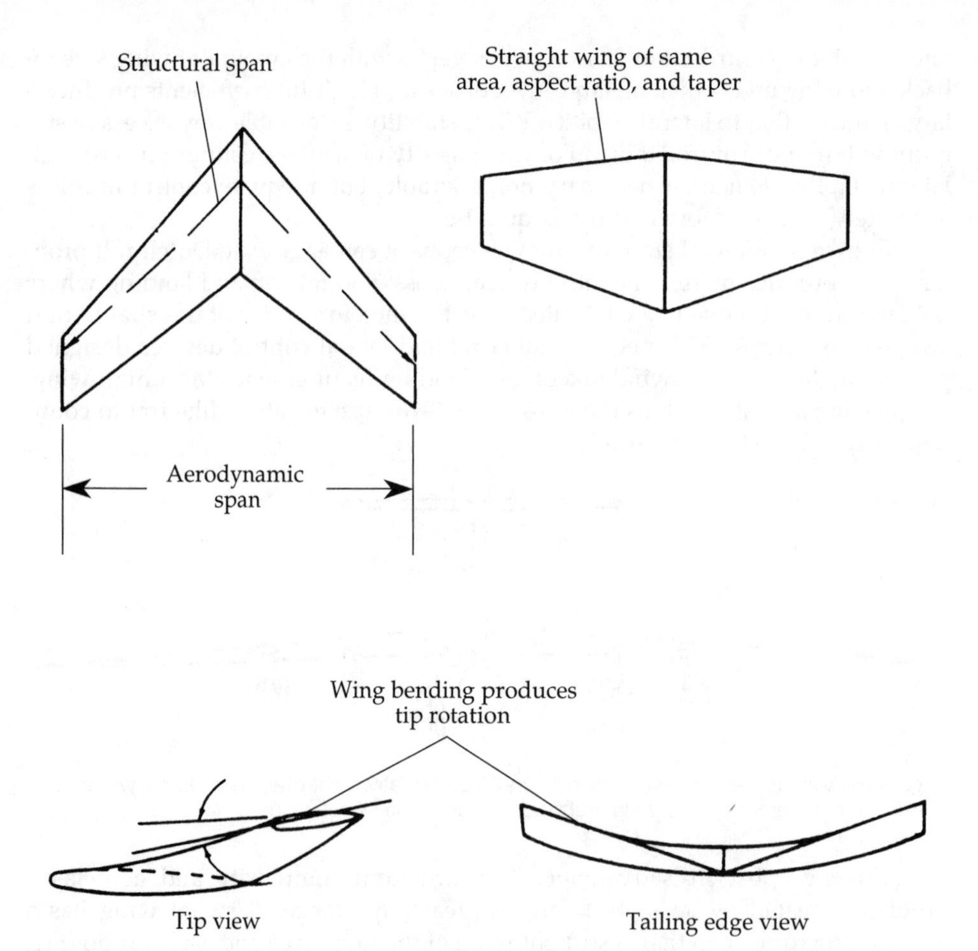

Fig. 8-17. Structural complications due to sweepback. As the wing bends upward, the tip angle of attack decreases.

apparent change in critical Mach number. This effect is shown for a typical 9-percent-thick symmetrical airfoil in the graph of Fig. 8-18.

Note that very low-aspect ratios are required to cause a significant increase in critical Mach number. Very-low-aspect ratios create the extremes of three-dimensional flow and a subsequent increase in free-stream speed to create local sonic flow. Actually, the extremely low-aspect ratios required to produce high critical Mach number are not too practical. Generally, the advantage of low-aspect ratio must be combined with sweepback and high-speed airfoil sections.

The thin rectangular wing in supersonic flow illustrates several important facts. As shown in Fig. 8-18, Mach cones form at the tips of the rectangular wing and affect the pressure distribution on the area within the cone. The vortex develops within the tip cone due to the pressure differential, and the resulting average pressure on the area within the cone is approximately one-half the pressure between the cones. Three-dimensional flow on the wing is then confined to the area

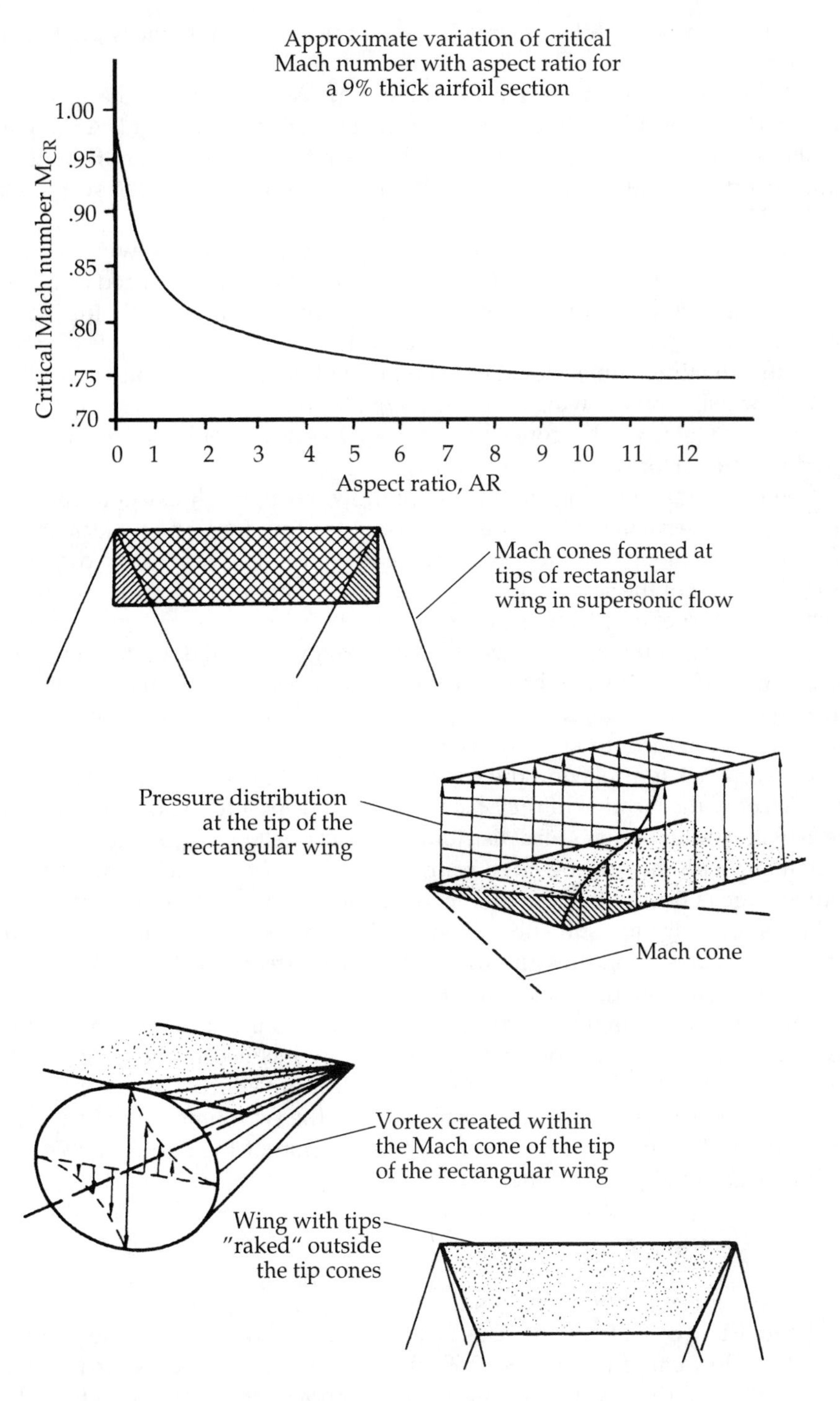

Fig. 8-18. General planform effects.

within the tip cones, while the area between the cones experiences pure two-dimensional flow.

It is important to realize that the three-dimensional flow on the rectangular wing in supersonic flight differs greatly from that of subsonic flight. A wing of finite aspect ratio in subsonic flight experiences a three-dimensional flow that includes the tip vortices, downwash behind the wing, upwash ahead of the wing, and local induced velocities along the span.

Recall that the local induced velocities along the span of the wing would incline the section lift aft relative to the free stream and result in induced drag (chapter 5). Such a flow condition cannot be directly correlated with the wing in supersonic flow. The flow pattern for the rectangular wing of Fig. 8-18 demonstrates that the three-dimensional flow is confined to the tip, and pure two-dimensional flow exists on the wing area between the tip cones. If the wingtips were to be "raked" outside the tip cones, the entire wing flow would correspond to the two-dimensional (or section) conditions.

Therefore, for the wing in supersonic flow, no upwash exists ahead of the wing, three-dimensional effects are confined to the tip cones, and no local induced velocities occur along the span between the tip cones. The supersonic drag due to lift is a function of the section and angle of attack. The subsonic induced drag is a function of lift coefficient and aspect ratio.

This comparison makes it obvious that supersonic flight does not demand the use of high-aspect-ratio planforms typical of low-speed aircraft. In fact, low-aspect ratios and high taper are favorable from the standpoint of structural considerations if very thin sections are used to minimize wave drag.

If sweepback is applied to the supersonic wing, the pressure distribution will be affected by the location of the Mach cone with respect to the leading edge. Figure 8-19 illustrates the pressure distribution for the delta wing planform in supersonic flight with the leading edge behind or ahead of the Mach cone. When the leading edge is behind the Mach cone, the components of velocity perpendicular to the leading edge are still subsonic even though the free-stream flow is supersonic, and the resulting pressure distribution will greatly resemble the subsonic pressure distribution for such a planform.

Tailoring the leading-edge shape and camber can minimize the components of the high leading edge suction pressure, which are inclined in the drag direction, and the drag due to lift can be reduced. If the leading edge is ahead of the Mach cone, the flow over this area will correspond to the two-dimensional supersonic flow and produce constant pressure for that portion of the surface between the leading edge and the Mach cone.

Control surfaces

The design of control surfaces for transonic and supersonic flight involves many important considerations. This fact is illustrated by the typical transonic and supersonic flow patterns of Fig. 8-19. Trailing-edge control surfaces can be affected adversely by the shock waves formed in flight above critical Mach number. If the

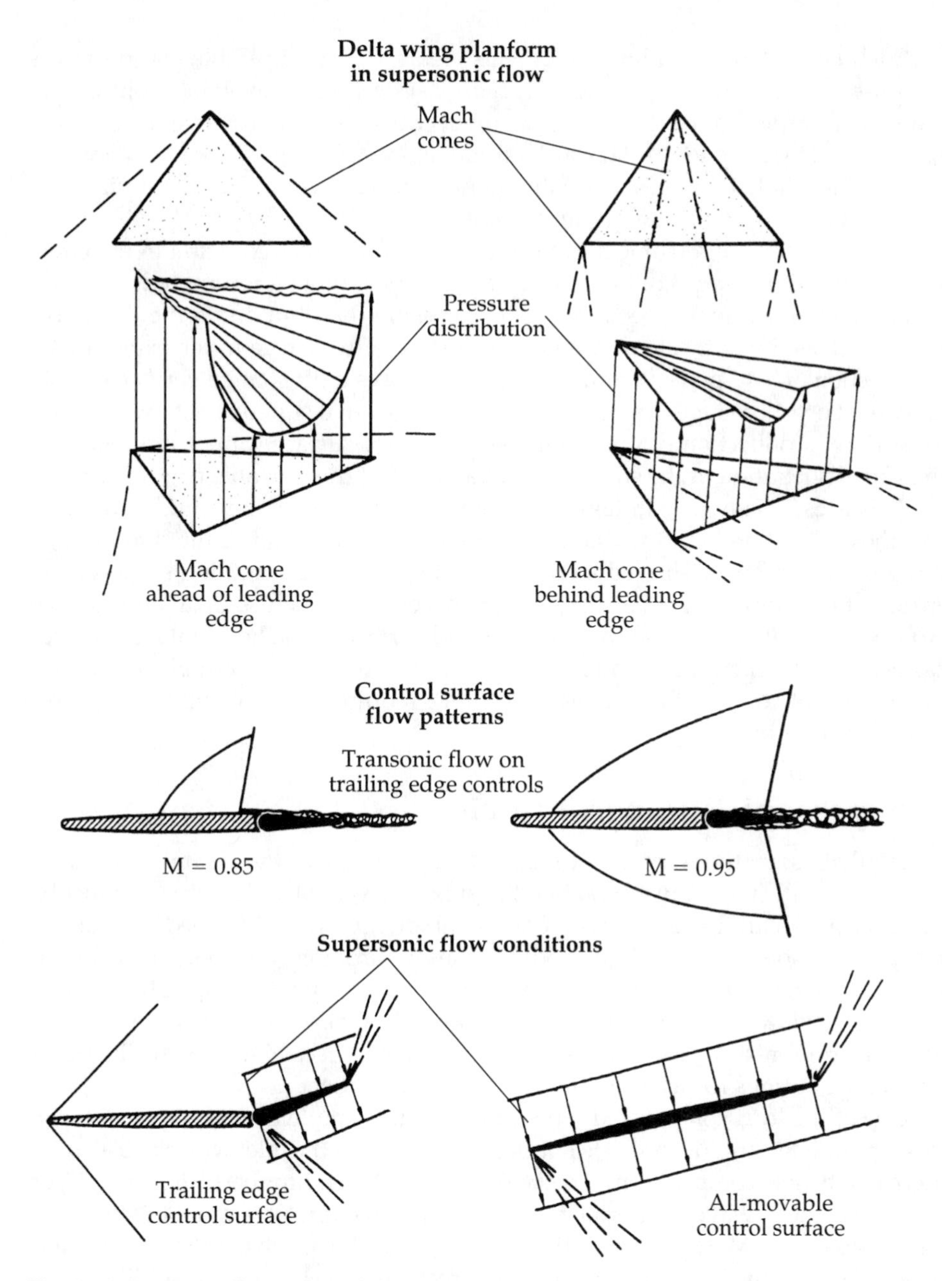

Fig. 8-19. Planform effects of a delta wing, and control surface flow patterns.

airflow is separated by the shock wave, the resulting buffet of the control surface can be very objectionable. In addition to the buffet of the surface, the change in the pressure distribution due to separation and the shock wave location can create very large changes in control surface hinge moments.

Such large changes in hinge moments create very undesirable control forces and present the need for an *irreversible control system.* An irreversible control system would employ powerful hydraulic or electric actuators to move the surfaces upon control by the pilot, and the airloads developed on the surface could not feed back to the pilot. Of course, suitable control forces would be synthesized by bungees, Q springs, bobweights, and the like.

Transonic and supersonic flight can cause a noticeable reduction in the effectiveness of trailing-edge control surfaces. The deflection of a trailing-edge control surface at slow subsonic speeds alters the pressure distribution on the fixed portion as well as the movable portion of the surface. This is true to the extent that a 1° deflection of a 40-percent-chord elevator produces a lift change very nearly the equivalent of a 1° change in stabilizer setting; however, if supersonic flow exists on the surface, a deflection of the trailing-edge control surface cannot influence the pressure distribution in the supersonic area ahead of the movable control surface.

This is especially true in high supersonic flight where supersonic flow exists over the entire chord and the change in pressure distribution is limited to the area of the control surface. The reduction in effectiveness of the trailing-edge control surface at transonic and supersonic speeds necessitates the use of an all-movable surface. Application of the all-movable control surface to the horizontal tail is typical because the increase in longitudinal stability in supersonic flight requires a high degree of control effectiveness to achieve required controllability for supersonic maneuvering.

Supersonic engine air inlets

Air that enters the compressor section of a jet engine or the combustion chamber of a ramjet usually must be slowed to subsonic velocity. This process must be accomplished with the least possible waste of energy. At flight speeds just above the speed of sound, only slight modifications to ordinary subsonic inlet design produce satisfactory performance; however, at supersonic flight speeds, the inlet design must slow the air with the weakest possible series or combination of shock waves to minimize energy losses and temperature rise. Figure 8-20 illustrates some of the various forms of supersonic inlets, or *diffusers*.

One of the least complicated types of inlet is the simple normal shock diffuser. This type of inlet employs a single normal shock wave at the inlet with a subsequent internal subsonic compression. At low supersonic Mach numbers, the strength of the normal shock wave is not too great, and this type of inlet is quite practical. At higher supersonic Mach numbers, the single normal shock wave is very strong and causes a large reduction in the total pressure recovered by the inlet. In addition, it is necessary to consider that the wasted energy of the airstream will appear as an additional undesirable rise in temperature of the captured inlet airflow.

If the supersonic airstream can be captured, the shock wave formations will be swallowed, and a gradual contraction will reduce the speed to just above sonic. The subsequent diverging flow section can then produce the normal shock wave, which slows the airstream to subsonic. Further expansion continues to slow the air to lower subsonic speeds. This is the convergent-divergent inlet shown in Fig. 8-20.

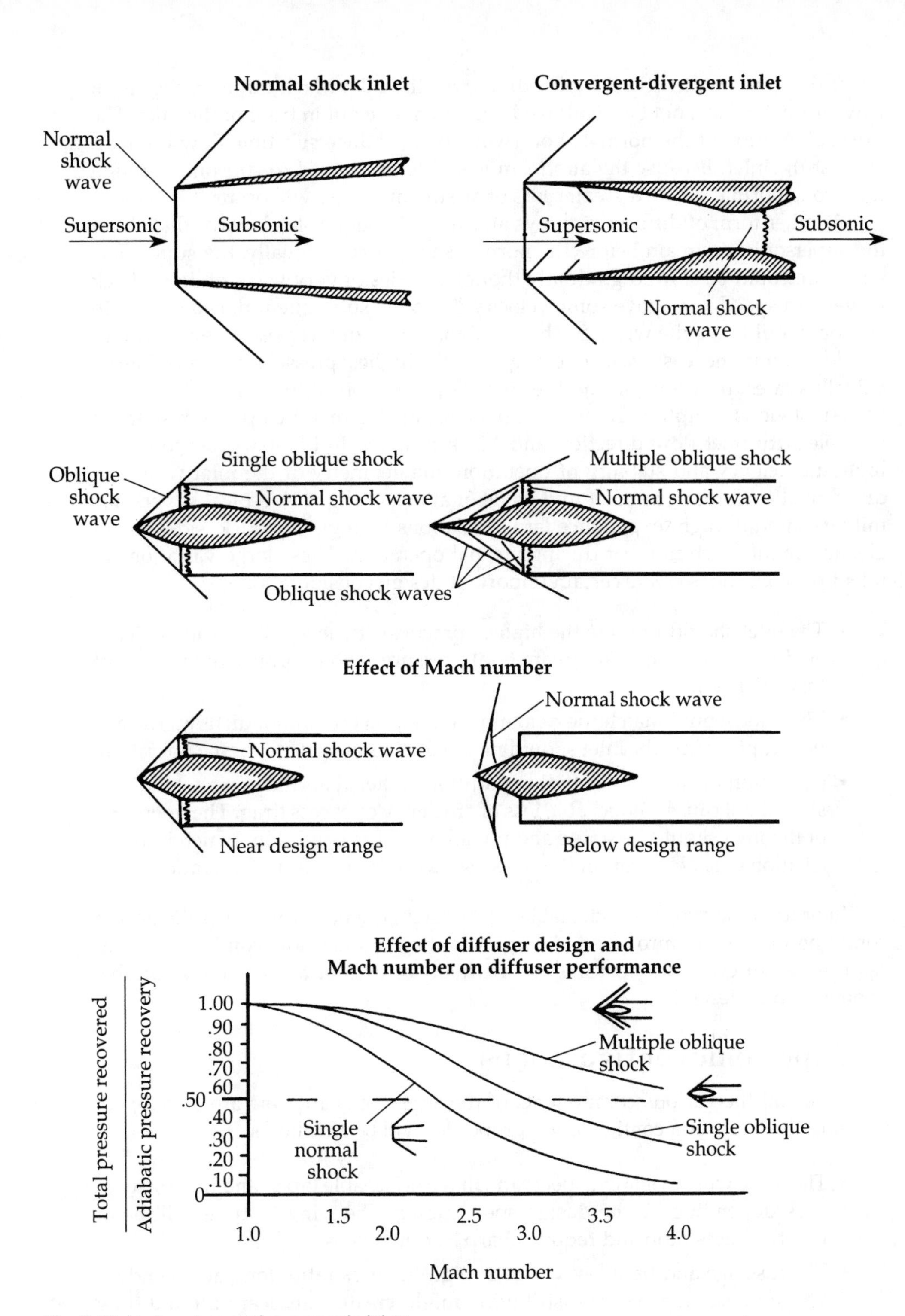

Fig. 8-20. Various types of supersonic inlets.

If the initial contraction is too extreme for the inlet Mach number, the shock wave formation will not be swallowed and will move out in front of the inlet. The external location of the normal shock wave will produce subsonic flow immediately at the inlet. Because the airstream is suddenly slowed to subsonic through the strong normal shock, a greater loss of airstream energy will occur.

Another form of diffuser employs an external oblique shock wave that slows the supersonic airstream before the normal shock occurs. Ideally, the supersonic airstream could be slowed gradually though a series of very weak oblique shock waves to a speed just above sonic velocity. Then the subsequent normal shock to subsonic could be quite weak. Such a combination of the weakest possible waves would result in the least waste of energy and the highest pressure recovery. Figure 8-20 illustrates this principle and the efficiency of various diffusers.

An obvious complication of the supersonic inlet is that the optimum shape is variable with inlet flow direction and Mach number. In other words, to derive highest efficiency and stability of operation, the geometry of the inlet would be different at each Mach number and angle of attack of flight. A typical supersonic military aircraft might experience large variations in angle of attack, sideslip angle, and flight Mach number during normal operation. These large variations in inlet flow conditions create certain important design considerations:

- The inlet should provide the highest practical efficiency. The ratio of recovered total pressure to airstream total pressure is an appropriate measure of this efficiency.
- The inlet should match the demands of the powerplant for airflow. The airflow captured by the inlet should match that necessary for engine operation.
- Operation of the inlet at flight conditions other than the design condition should not cause a noticeable loss of efficiency or excess drag. The operation of the inlet should be stable and not allow a *buzz* condition, which is an oscillation of shock location that is possible during off-design operation.

In order to develop a good, stable inlet design, the performance at the design condition may be compromised. A large variation of inlet flow conditions might require special geometric features for the inlet surfaces or a completely variable geometry inlet design.

Supersonic configurations

When all the various components of the supersonic airplane are developed, the most likely general configuration properties will be as follows:

- The wing will be of low-aspect ratio, have noticeable taper, and have sweepback depending on the design speed range. The wing sections will be of low-thickness ratio and require sharp leading edges.
- The fuselage and nacelles will be of high-fineness ratio (long and slender). The supersonic pressure distribution might create significant lift and drag and require consideration of the stability contribution of these surfaces.

- The tail surfaces will be similar to the wing: low-aspect ratio, tapered, swept, and of thin section with sharp leading edge. The controls will be fully powered and irreversible with all-movable surfaces the most likely configuration.
- In order to reduce interference drag in transonic and supersonic flight, the gross cross section of the aircraft can be "area-ruled" to approach that of some optimum high-speed shape.

One of the most important qualities of high-speed configurations will be the low-speed flight characteristics. The low-aspect-ratio, swept-wing planform has the characteristic of high induced drag at slow flight speeds. Steep turns, excessively slow airspeeds, and steep, power-off approaches can then produce extremely high rates of descent during landing. Sweepback and low-aspect ratio can cause severe deterioration of handling qualities at speeds below those recommended for takeoff and landing.

On the other hand, thin, swept wings at high wing loading will have relatively fast landing speeds. Any excess of this basically fast airspeed can create an impossible requirement for brakes, tires, and arresting gear. These characteristics require that the pilot account for the variation of optimum speeds with weight changes and adhere to the procedures and techniques outlined in the flight handbook.

Figure 8-21 shows the distinctive wing planforms and fuselage shapes depending on the Mach number and purpose of the airplane.

AERODYNAMIC HEATING

When air flows over any aerodynamic surface, certain reductions in velocity occur with corresponding increases in temperature. The greatest reduction in velocity and increase in temperature will occur at the various stagnation points on the aircraft. Of course, similar changes occur at other points on the aircraft, but these temperatures can be related to the ram temperature rise at the stagnation point.

Subsonic flight does not produce temperatures of any real concern, but supersonic flight can produce temperatures high enough to be of major importance to the airframe and powerplant structure. A graph in chapter 10 (Fig. 10-6) illustrates the variation of ram temperature rise with airspeed in the standard atmosphere. The ram temperature rise is independent of altitude and is a function of true airspeed. Actual temperatures would be the sum of the temperature rise and the ambient air temperature; thus, low altitude flight at high Mach numbers will produce the highest temperatures.

In addition to the effect on the crew's environment, aerodynamic heating creates special problems for the airplane structure and the powerplant. Higher temperatures produce definite reductions in the strength of aluminum alloy and require the use of titanium alloys, stainless steels, and the like, at very high temperatures. Continued exposure at elevated temperatures causes further reductions of strength and magnifies the problems of "creep" failure and structural stiffness.

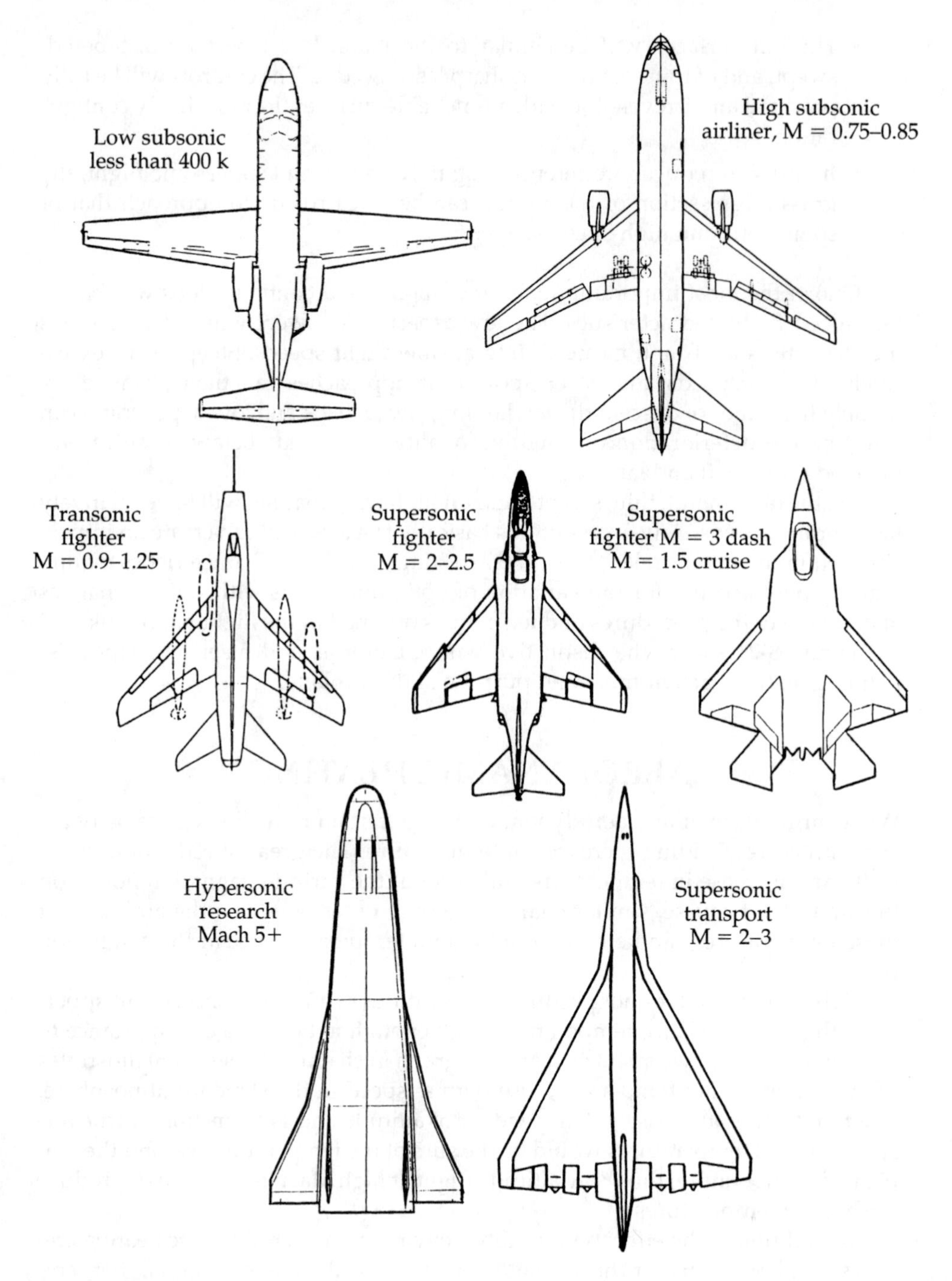

Fig. 8-21. The shape of the high-speed airplane depending on speed and mission requirements.

The turbojet engine is adversely affected by high compressor inlet air temperatures. Because the thrust output of the turbojet is some function of the fuel flow, high compressor inlet air temperatures reduce the fuel flow that can be used within turbine operating temperature limits. The reduction in performance of the turbojet engines with high compressor inlet air temperatures requires that the inlet design produce the highest practical efficiency and minimize the temperature rise of the air delivered to the compressor face.

Faster flight speeds and compressible flow dictate airplane configurations that are much different from the ordinary subsonic airplane. Subsequent chapters and appendices examine the design and development process that found solutions to the problems of high-speed flight.

9

Supersonic transports

THE 1950s AND EARLY 1960s WERE HEADY TIMES for the aerospace industry. The sound barrier was hurdled. The Century Series of fighters routinely flew supersonically. The Boeing 707 and Douglas DC-8 ushered in the jet age in air transport with phenomenal speeds of over 500 mph. Sputnik, Mercury, and Gemini were orbiting the Earth. Project Apollo was implemented to land men on the moon. The United States and Russia were engaged in a space race as well as a technological race. The British-French Concorde and the Russian TU-144 supersonic transports were being developed.

On June 5, 1963, in a speech before the graduating class of the United States Air Force Academy, President Kennedy committed this nation to "develop at the earliest practical date, the prototype of a commercially successful supersonic transport, superior to that being built in any other country in the world . . . ."

What lay ahead was years of development, competition, controversy, and ultimately rejection of the supersonic transport (SST) by the United States in 1971. The project did not proceed to the flying prototype stage. The British-French Concorde (Fig. 9-1) and the Russian TU-144 designs, however, progressed to the flight stage. After a disastrous crash at the Paris airshow and other development problems, the TU-144 project was abandoned by the Russians. Only the Concorde became airline operational, but only in small numbers, and it proved to be commercially uneconomical.

NASA did considerable work, starting in 1959, on basic configurations for the SST. Four basic types of layout evolved that were developed further by private industry. Douglas, North American Aviation, Boeing, and Lockheed all responded to the Request for Proposal (RFP) issued by the U.S. government through the Federal Aviation Administration in response to President Kennedy's commitment. After deliberation by the FAA, by the president's advisory committee, and by the president himself, preliminary design contracts were awarded to Boeing and Lockheed for the airframe and to General Electric and Pratt & Whitney for the engines. Douglas and North American Aviation were eliminated or dropped out.

One problem associated with an SST is the tendency of the nose to pitch down as it flies from subsonic to supersonic speed due to the rearward travel of the cen-

Fig. 9-1. The British-French Concorde supersonic transport. Note the striking difference between the Mach 2.2 Concorde and the subsonic Mach 0.8 Boeing 747 in the background.

ter of lift (refer to chapters 4 and 8). Boeing proposed the *swing-wing* to maintain the airplane's balance and counteract the pitch-down motion. Lockheed's original fixed-wing design needed *canards,* which are small wings placed forward of the wing to counteract the pitch-down tendency (Fig. 9-2).

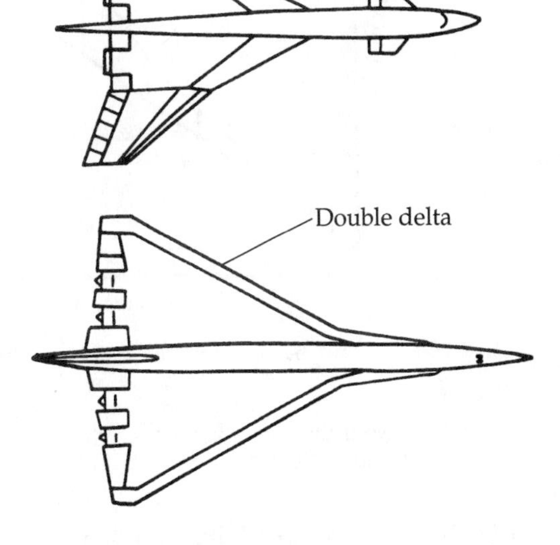

Fig. 9-2. The original Lockheed SST design on the top used canards to prevent a pitch down tendency during transonic flight. The final double-delta configuration on the bottom negated the pitchdown tendency because the forward delta generates lift supersonically.

Eventually, the Lockheed design used a double-delta configuration (Fig. 9-2), and the canards were no longer needed. This design proved to have many exciting aerodynamic advantages. The forward delta begins to generate lift supersonically (negating pitch down). At slow speeds, the vortices trailing from the leading edge of the double-delta (Fig. 9-3A) increase lift as shown in Fig. 9-3B. This means that flaps and slats could be reduced or done away with entirely, resulting in a simple wing design.

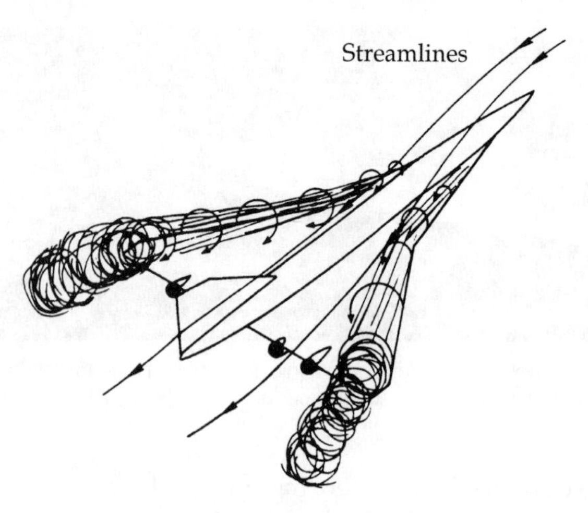

A. Vortices on double delta wing

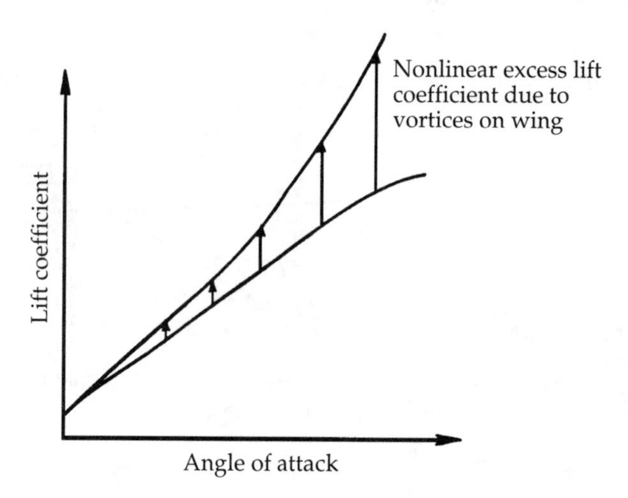

B. Lift coefficient increase due to vortices

Fig. 9-3. At low speeds, the vortices trailing from the forward delta increase lift resulting in a simple wing design without flaps or slots.

In landing, the double-delta experiences a ground-cushion effect, which allows for slower landing speeds. Figure 9-4 shows the British-French Concorde and the Russian TU-144 prototypes. They used a variation of the double-delta wing called the *ogee wing*. It, too, uses the vortex-lift concept for improvement in low-speed subsonic flight.

The British-French Concorde and Russian TU-144 cruised at Mach 2.2–2.4, whereas the United States proposed SST was designed to cruise at Mach 2.7–3.0. Speeds much over Mach 2.0 provide a temperature rise at which aluminum alloys

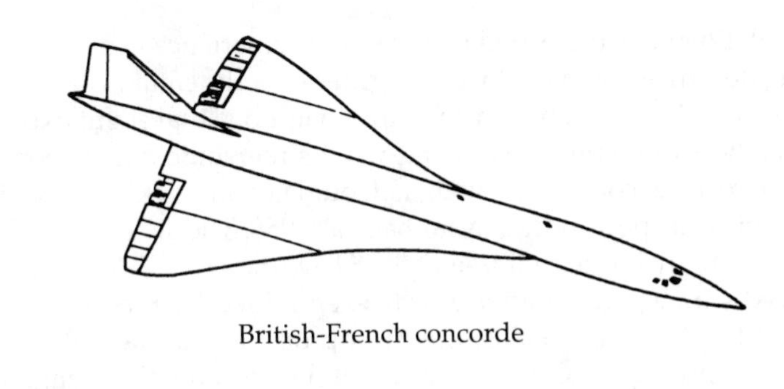

British-French concorde

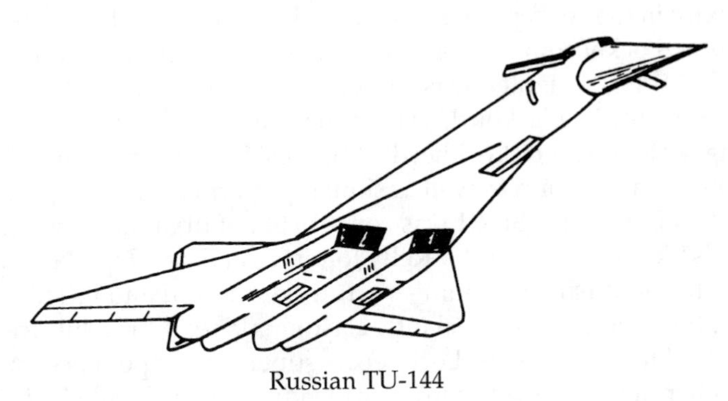

Russian TU-144

Fig. 9-4. The British-French Concorde and Russian TU-144 SST designs. The TU-144 flew as a prototype only and was eventually abandoned. The Concorde is the only operational SST.

begin to lose strength. The speeds of the Concorde and TU-144 were just below the temperature limit of aluminum alloys; therefore, they were constructed of these familiar materials. America's SST cruised at speeds that provided structural temperatures requiring titanium alloys as the basic material.

Government management of the United States' SST competition was an interesting case of politics at work. The Pentagon had a long established system for evaluating complex new projects, but the Pentagon was a military organization and the SST was a civil aircraft; therefore, the SST program was assigned to the Federal Aviation Administration (FAA) of the Department of Transportation. The FAA, however, had no background in programs of this complexity.

Since the SST competition was now between Boeing and Lockheed, a review of the background of these two aerospace giants shows that both had equal but different qualifications for coming up with a successful supersonic transport.

Boeing had practically cornered the market for jet airliners with the 707, 727, and 737. The 747 was under development. In addition, its military division had produced the six-engine B-47 and eight-engine B-52 jet bombers, as well as the KC-135 tanker. Boeing's transport experience and ability to develop and produce large jet airliners was unquestionable; however, Boeing had never produced a supersonic airplane. Its jet transports and bombers were all fast subsonic airplanes.

Although Boeing had gleaned considerable experience in supersonics with its swing-wing design entry in the tactical fighter experimental (TFX) competition, it lost out to General Dynamics, and therefore had no actual flight experience. (The TFX eventually became the F-111 swing-wing supersonic fighter-bomber.) Along with other aerospace companies such as Douglas and Lockheed, Boeing had begun supersonic transport studies as far back as 1958. The swing-wing F-111 design studies undoubtedly influenced Boeing's SST work.

Lockheed's transport experience was exemplified by the piston-engine Constellations of the 1940s and 1950s, the turboprop Electra, the military turboprop Hercules, and four-engine C-141 Starlifter jet. Lockheed's three-engine TriStar jet airliner was only in the conceptual stage in 1964; however, at the beginning of the SST competition, Lockheed had considerable supersonic flight experience.

The Mach 2.0 F-104 fighter was in production. The Mach 3.0 A-11/YF-12, which became the SR-71 Blackbird, was flying. The Blackbird airplane was, in a sense, a flying scale-model of Lockheed's SST design. Blackbird and SST designs had the double delta planform as well as similar performance of Mach 3.0 at 70,000 feet altitude. In addition, the SR-71 was constructed of titanium, just like the SST.

Lockheed's Skunk Works, under Kelly Johnson, had developed the techniques and processes for producing titanium structures. Lockheed and its subcontractors came up with high-temperature sealants, elastomeric seals, oils, lubricants, glass, hydraulic oil, and hydraulic seals. Lockheed's supersonic experience far exceeded that of any other aerospace company. (Appendix C is a detailed discussion of Lockheed's work on an SST.)

Initial work on the SST was either company-funded or consisted of preliminary research studies for government agencies, including the Supersonic Commercial Air Transport (SCAT) feasibility studies for the National Aeronautics and Space Administration (NASA). In August 1963, the U.S. government officially launched America's supersonic transport program by issuing a request for proposals (RFP) through the FAA.

Responses to the RFP were designated Phase I of what was essentially an FAA design competition. Three airframe manufacturers and three engine manufacturers submitted proposals. As stated previously, after deliberation by the FAA, by a presidential advisory committee, and by the president himself, Phase IIA contracts were awarded to Boeing and Lockheed for the airframe, and to General Electric and Pratt & Whitney for the engines.

The feasibility of the SST was considered firmly established in Phase IIA. Phase IIB was initiated in early 1965 with the same airframe and engine manufacturers participating. Phase IIC began in July 1965 and was scheduled to end in December 1966. It was anticipated that one airframe manufacturer and one engine manufacturer would be selected in late 1966 or early 1967 to proceed with Phase III of the U.S. SST program. Phase III would involve building prototype and production airplanes and final FAA certification before the SST was scheduled to enter airline service in approximately 1974.

The challenge of the SST was severe, the problems were difficult, and the stakes were high. The process of evolving a final design involved trade-off studies, optimization studies, and parametric analyses to ensure the establishment of the best possible design.

Ultimately, in December 1966 Boeing's swing-wing design was selected as the winner of the SST competition. This came after approximately 8 years of research by both Boeing and Lockheed. General Electric was selected over Pratt & Whitney for SST engine development. The government-industry program was administrated by the FAA. Phase III of the SST program had begun.

At this time, it was estimated that the program would cost approximately $1.5 billion. Of this, the government was to advance 90 percent minus approximately $60,000,000 risk money from the airlines that had ordered 122 SSTs, plus additional money contributed by the aircraft industry. It was expected that the government would be repaid all of its invested money plus a return of about $1 billion on a minimum sale of 500 supersonic transports. These were euphoric times for the aerospace industry, the airlines, and the FAA!

The euphoric times did not last long. As the SST program progressed, many technical problems surfaced. The size of the airplane grew to meet airline payload requirements. Major design changes were incorporated into the original Boeing 2702-100 design. The supersonic cruise lift-drag ratio increased from 6.75 to 8.2. The engines were moved further aft to alleviate the exhaust impinging on the rear tail surfaces.

Despite the advantages previously quoted for a swing-wing concept, technological advances in construction did not appear in time. Because of the swing-wing mechanisms and beefed-up structure due to engine placement, incurable problems in reduction of payload resulted. Boeing had no recourse but to adopt a fixed-wing concept (Fig. 9-5).

Technical problems were not the only ones plaguing the SST. The high development costs associated with the SST led to political problems. Environmental concerns surfaced due to sonic boom effects, noise, and claims of ozone layer depletion. The continued bad press had its effect on the potential airline customers for the SST. Pan American and TWA in the United States were the major transoceanic carriers; therefore they were expected to be the major SST operators. TWA was losing interest, and Pan Am remained committed, but it was becoming skeptical, too.

The SST controversy became overwhelming. Eventually, Congress canceled the SST project in 1972, well before a flying prototype could be built. (That left the British-French Concorde as the only operational supersonic transport.)

HIGH-SPEED CIVIL TRANSPORT

SST efforts of the late 1960s and early 1970s did not result in a fleet of superfast airplanes traversing the globe. Except for a handful of British-French Concordes, all airline flying until the end of the twentieth century and well into the twenty-first century will be at speeds slower than Mach 1. Supersonic transport research is continuing under the direction of NASA's Langley Research Center; the *high-speed research* (HSR) is supposed to lead toward the *high-speed civil transport* (HSCT).

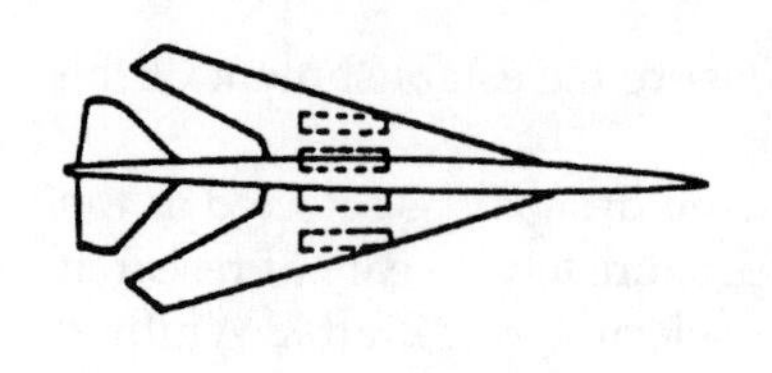

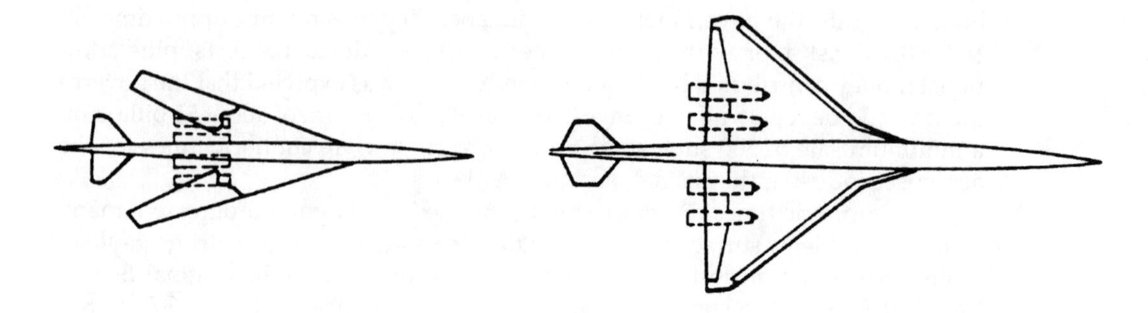

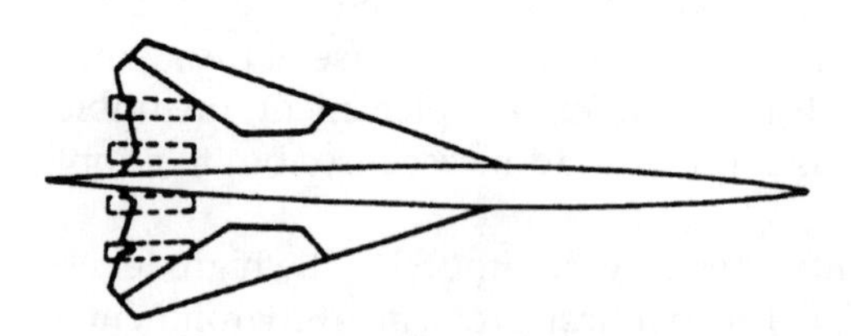

Final version model 2707-300

Fig. 9-5. The evolution of the Boeing SST design from the original, swing concept to the final fixed-wing configuration. The program was canceled before a flyable prototype was built.

Figures 9-6 and 9-7 show two different HSCT configurations in the NASA Langley low-speed wind tunnel.

Studies indicate that there will be a substantial opportunity for a future HSCT aircraft to satisfy the rapidly growing long-haul market, especially for travel across the Pacific. Estimates show that during the period from 2005 to 2015, this market could support 500–1,000 HSCT aircraft, creating a multibillion dollar sales opportunity for its producers.

If the United States builds such an aircraft, it alone could be worth more than $15 billion per year to the country's aerospace industry and provide more than 140,000 direct high-value added jobs. Although today's technology would not support a successful HSCT, studies further show that aggressive technology development and application could let the industry produce an economically viable and environmentally compatible supersonic airliner.

In Europe, the United Kingdom's British Aerospace and France's Aerospatiale are jointly conducting studies of a supersonic transport successor to the Concorde.

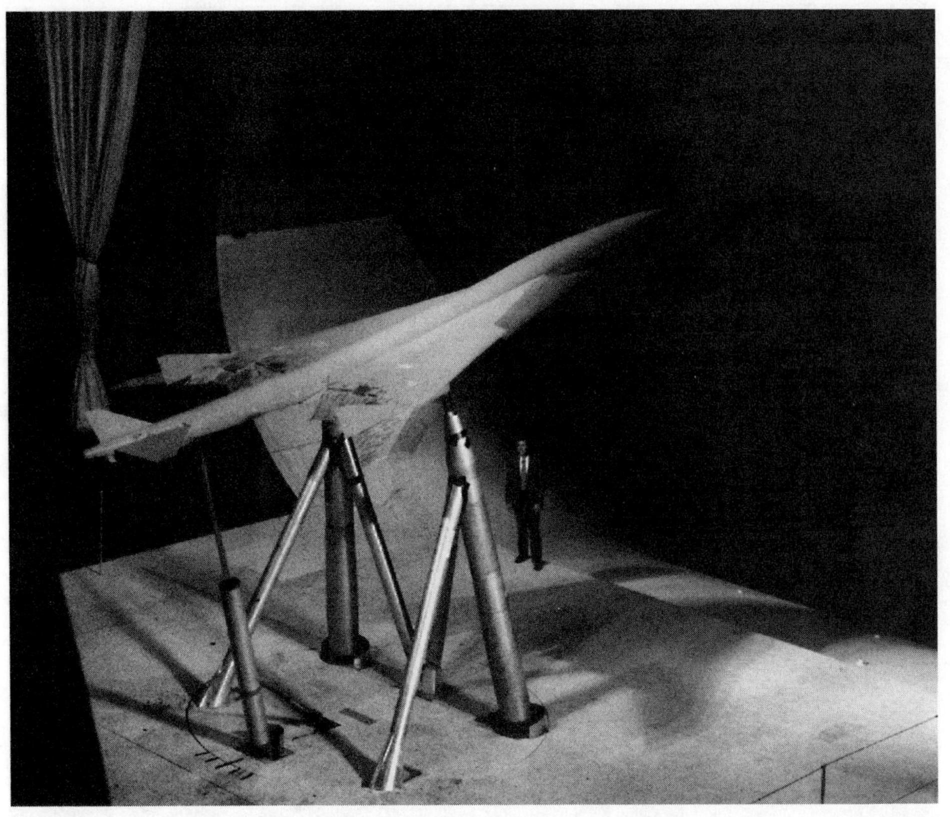

Fig. 9-6. A supersonic transport model in the NASA Langley full-scale wind tunnel for low-speed testing.

Similarly, the British Rolls-Royce Company and Snecma of France are working together on new engine concepts for tomorrow's advanced supersonic transport. Japan, through its Ministry of International Trade and Industry, has started a supersonic/hypersonic technology program, initially focused on a high-speed aircraft propulsion system.

These examples of investment in high-speed transport research underline the interest among aircraft manufacturers of foreign nations in getting an early jump on what will be the next plateau for international aviation competition: the long-range, economical, environmentally acceptable supersonic passenger transport. Aviation experts feel that with sufficient technological development, high-speed transports can become competitive with subsonic jetliners and capture a significant portion of the growing long-haul intercontinental market, especially in transpacific service, where passenger traffic is expected to quadruple by the start of the new century.

In the interest of maintaining the United States' world leadership in commercial aviation, and in recognition of the economic potential of the multibillion dollar transport market, NASA is conducting the HSR program to help U.S. manufacturers prepare for the coming competition. The research program is

Fig. 9-7. A one-tenth-scale blended-wing supersonic transport model in the NASA Langley full-scale wind tunnel for low-speed testing.

addressing the key technologies essential to resolving environmental and economic barriers to supersonic flight.

HSR is a follow-on to a first-phase effort that involved high-speed civil transport studies conducted for NASA by Boeing and Douglas, the nation's two leading jetliner producers. Figures 9-8 and 9-9 show artists' concepts for a second-generation supersonic transport resulting from this study.

The companies identified technological advances that should be possible by the early twenty-first century developments that could make supersonic airliners more efficient, which would permit fare levels competitive with subsonic fares; however, the studies cautioned that demand for an advanced supersonic transport will materialize only if, in addition to being operationally efficient, the airplane also can meet allowable noise standards and demonstrate that it will have no harmful effects on the atmosphere.

United States research efforts consider sonic booms a significant part of the HSCT program, even though NASA researchers are skeptical that populated areas would ever put up with even softer sonic booms. After flying the supersonic Concorde operationally since January 1976, the British and French are not concentrat-

Fig. 9-8. Artist's concept of a McDonnell Douglas Mach 2.4 high-speed civil transport resulting from a NASA-funded first-phase study program.

Fig. 9-9. Artist's concept of a Boeing Mach 2.4 high-speed civil transport resulting from a NASA-funded first-phase study program.

ing on sonic booms in their next-generation supersonic transport research. Airline operators assume supersonic transports will always be restricted to supersonic flight over water and fly subsonically over land, like the Concorde.

The British-French Concorde was designed to fly at Mach 2.2–2.4, just below the aerodynamic heating temperature rise at which aluminum alloys begin to lose strength; thus, the Concordes were constructed of these familiar materials. The United States' SST of the 1960s was designed to fly Mach 2.7–3.0, which would have required titanium structures. The structures would have been much heavier, more expensive, and more difficult to fabricate than aluminum.

The HSCT has no speed limit at this point in time; NASA is quoting a speed range of Mach 2.0–3.0 with a probable cruise speed of Mach 2.4, which would mean that a heavy and expensive all-titanium structure might not be necessary. Advanced materials and structures technology could provide a 30–40 percent weight saving in meeting HSCT requirements of 60,000 hours durability at up to a 350°F operating temperature.

The planned program addresses development, scale-up, and verification of high-temperature polymer matrix composites and aluminum alloys, including fabrication and combined load testing of large wing and fuselage structures.

CONCLUSION

The high-speed research program provides needed high-risk, long-term research and technology development that is currently beyond industry's investment reach. The potential payoffs for the country are substantial: positive balance-of-trade ($200–$350 billion in sales), creation of about 140,000 high-value jobs, and preserving the general strength of the U.S. aerospace industrial base for civil and military applications.

The required investment for HSCT technology development is large. An estimated $4.5 billion would be necessary prior to an industry go-ahead decision leading to certification and production. Current technical, environmental, economic, and political uncertainties are too great for industry to make major funding investments in the very early stages of technology development.

Industry has indicated that it will provide additional parallel investments along a schedule consistent with the progress and success achieved in the NASA program. It is a commitment that reflects the uncertainties.

A national team effort is an integral part of the high-speed research program approach. For example, NASA and the nation's two leading engine manufacturers, General Electric and Pratt & Whitney, have formed a strong team with a broad base of other engine manufacturers, material suppliers, and academia to develop a new class of high-temperature composite material for HSCT propulsion systems.

To assure that the program is market-driven with industry's needs being met in the most cost-effective manner, future plans, system design, and economic analysis will be comprehensively reviewed on a quarterly basis with NASA and industry management. Periodic reviews with the NASA aeronautics advisory committee and the National Academy's aeronautics and space engineering board will also be conducted.

The goals in each technology area are very challenging and high-risk, but attainable. If it appears that it is not possible to achieve the environmental or economic success requirements, the program will be reassessed and redirected or canceled as appropriate.

The HSR program represents a realistic estimate of the total technology development required and a national team approach with shared government and industry responsibility.

Don't expect to fly supersonically until early in the twenty-first century, unless you fly the Concorde!

10

Beyond supersonic toward hypersonic flight

HYPERSONIC FLIGHT IS ARBITRARILY DEFINED as flight at speeds beyond Mach 5.0, although no drastic flow changes are evident to define this. To date, short-time speeds of this magnitude have been achieved only by rockets, spacecraft, and the NASA X-15 research airplane. Several formidable problems are encountered at these speeds.

First, the shock waves generated by a body trail back at such a high angle that they might seriously interact with the boundary layers about the body. For the most part, these boundary layers are highly turbulent in nature.

Secondly, across the strong shocks, the air undergoes a drastic temperature rise up to 2,000°F. Aerodynamic heating of the body is therefore a major problem. For sustained hypersonic flight, most normal metals used in today's airplanes would quickly melt; therefore, new materials or methods that can withstand the high-temperature effects are required. Wing leading edge temperatures can be reduced by using a high degree of sweepback.

Control surfaces for hypersonic flight must be strategically placed so that they encounter sufficient dynamic pressure about them to operate. Otherwise, if shielded from the approaching flow by the fuselage, for example, they will be ineffective.

Propulsion is another major problem at hypersonic speeds. Economically, the most promising prospect is the ramjet engine. The ramjet engine works on the principle that at high Mach numbers the shock waves compress the air for combustion in the engine. This does away with any moving parts and represents an efficient propulsion method.

(Much of the material in this chapter was excerpted from an article in *Lockheed Horizons* magazine by G. Daniel Brewer, manager of airbreathing hypersonic vehicle systems, and Milford G. Childers, senior structures research and development engineer. The information is based on research conducted by Lockheed during the 1960s.)

LIFTING BODIES—THE SHAPE OF HYPERSONICS

Since the late 1950s, NASA has been designing aircraft that produce more lift than drag and yet resemble spacecraft. They are called *lifting bodies* because they have no wings or small wings and obtain most of their lift because of their body shapes. Some of the early designs are shown in Fig. 10-1; none were developed into flight articles. Figure 10-2 shows four of the shapes that were flight tested to explore the subsonic and low supersonic speed ranges and evaluate the handling characteristics and flight qualities of this unusual concept.

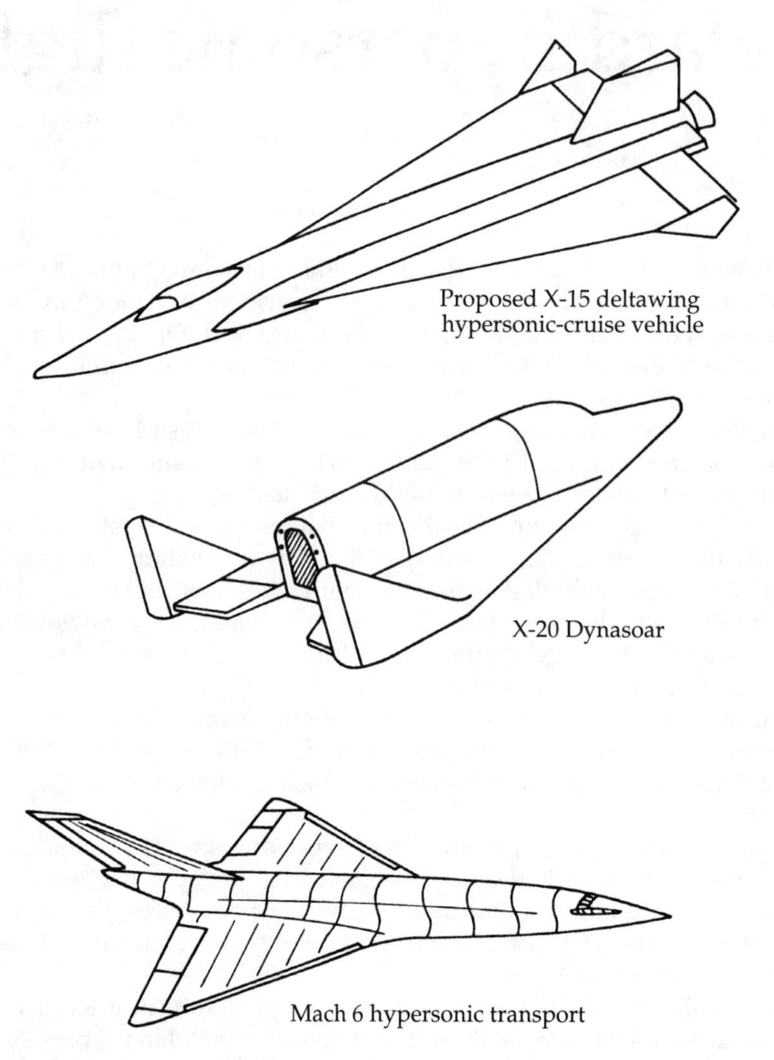

Fig. 10-1. Examples of early hypersonic vehicle designs. None were developed into flight articles.

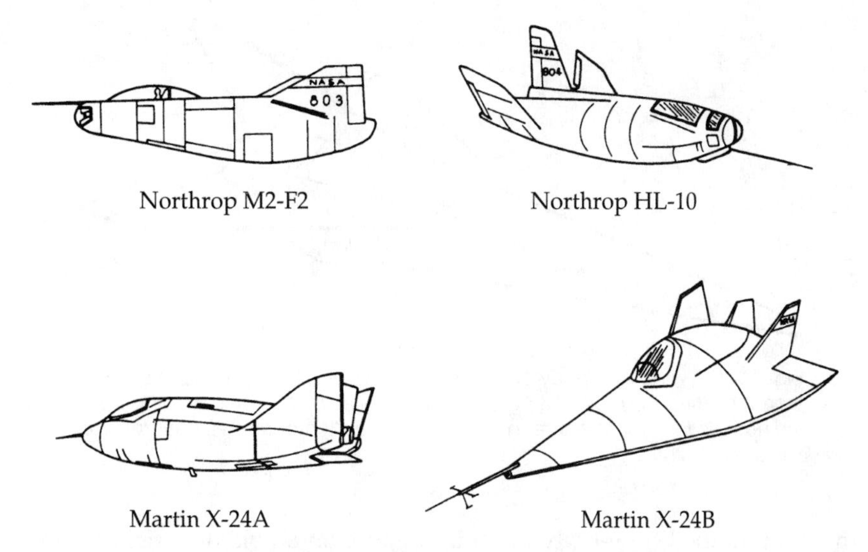

Fig. 10-2. These four lifting body shapes were built and flight tested to explore subsonic speed ranges and evaluate handling characteristics.

The vehicle shown in Fig. 10-3 is a typical lifting-body type that evolved from the NASA studies. The lifting-body configuration provides a considerable performance advantage over the more conventional wing-and-body shape shown in Fig. 10-4. This advantage primarily stems from the unit weight of the wing in the wing-and-body configuration. The weight would be excessive if the wing were to have the necessary structural strength at the elevated temperatures generated in hypersonic flight.

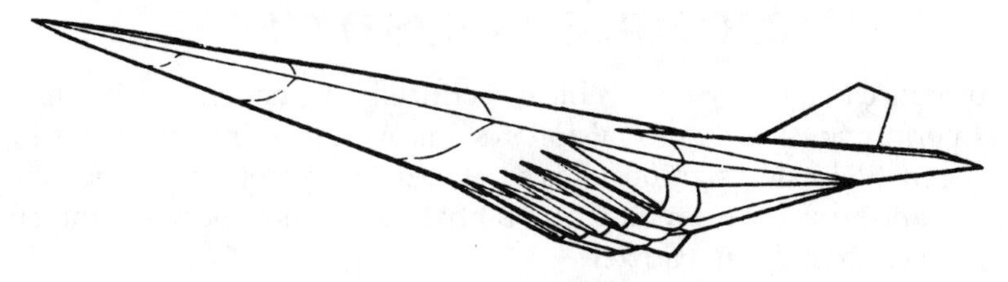

Fig. 10-3. A typical lifting body hypersonic design that has evolved from NASA and industry studies.

Although the wing of the wing-and-body vehicle is more efficient as an aerodynamic lifting surface, the lifting-body design represents a better integration of vehicle elements and results in a better vehicle structural weight fraction and improved flight performance. The *structural weight fraction* of a typical lifting-body configuration is calculated to be approximately 17 percent lower than that of the corresponding wing-and-body vehicle.

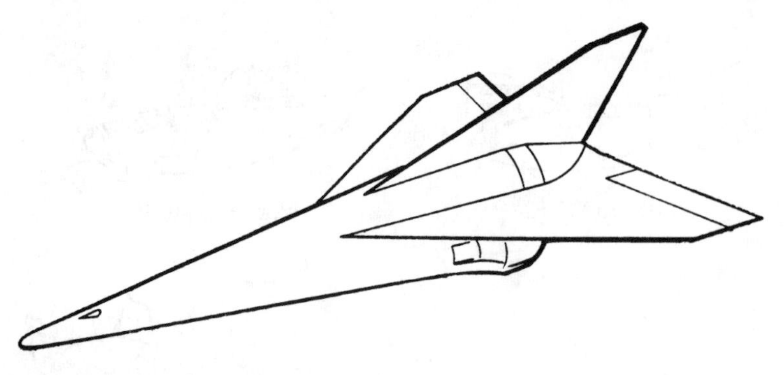

Fig. 10-4. A wing-and-body configuration, although better aerodynamically than the lifting body, would be excessively heavy if the wing were to have the necessary structural strength at the elevated temperatures generated in hypersonic flight.

In addition, the lifting-body provides a more favorable flow field for the air-breathing propulsion system. The entire forebody of the vehicle forms a surface for precompression of the air entering the engine inlets, which permits a reduction in the size and weight of the air-breathing engines over those required for the wing-and-body configuration.

The lifting-body configuration provides various desirable features: a thick center section for storage of fuel, relatively sharp leading edges for efficient aerodynamic performance, and a bottom surface bulged along the center line to force concentration of the boundary layer along the outer edges of the delta planform. This last feature is desirable to achieve a thinner boundary layer at the inlet of the engines, which minimizes the ingestion of low-energy-level air.

PROPULSION SYSTEMS

The performance of a hypersonic vehicle is critically dependent on the characteristics and performance of its propulsion system. Air-breathing engines using liquid hydrogen fuel offer significant advantages for hypersonic propulsion. A primary advantage stems from their capability to use atmospheric air, thereby obviating the need to carry an oxidizer.

Of the various air-breathing engines, *supersonic combustion ramjets* (*scramjets*) are more efficient than the conventional ramjets at hypersonic speeds. In a conventional ramjet, incoming air is slowed to subsonic speeds in the inlet; fuel is introduced, mixed, and burned; and the combustion products are accelerated in a convergent-divergent nozzle.

In a scramjet, incoming air is simply slowed to supersonic speed, and fuel is introduced and burned so that the flow remains supersonic. The combustion products are accelerated to even higher velocities in a simple divergent duct. Neither ramjets nor scramjets are self-accelerating propulsion systems from static conditions, and both require acceleration to an appreciable forward velocity by a boost system before they are capable of providing net positive thrust.

One of the most crucial components of an air-breathing propulsion system is the inlet. It must convey air from the atmosphere to the engine of the aircraft with the least possible disturbance to the external flow. For subsonic combustion engines, the air induction system is called upon to provide air at the maximum pressure and with the least possible drag and interference. Inlet flow is compressed to near-sonic velocities at the inlet throat and shocked down to subsonic velocities. A subsonic diffuser is employed to effect maximum recovery of pressure and temperature.

For scramjet operation, the incoming air is compressed only to a lower velocity relative to the free-stream in the inlet; thus, scramjets operate at lower combustion-chamber static temperatures and pressures than ramjets. Scramjets have lower combustion efficiency than ramjets but gain their net increase in thrust from decreased losses in the inlet section.

Ideally, inlets for scramjets should have efficient flow delivery with good compression. Additionally, scramjet inlets should have provisions for spillage to encourage low-Mach-number starting, or have provisions for noninterference spillage from adjacent inlets in the event of inlet choking, often referred to as an "unstart." Exposure to the airstream must be sufficient to provide radiation cooling of as much of the inlet structure as possible. Fixed geometry structure is desirable to minimize weight.

The integration of the inlet and the scramjet propulsion system with the vehicle is a major problem. Design of inlet and exhaust system surfaces on a scramjet-powered hypersonic vehicle represents one of the major tasks that must be undertaken before successful development of such a vehicle is possible.

TEMPERATURE ENVIRONMENT

As flight speeds increase beyond supersonic, beyond 3,500 miles per hour into the hypersonic speed range, engineers encounter increasingly severe material and structural design requirements. Greater speed results in vastly increased temperatures on a vehicle's surfaces by aerodynamic heating. Higher speed also results in higher pressure loads on the surface of the craft, and heavier structure is required to resist the forces generated. Both factors, temperature and pressure, can be greatly reduced by flying higher in the atmosphere; therefore, aerospace vehicles are generally designed to fly at higher altitudes when flying at higher speeds.

Figure 10-5 illustrates the effect of velocity and altitude on the temperature at the nose of the aerospace vehicle. Figure 10-5 shows only general conditions; local areas on the vehicle might be even more severely taxed by high pressures combined with high temperatures. Typical of such local areas are the inlet ducts for air-breathing engines. Not shown in Fig. 10-5, but of major importance in the selection of materials and in the design of the structure of hypersonic vehicles, is the time during which the materials are subjected to high temperatures and the requirement for reusability of the vehicle.

In spite of these increasing demands, materials and structures engineers are expected to provide a design with ratios of structural weight to gross operating weight (structural weight fractions), comparable with or better than those experienced in slower vehicles. Fortunately, the aerodynamic demands of a hypersonic

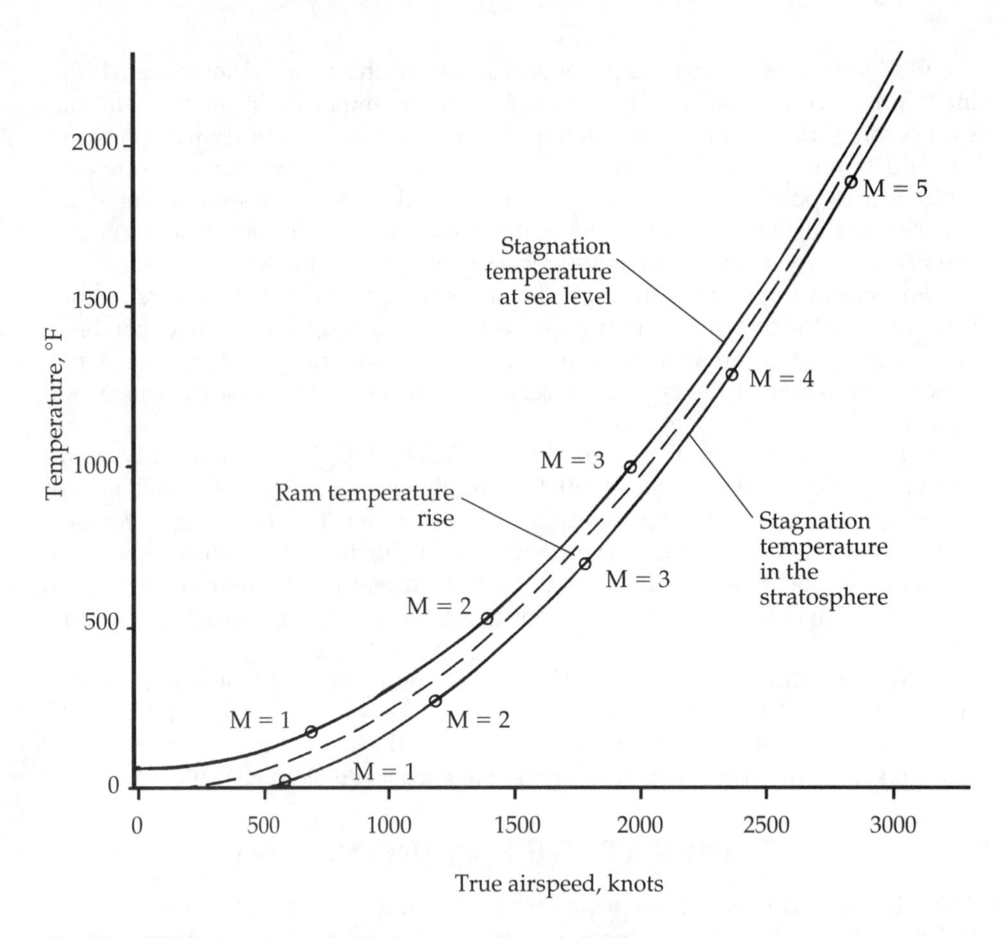

Fig. 10-5. Effect of speed and altitude on aerodynamic heating.

vehicle tend toward design configurations with short-span wings and lifting bodies that result in a compact airframe. On the other hand, the use of hydrogen as a fuel means that the fuel tanks are very large since the density of hydrogen fuel is very much less than the density of conventional jet fuel.

This essentially outlines the major problems and goals of the materials engineer and structures engineer, who must blend their respective technologies to provide a vehicle structure of exceptional efficiency for operation in an alien environment.

(Note: Although the basic theme of this book concentrates on aerodynamics and propulsion, due to the severe aerodynamic heating environment of the hypersonic vehicle, some discussion of structural implications is in order.)

MATERIALS SELECTION CRITERIA

Temperature, the most obvious factor that influences material selection in hypersonic flight, can be considered as a function of location on the vehicle and the structural concept employed, whether the structure is protected or not. Figure

10-6 is typical of the surface-temperature variation on a hypersonic vehicle at cruise altitude and velocity. Different parts of the vehicle experience a wide range of temperatures; therefore, it is necessary to choose a material that is compatible with the specific temperature environment while also meeting the requirements for strength, fabricability, and reliability.

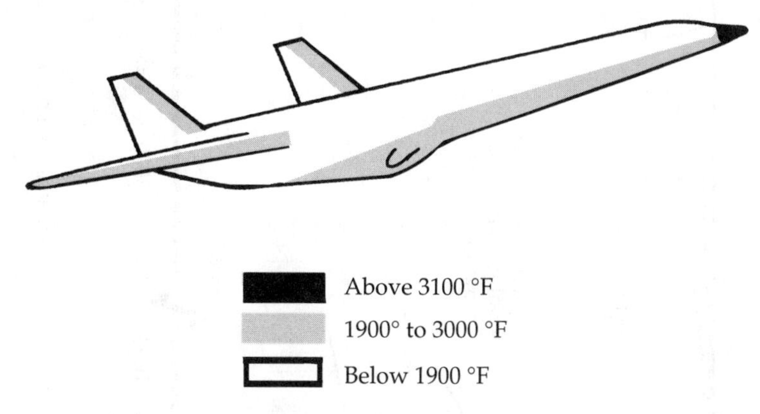

Fig. 10-6. Surface temperature variation on a typical hypersonic vehicle.

The most common structural materials are usually classified according to the degree of heat that they are able to withstand. Figure 10-7 shows the temperature ranges for which different classes of material have useful structural properties. It should be noted that time of exposure has an appreciable effect on temperature capability. From this chart it is seen that the standard aircraft aluminum alloy must not be operated at temperatures above 250°F for extended periods of time. The class of materials with moderate temperature capabilities, represented by stainless steel and titanium, are restricted to about 900°F. Superalloy materials have potential usefulness up to 1,900°F. Between 1,900°F and 3,100°F, it is necessary to use *refractory metals*, which are metals with melting points above 4,000°F that also have the ability to retain their strength at very high temperatures. Graphite and the ceramics occupy the temperature regime above 3,100°F.

Based upon information in Fig. 10-6 and the use of materials for specific heat ranges, it could be generally concluded that the upper surface of a hypersonic vehicle can be made from superalloys and the bottom surface and leading edges from refractory metals. The nose cap requires the equivalent of a ceramic material.

Although Fig. 10-7 is useful in determining the class of material required for a specific temperature-time combination, other considerations can determine the temperature limitation on a specific material. One of the most important factors influencing the use of a specific material is its oxidation resistance. For instance, the superalloys are generally considered to be resistant to oxidation at temperatures up to 2,000°F; however, Rene 41, one of the higher strength nickel-base super-alloys, exhibits an intergranular oxidation at temperatures of 1,500°F and above.

Going beyond the problem of oxidation, the selection of a material is influenced by its mechanical, physical, metallurgical, and chemical properties, and by

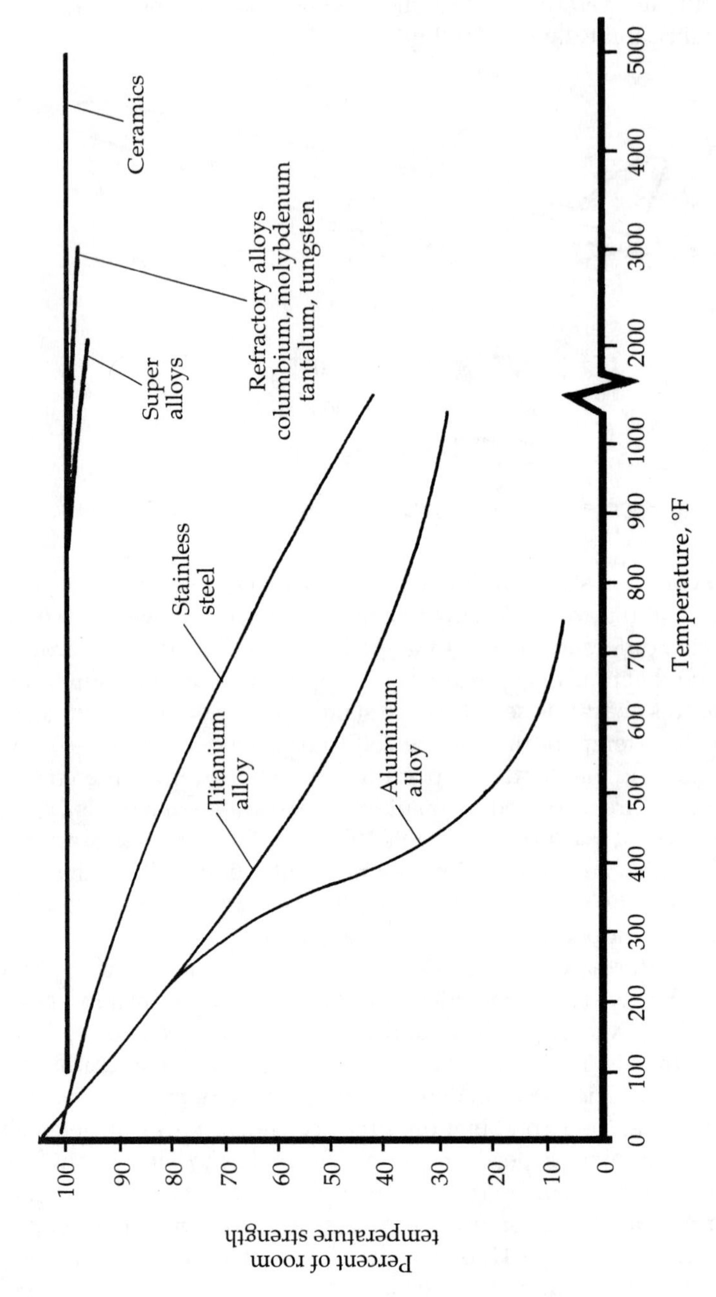

Fig. 10-7. Approximate effect of temperature on tensile ultimate strength of various structural materials (½-hour exposure).

the ease with which parts can be made from the material. The relationship between the material characteristics and structural design is so intimate that they must be considered simultaneously.

STRUCTURAL DESIGN REQUIREMENTS

The environmental conditions encountered by a vehicle traveling at hypersonic velocities are so severe that the structural concept chosen must be carefully tailored to the specific application. Although temperature is perhaps the major factor to be considered, other factors of importance include:

- Load/temperature time cycle
- Structure surface pressures
- Vehicle reusability requirement
- Exterior surface smoothness
- Manned or unmanned vehicle
- Type of fuel

INTERIOR ENVIRONMENT REQUIREMENTS

The structural concept chosen for a given application must satisfy the basic requirement of structural integrity. This means that structure must be capable of supporting the operating loads without failure and without excessive deflection for the missions that are contemplated. This requirement implies the ability of the structures engineer to predict the applied loads, the external and internal distribution of these loads, and to determine the strength of the structure. Structural integrity is normally demonstrated by a combination of ground-load tests and flight tests.

The requirement for structural efficiency means that the structure must perform its function with minimum weight. For structural success, it is necessary to integrate the materials and structure, utilizing the best available technologies in the manufacturing and processing of these materials. It is also necessary to be on the alert to take advantage of advancing technologies by projecting the design into the future.

Another important requirement is reliability of the structure. For purposes of this discussion, reliability is defined as the ability of the airframe to perform its mission repeatedly and safely without an excessive amount of maintenance. Problems of metal fatigue, corrosion, and oxidation must be taken into account in designing and producing a reliable structure.

Generally referred to as fail-safe or safe-life concepts of design, the structural design must also be such that inadvertent damage must be anticipated and provided for to assure structural integrity. For hypersonic vehicles, the added complications that result from the thermal environment make design for reliability even more difficult to achieve.

STRUCTURAL CONCEPTS

With these ground rules in mind, it is now possible to examine various structural concepts to determine which of these best meet the high-temperature requirements of hypersonic flight. There are three general categories of structures that can be considered for hypersonic vehicles: radiation-cooled structure, insulated structure, and internally cooled structure.

The radiation-cooled structure depends on the radiation of the thermal energy away from the structure to maintain a balance with the thermal input. For a given thermal input, the temperature of the structure reaches an equilibrium value that is dependent on the emissivity of the surface, the temperature of the surrounding environment, and the amount, disposition, and thermal characteristics of the structural material.

Realize that although the structure is radiatively cooled, the cooling is only relative to the high thermal input created by the environment in which the vehicle operates. Because the amount of energy radiated depends also on the surface temperature of the structure, these radiation-cooled structures will be very hot in order to provide the desired thermal balance.

The presence of concentrated masses of materials, such as stiffeners or beam caps, is responsible for an uneven distribution of temperature throughout the structure. These temperature variations result in nonuniform thermal expansions, which in turn produce stresses in addition to those imposed by the external loads. Highest stress levels are generally encountered during the initial and terminal phases of the reentry heating cycle when the temperature gradients are at a maximum.

In the case of insulated structure, the primary structural elements are protected from the direct effect of the hot environment by a shield. This shield might take one of two forms or a combination of both. The first part of these "passive" systems consists simply of a layer of insulation that by virtue of its insulative properties slows down the flow of heat into the structure, which retards the heat absorbed by the interior structure.

If the heat exposure time is short enough, the maximum temperature of the structure will be lower than it should be if exposed directly to the hot environment. For this concept, the outside of the insulative layer (the protective shield) might be even hotter than the equivalent hot radiative structure because the transfer of thermal energy to the interior is retarded by the insulation.

It is apparent that if the structure is exposed to the high-heat environment for a long enough period of time, it could become as hot as an uninsulated structure. But heating will take place much more slowly, reducing the temperature variations within the structure, which reduces the thermal stresses.

The second type of shield is an "active" type, where the metallic surface to be protected is covered by a layer of material, generally a plastic. When suddenly subjected to the high temperatures encountered during reentry, this material essentially changes from the solid to the gaseous state. The conversion to and from the liquid transitional state occurs so rapidly that it is insignificant in the process. Converting the solid material into a gaseous material absorbs most of the heat that otherwise would be available for heating the structure. This conversion process is called *ablation*. Material between the ablating surface and the structure provides an insulative layer that slows down the transfer to the structure of the heat not absorbed by the ablation process.

The effectiveness of the ablation system depends on the total amount of thermal energy to be absorbed and the time required. The ablation system involves the consumption of material; hence, to achieve reusability, it requires refurbishment or replacement of the material prior to each flight. Ballistic reentry vehicles such as Mercury, Gemini, and Apollo, which were characterized by a high-heat pulse for a short period of time, employed very effective ablative systems on the shield that faced directly into the flight path during reentry.

It becomes difficult to justify the pure ablation system for vehicles that have a sizable lift during reentry, and combinations of ablative and insulative systems have been proposed. For vehicles that have a ratio of lift-to-drag in the neighborhood of 3, the weight and complexity of this system tend to become prohibitive, particularly when reusability is a requirement.

The third concept is the internally cooled structure. There are a number of variations, depending on the method of cooling. The most highly developed system employs a coolant that circulates through the structure and is either recovered or jettisoned after use.

For the active-cooling concept there is obviously a trade-off between the weight of insulation and the coolant plus the system required to circulate it. For some applications where the fuel is used as a coolant, enough is available so that no insulation is required. An advantage of this system is that the structure operates at lower temperatures, and because structural materials have higher strength at lower temperatures, advantage can be taken of this property.

Circulating-type cooling, including internal environmental control systems, will be necessary for all vehicles carrying a crew, possible passengers, and temperature-sensitive equipment. Complexity of the system with its tubes, pumps, heat exchangers, and the like is a factor on the debit side and one for serious consideration; so is the reliability of such a system. In comparison, the reliability of the oxidation coating for a refractory heat shield is not so critical because it is possible to design the shield to be expendable.

A fourth concept should only be mentioned in passing since it has a limited application where weight is not a prime consideration. This is the *heat sink* structure that utilizes a mass of material large enough to absorb the excess heat not radiated away to space without increasing the temperature beyond acceptable values.

STRUCTURAL CONFIGURATION

The next step in the chain of considerations leading to a structural design is an investigation of the structural configuration of the vehicle. A designer is again faced with several possible choices. The designer can choose to carry the primary bending loads in heavy members or longerons running lengthwise down the vehicle. These members are supported at intervals by rings that are stiff enough to prevent general instability failure of the structure. For this type of structure the primary function of the surface covering is to transfer pressure loads to the longerons and rings and to support the shear loads.

The designer also has the option of carrying the primary bending loads in a semimonocoque surface structure. The heavy members are eliminated in this case, and the surface is stiffened to carry the compressive loads in addition to the pressure and shear. As in the previous case, stiff rings are required at appropriate intervals to prevent general collapse of the structure.

Among surface structures, there are three general types that can be used: sandwich structure, corrugation-skin surfaces, and stiffener-skin surfaces. Sandwich structures incorporate two flat-face sheets that are separated and held in position by a core or interface structure. The two most commonly used are a honeycomb semistructural core and a structural corrugated core. The sandwich structure surface has the advantage of being rigid, and with proper support it is able to sustain high stresses efficiently.

The corrugation-skin structure is also efficient, particularly for applications where the loading is predominantly in one direction. It also has the advantage of being relatively easy to build because the assembly of the corrugations to the skin is by open spot welding or riveting. A variation of this concept results when the inner face provides stiffening in two directions, as with a stamped waffle-pattern inner sheet.

The stiffener-skin surface, while not as inherently rigid as the other two, does provide the advantage of simplicity. The stiffeners are generally arranged to run in one direction only. They can be attached with either rivets or spotwelds, or else machined or fusion welded to make them integral with the surface skin. The grid-stiffened sheet is a variation that complicates the concept but does provide rigidity in all directions.

Selecting the configuration to be used for a specific application requires considering the environment, rigidity needs, fabricability, smoothness, and reliability; hence, a detailed analysis is required to provide a valid basis for such a selection.

Practical application

These concepts and structural configurations provide a basis for a qualitative evaluation of a structure for a specific application. The application chosen for discussion is a vehicle that has a high hypersonic lift over drag ratio. In general, this vehicle is characterized by a low-level heating rate but with a long heating time, resulting in a large total amount of heat to be absorbed or rejected.

The maximum temperature experienced by the external surface at one designated point is 2,200°F. If it is assumed that temperatures above 1,500°F require the use of coated refractory material, then this material is exposed to temperatures above this value for a period of 4,500 seconds, or 1.25 hours. This temperature-time cycle is within the capability of coated columbium, which is reusable.

A radiation-cooled structure can be used for this particular hypersonic vehicle; however, if an insulated structure is considered, the amount of insulation required will depend on the allowable temperature of the structure. The use of an aluminum structure would require no less than 4½ inches of insulation. This thickness would seriously reduce the interior volume of the vehicle and might result in a thickness greater than the permissible design total for a wing or fin. The use of a superalloy with an allowable peak temperature of 1,500°F reduces the thickness requirement to approximately 1½ inches, and merits favorable consideration.

The use of an ablative heat shield for vehicles with high-heating cycles is considered impractical due to the required weight. Other disadvantages accrue, such as a change in shape or contour that would affect the aerodynamics of the vehicle. This type of thermal protection system is not suitable for this vehicle.

Finally, the insulated and internally cooled structure presents some attractive features. Studies have shown that for high-temperature time cycles in the neighborhood of 30 minutes, such a system is competitive weightwise with a purely radiative system; however, as more time at high temperatures is required, the amount of coolant required also increases. If the coolant is carried for this purpose alone, weight might become an important factor.

If the coolant is also the fuel, a special consideration arises for those vehicles where engine operation is not required during certain portions of a flight trajectory where temperatures are high and cooling is necessary. In any case, special design provisions must be made to cope with problems of this kind. Certainly the use of the internally cooled structure system requires a detailed analysis of the factors involved.

It is clear from the previous discussion that the selection of the material and the design of a structure for vehicles that fly at hypersonic speeds is a very complex problem. Each vehicle and each part of the vehicle must be examined and analyzed to provide the best possible structure for each particular application. There is very little flight experience to provide a basis for analysis, and what experience does exist is mostly on vehicles totally unlike the ones being considered for this flight regime.

It is therefore necessary to rely on laboratory tests alone to substantiate structural integrity. Unfortunately, it is difficult to simulate all the necessary environmental and design conditions in the laboratory; therefore, complete substantiation cannot be obtained without the actual flight testing of research vehicles.

NATIONAL AERO SPACE PLANE (NASP)

As a presidentially directed joint project of the Department of Defense (DOD) and NASA, the NASP program objective is to develop and demonstrate hypersonic technologies with the ultimate goal of a single-stage-to-orbit vehicle. These vehicles would be capable of horizontal takeoff and landing and long-range hypersonic flight (Mach 5.0 and above) within the atmosphere.

The program might result in demonstration and flight test of new technologies for hypersonic (in excess of 4,000 mph) operations, using an experimental vehicle designated the X-30. If successful, flight tests will prove the viability of possible, future families of vehicles performing a variety of missions in the atmosphere and in space.

Both DOD and NASA have been performing research on hypersonics, and this program unifies the separate research efforts. Recent advances in hypersonic propulsion, materials and structures, and computational fluid dynamics resulting from these efforts have contributed to the conclusion that a space plane might be possible sometime early in the twenty-first century. Technological advances are supporting this goal: tests of supersonic combustion phenomena; development of high-strength, lightweight, and high-temperature materials; and the availability of supercomputers for engine/airframe design integration.

The concept of a NASP research vehicle is the result of studies performed in 1984 and 1985 by DOD and NASA with widespread industry participation. The concept centers on a manned, hydrogen-powered, single-stage-to-orbit vehicle capable of horizontal takeoff, reaching orbital speeds, cruising for sustained periods at hypersonic speeds within the atmosphere, and landing. The vehicle will be used to demonstrate and flight-test technologies now under development in the program.

Basic configuration

The National Aero Space Plane employs the lifting-body concept pioneered by NASA studies. Figures 10-8 and 10-9 show the basic configuration of the vehicle, which is approximately the size of a Boeing 727. The X-30 is expected to have a takeoff weight of 250,000–300,000 pounds and a length of 150–200 feet.

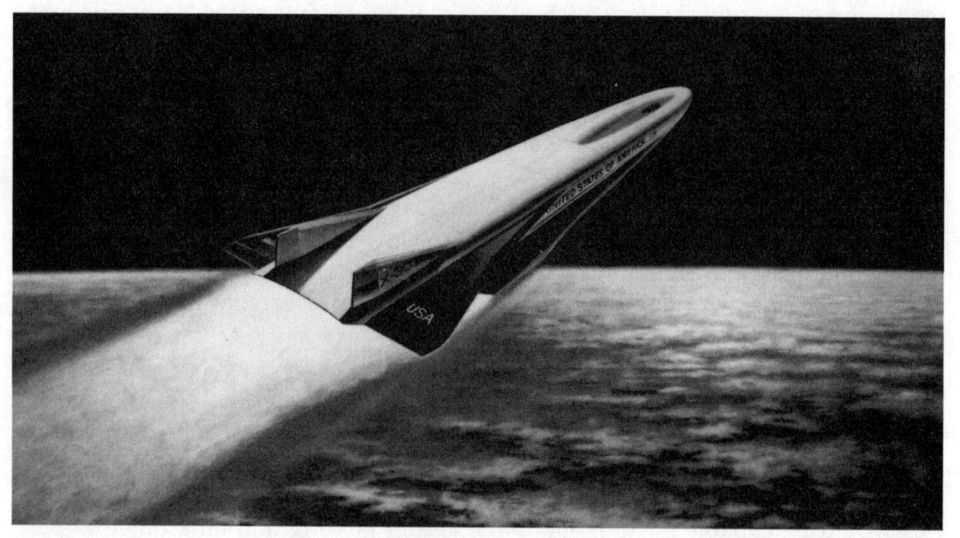

Fig. 10-8. Artist's concept of the joint DOD/NASA National Aero Space Plane that would fly at hypersonic speeds of Mach 5.0, faster than 4,000 mph.

The X-30 will be the ultimate X-plane. It will be a flight test vehicle that will take off horizontally, fly into orbit using air-breathing engines as its primary propulsion, then return through the atmosphere to land on a runway.

The lifting-body shape of the X-30 will produce aerodynamic lift as the space plane rushes through the air. This lets the wings be smaller than would be possible

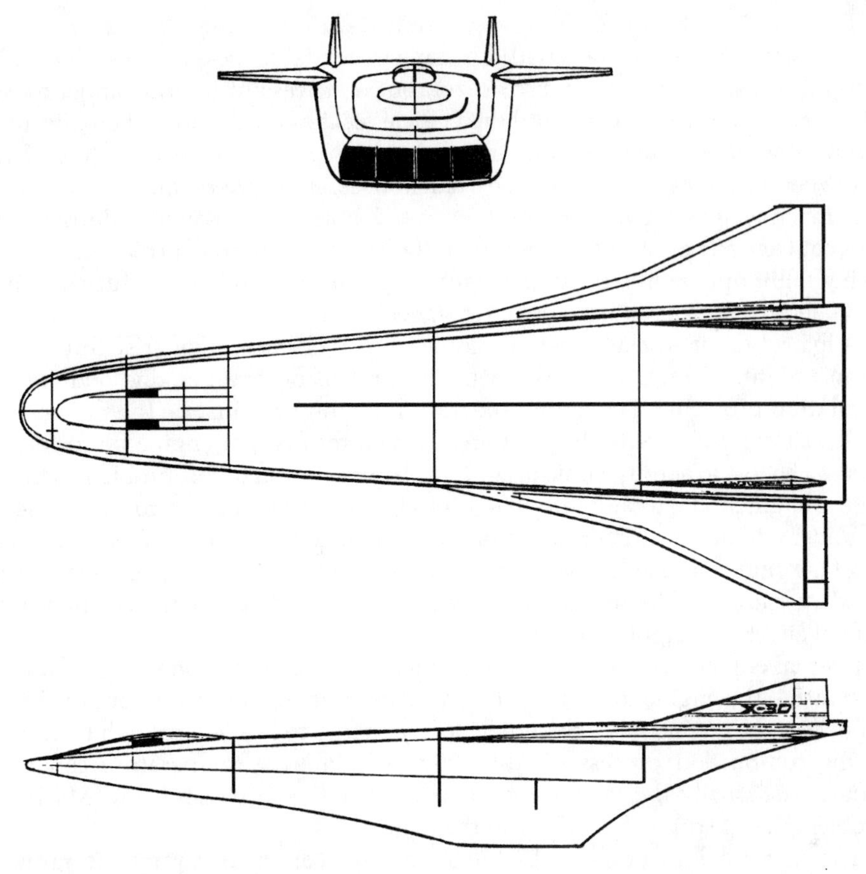

Fig. 10-9. The lifting-body space plane is designed to fly at Mach 5.0. The vehicle's engine inlets are used to compress inlet air to the scramjet engines that are integrated into the structure.

with a more conventional design. Twin vertical tails will give the X-30 good stability and control during the atmospheric part of its flight path. Two pilots will fly the X-30 from a cockpit in a blister near the nose.

To meet its goals, the X-30 must use lightweight materials in its airframe and engines. Supercold hydrogen fuel will circulate in many of its structures to keep the materials within temperature limits as air friction heats the vehicle during ascent and reentry. The hydrogen will be carried as slush so that the X-30's fuel tanks can be smaller than those needed for liquid hydrogen.

Much of the X-30 will be made of titanium-based alloys, which are strong, tough, and resistant to heat. Special coatings will be used to protect the materials because most titanium alloys would chemically react with the onboard hydrogen propellant. The hottest areas of the X-30's airframe will be protected with advanced carbon-carbon composite panels similar to those used on the nose and wings of the space shuttle orbiters.

The NASP will be powered by hybrid ramjet-rocket engines allowing it to take off from conventional runways without using expendable rocket stages. Currently, spacecraft must carry their own oxidizer, but a successful hybrid ramjet-rocket will use air in the atmosphere until the vehicle reaches an altitude where the air is too thin, whereupon the rocket engine will begin to operate by using a store of liquid oxygen to attain orbit. Since the vehicle uses atmospheric air for part of its flight, much weight in oxidizer would be saved, making it possible to design a vehicle that can take off and land horizontally like conventional airplanes. Besides making flight operation more convenient, such a design will also reduce expendable materials, such as jettisoned rocket stages.

A hybrid ramjet-rocket must overcome several obstacles that arise during hypersonic flight. At high Mach numbers, the protrusion of an engine pod creates critical drag problems. The higher the speed, the higher must be the ratio of the nozzle exit area relative to the inlet area, which means that engine performance gains can be lost to aerodynamic drag. The NASP solution to this problem is to design the engine to be totally integrated into the airframe in order to save weight. The underside of the aircraft is contoured to help compress the air, which forms part of the engine intake. The airframe behind the engine flares out to form part of the exhaust nozzle. The requirements of hypersonic operation are satisfied with minimal effect on weight and drag.

Internal engine heat is another challenge. Air for combustion must be slowed down inside the engine to prevent the flame from being blown out, but the braking effect of the engine inlet converts velocity energy to heat energy, which can affect the combustion process. Additionally, the internal temperature can melt engine parts. Finally, the shock wave created when the air passes below Mach 1 is a problem that disrupts the airflow in the engine.

The NASP will be powered by the supersonic-combustion ramjet (scramjet), which is designed to overcome the problems of internal heat and shock. The scramjet will use hydrogen as a fuel because hydrogen can burn within a fast airstream; therefore, the air will not have to be slowed as much as it would be if conventional fuels were used. New materials, aerodynamic concepts, and variable airflow geometry controlled by high-speed computers are other factors that will go into the making of a scramjet.

Recall that ramjets need considerable forward speed before the inlets can compress the air enough to sustain combustion. Any ramjet design will need another form of propulsion to enable it to reach ramjet speeds; therefore, the NASP will use a rocket engine or engines that will allow the aircraft to take off and accelerate to ramjet speeds as well as operate at slow speeds for approach and landing.

The NASP would take off with its rocket engine, switch to the scramjet at high speeds, then at high altitudes switch back to the rocket engine to attain orbit. The scramjet-rocket would reduce some of the need for liquid oxygen. Except for taking off and landing, the scramjet will be used for long-range hypersonic flight within the atmosphere. Again, the rocket engine will provide slow-speed propulsion.

Actual construction of the piloted X-30 is not assured. A decision on the X-30 will be made in the late 1990s or early twenty-first century. This will depend on results of the continuing research program that includes flight tests of key NASP aeropropulsion technologies (engine and forebody/inlet flow at Mach 12–15) with a series of subscale, unpiloted vehicles mounted on surplus ICBM boosters. The flight tests will be anchored by focused ground tests. These experiments will advance the hypersonic technology database for high Mach numbers in three flight stages:

- Boundary-layer transition
- Propulsion performance
- Hypersonic stability and controllability, and airframe/propulsion integration

The value of NASP

The program is invigorating the United States' hypersonic technology base by pushing the state of the art in materials, propulsion, and systems integration, which are all crucial technology areas. It is also upgrading the nation's hypersonic wind tunnels and maintaining the technical expertise of our scientists and engineers.

The NASP program already has achieved significant results in technology transition to industry, especially in areas such as materials, materials processing, and computational methods. The NASP team continues to set an outstanding example of industry/government cooperation on high-technology challenges.

Specific technological advancements by the NASP program include:

- A broad experimental space-plane capability for hypersonic flows around slender-body shapes.
- Quantum leaps in computer science for internal and external flows, even in extreme conditions of heat and speed.
- Greatly advanced United States technology base for very-high-speed, air-breathing propulsion (scramjets).
- Technology for slush hydrogen, which is a denser liquid/solid mixture that allows for smaller, more efficient vehicles.
- Lightweight materials with high strength at high temperatures and other properties such as corrosion resistance.
- Large-scale structures of advanced material demonstrated in the extremes of the NASP flight environment.

Figures 10-10 and 10-11 show various wind tunnel tests conducted on models of hypersonic vehicles configurations.

JUDGMENT

It is obvious that developing an airplane to fly hypersonically within the earth's atmosphere is a formidable undertaking. One can conclude that hypersonic flight within the earth's atmosphere will always be too costly and not worth the effort.

Fig. 10-10. This NASA wind-tunnel test from the 1970s is one of a series of aerodynamic tests that investigated the concept of engine-body blending for hypersonic aircraft. Mounting a row of scramjet engines to the underside of this research model keeps the engines within the aircraft's bow shock wave, increasing aerodynamic pressure for better engine performance. In this photo mosaic, schlieren photography was used to help researchers visualize the complex airflow around the configuration. The 20-inch Mach 6.0 wind tunnel is located at NASA's Langley Research Center, Hampton, Va.

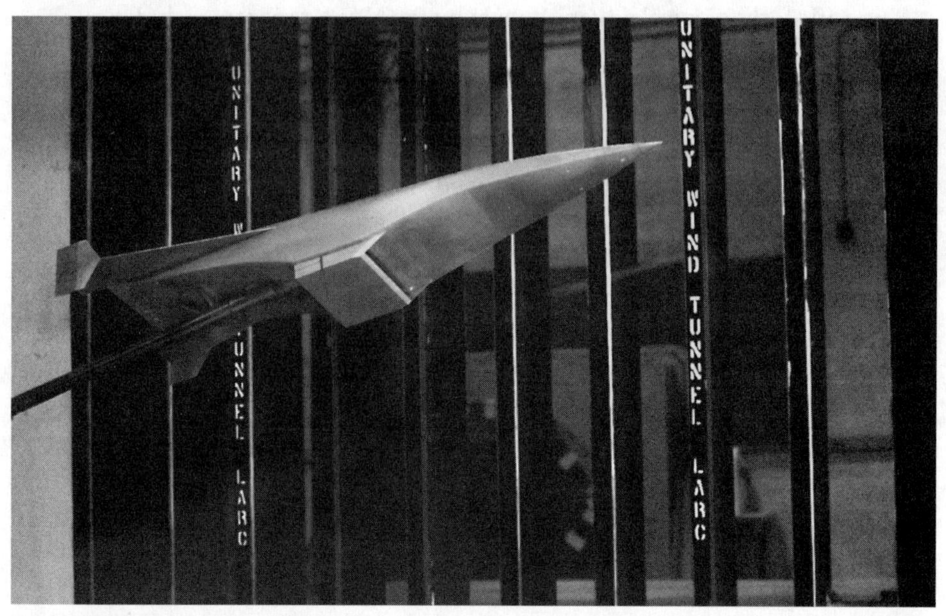

Fig. 10-11. The NASP Langley Test Technique Demonstrator (TTD) model in the unitary plan wind tunnel at NASA Langley Research Center. The NASP TTD model was designed to pave the way for future investigations of NASP configurations. Both static and dynamic aerodynamic characteristics were measured to determine stability, control and performance at Mach 1.5–4.6 (approximately 1,000–3,000 mph). These results are compared with similar data generated on advanced computers using computational fluid dynamic codes.

11

Quest for the fuel-efficient airliner

(This chapter includes excerpts from an article in *Lockheed Horizons* magazine, spring 1982, with permission of the Lockheed Corporation. The article was written by William E. Arndt, manager of Lockheed's propulsion and acoustic department.)

FROM ITS INCEPTION, AVIATION HAS BEEN DRIVEN by the consistent desire to fly faster, farther, and higher. Achieving supersonic flight in the 1950s was viewed by many as the triumph of the century, a symbol of United States technological superiority. But as the 1970s arrived, that perspective was altered. Environmentalists began raising concerns about air and noise pollution.

A then unknown force called the Organization of Petroleum Exporting Countries (OPEC) made Americans face the fact that energy sources were not unlimited and would not remain cheap. Lines appeared at fuel pumps and the compact, fuel-efficient car came into high demand. People learned to cope by traveling less, driving slower, carpooling, and using their air conditioners and heaters more conservatively.

The available solutions for airlines, however, were more severe. Cutting back on flights and increasing ticket prices meant losing business. The aircraft industry faced the possibility of surrendering its position as a world leader in the transport aircraft market.

At the direction of Congress, NASA began exploring solutions to the aircraft fuel problem. A fuel conservation task force headed by NASA was formed in 1975 to study every potential fuel-saving concept that aviation technology could produce. The result was the Aircraft Energy Efficiency (ACEE) Program targeted for implementation in fiscal year 1975. Six major technological elements of the program were related to the subsonic transport.

Three were related to airframes:

- composite structures
- active controls
- aerodynamic improvements

Three were propulsion improvements:

- improved then current high-bypass ratio turbofan engines
- advanced high-bypass ratio turbofan engines
- development of the advanced turboprop

All of these technical concepts were being actively pursued by NASA and the majority of the aircraft industry by the early 1980s. The optimistically predicted fuel savings for each concept are shown in Fig. 11-1. In 1982, airline fuel costs were predicted as shown in Fig. 11-2. Figure 11-3 indicated the projected percentage of fuel direct operating costs.

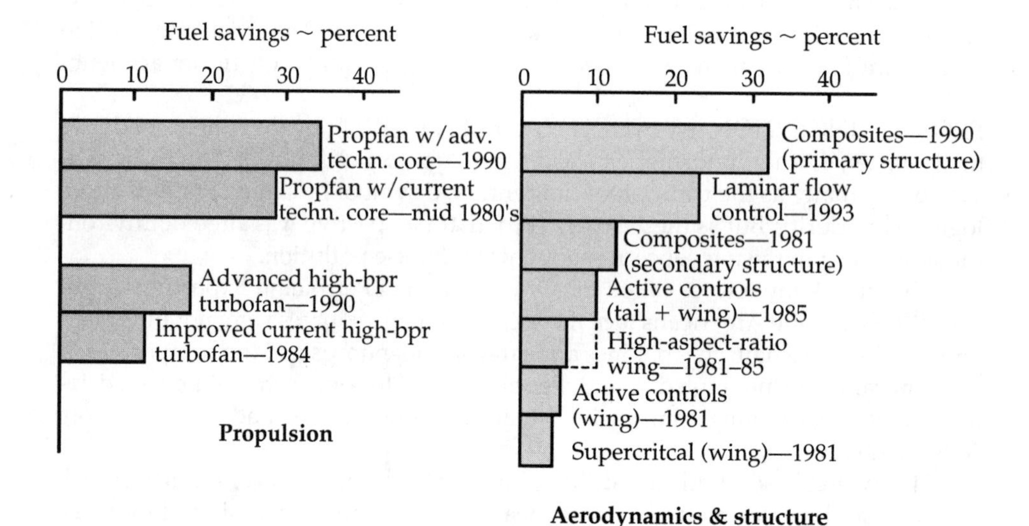

Fig. 11-1. New-technology benefits as predicted in 1981. These benefits were changed or revised downward as experience was gained through the 1990s.

Fuel prices at $3 per gallon did not materialize by the 1990s, although at $1 per gallon, fuel prices were still significantly higher than the 12–13 cents per gallon of the 1950s and 1960s. For this reason, as well as other considerations, the six major technological elements previously indicated were not fully implemented in the 1990s airline aircraft.

Aircraft geometry and systems affecting fuel consumption in transports designed in the 1950s and 1960s received only their proper due, considering the fact that fuel represented less than 10 percent of total operating cost. This was the era

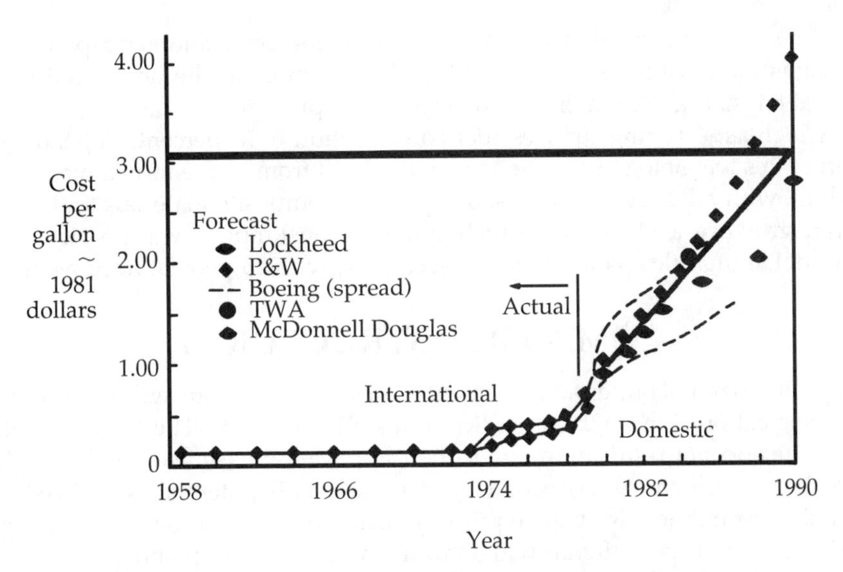

Fig. 11-2. Cost per gallon of jet fuel as predicted by aircraft and engine companies and airlines. These costs were not reached by 1990, which severely dampened enthusiasm for various technological improvements.

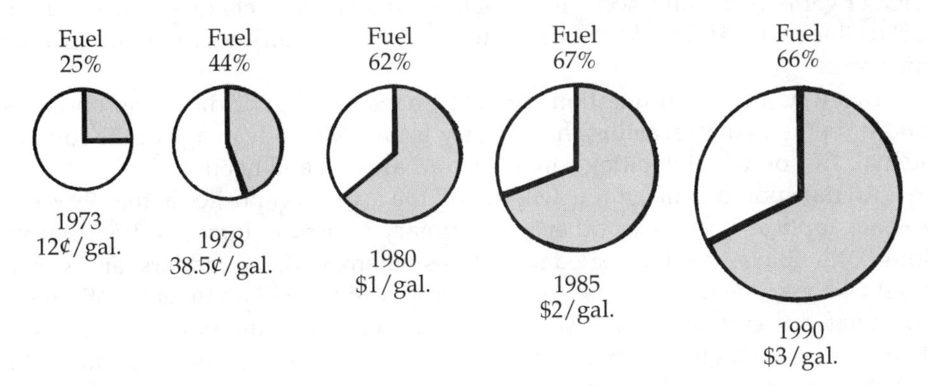

Fig. 11-3. Fuel as a percentage of direct operating costs as predicted in 1981. The $3 per gallon was not realized by 1990.

of the first-generation jets that revolutionized air travel: the Boeing 707, 727, early 737, and early 747; the McDonnell Douglas DC-8 and DC-9.

BENEFITS OF AIRFRAME-RELATED TECHNOLOGIES

A 1980s perspective of the fuel-saving benefits of various airframe related technologies is shown in Fig. 11-1. A supercritical wing gives about a 3-percent fuel reduction. Active controls for wing load reduction offer another 5-percent improvement, and when combined with active controls on the horizontal tail for reduced tail size, a total saving of 10 percent was projected.

A high-aspect wing reduces fuel consumption 6–10 percent, depending on whether this technology is applied to a wing built from conventional materials or one built with advanced materials. By applying composite materials to the major aircraft structure, a 32-percent reduction in fuel consumption was predicted to be realized. Laminar flow control was projected to give a 23-percent improvement.

COMPOSITE STRUCTURES

Composite materials offer advancements in fuel savings, but will not provide a technological breakthrough as predicted in 1981 (Fig. 11-1). The large benefits of composites do not result from either their application as substitute materials or from their employment in secondary structure. Full potential is realized only when the commitment is made to primary structure. Application of composites in the design of transport aircraft will occur in an evolutionary manner.

No commercial transport manufacturer can make a commitment to primary structure without obtaining more facts concerning durability. Extensive testing of composites will be conducted by the manufacturers in order to be ready for the greater gains that could occur in the late 1990s. Material characteristics will be determined so that it will be possible to guarantee the airlines at least a 20-year airplane.

Many nontechnical questions remain unanswered and, therefore, obstacles still lie in the path preventing the industry from taking advantage of the full potential. Two of these inhibiting considerations are cost and liability.

All transport manufacturers will extend the use of composites in the 1990s but restrict application to areas other than primary structure. Interiors, floor beams and posts, fillets, and trailing-edge surfaces (ailerons, flaps, rudders, and elevators) will be made with composites. These applications will occur as retrofit items to aircraft currently in service, substitution for incorporation into the production lines of the current transports, or in new aircraft in the same areas. Generally, the substitution can be made at a cost increase within the boundaries acceptable for reduction in weight.

The fuel reduction to be achieved through this type of application is, however, rather limited. On aircraft such as the L-1011 and DC-10, a weight reduction of approximately 5,000 pounds must occur to save 1 percent in fuel; thus, since the use of composites, in round numbers, reduces the weight of a particular piece of hardware by about 20 percent, more than 25,000 pounds of structure must be altered in order to effect a 1-percent fuel saving. A 1–2 percent fuel saving should not be considered discouraging; it is an encouraging portion of a cumulative effort.

These applications must be made as part of the learning cycle so that eventually composites can be employed in applications where they will really pay off. Until the characteristics of the material with respect to strength, durability, and cost are resolved, major application must wait. Cost of the material is still a problem, and the total question regarding cost is a "chicken and egg" problem. Until the material is used in large quantities, the price will not be reduced dramatically, and not until the price is reduced will it be used in large quantities. Military applications will probably pave the way for extended commercial use.

ACTIVE CONTROLS/CONTROL-CONFIGURED VEHICLES

The most dramatic flight control technology benefits in airplane performance economies are those resulting from what is termed as *control configured vehicle* (CCV) techniques or *active controls technology* (ACT). These terms are usually considered to have the same meaning in that they both imply the use of automatic feedback control systems that sense aircraft motion and provide signals to a control surface, such as an elevator, aileron, rudder, or spoiler, to reduce vehicle weight or to improve airplane performance.

The basic airplane configuration can be altered by incorporating active controls; for example, the wing structural weight can be reduced or a smaller tail assembly can be employed. Some of the functions considered to be active controls are augmented stability, including relaxed static stability, structural wing maneuver and gust-load alleviation, ride control, flutter or other dynamic elastic mode control, and structural fatigue reduction.

All of these active control functions, with the possible exception of ride control, are very beneficial in reducing aircraft weight and/or drag for better fuel economy. Ride control would indeed enhance passenger comfort on commercial aircraft but would normally be used only to reduce crew fatigue, which would improve the mission effectiveness on subsonic military transports.

Active wing load alleviation system

Active controls can provide an answer to increased span without wing redesign. Employment of this technique to today's aircraft provides significant reductions in fuel costs while incurring only a small development cost. Figure 11-4 shows an active aileron concept in which the outboard ailerons are automatically deflected symmetrically, responding to a maneuver or gust load.

With this system, which operates independently and in addition to the normal aileron control inputs, the center of the air load is forced to move inboard. This is accomplished by unloading the outboard wing with the ailerons and requiring the airplane to rotate to a slightly higher angle of attack in order to compensate for the loss of outboard lift, which increases the loading of the inboard wing. Design loads (a wing's bending moments) are reduced.

The airplane can now incorporate higher design weights or do something more clever. Because of fuel prices, the more clever approach is to add wing span in an amount that just restores the bending moments to their original design value; thus, an induced drag reduction and a fuel reduction occurs, and no structural change is required to the wing.

A typical *active control system* (ACS) configuration for reducing wing loads is shown in Fig. 11-5 for the Lockheed L-1011-500 transport. It consists of sensitive

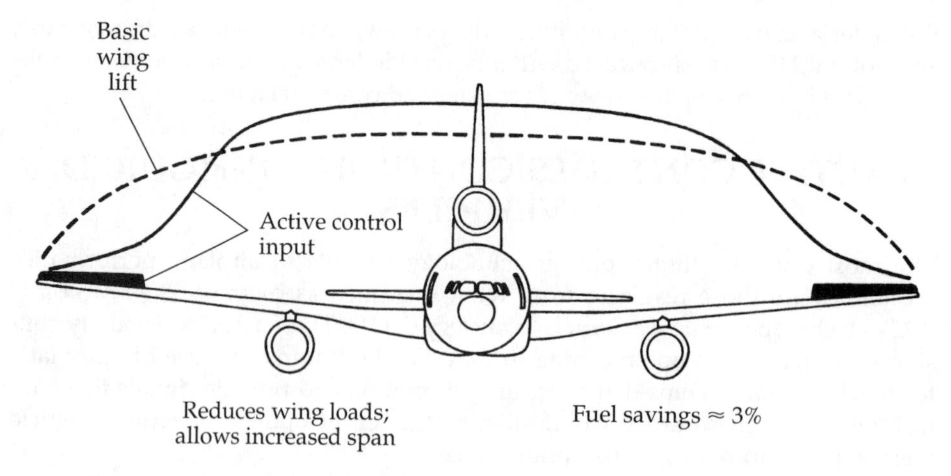

Fig. 11-4. Active ailerons automatically deflect upward in response to maneuver or gust loads, which reduces wing bending moments. The airplane can now incorporate higher design weights or add more wing span, which reduces induced drag.

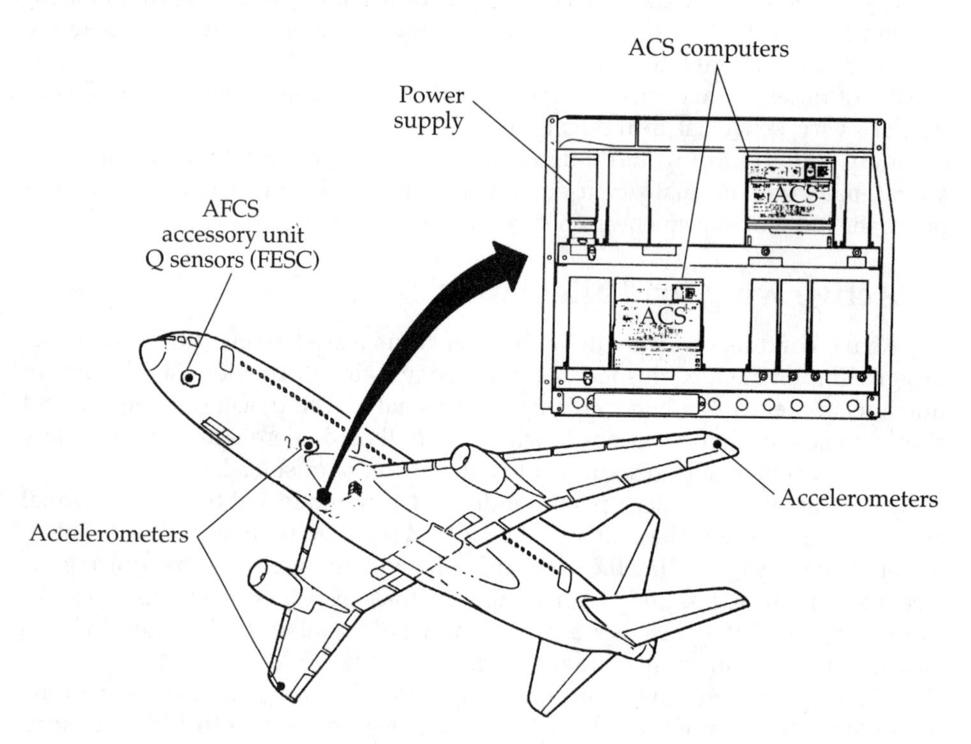

Fig. 11-5. The active control system incorporates accelerometers and sensors that control the aircraft through the active control system computers and the autopilot.

accelerometers mounted at each wingtip and at the aircraft center of gravity. Together with an airspeed sensor (Q sensor), these instruments detect accelerations of the aircraft due to maneuver or gust loads. These signals are transmitted to the ACS computers that send signals to the automatic flight control system (AFCS) to operate the flight controls accordingly, and entirely independent of pilot input.

LONGITUDINAL STABILITY AUGMENTATION

The center of gravity is ahead of the aerodynamic center (center of lift), and a downward load on the tail (horizontal stabilizer) provides a force balance for the inherently stable airplane (Fig. 11-6). This is the classic longitudinal stability configuration of forces that provide satisfactory handling qualities.

This standard force balance concept has the disadvantage of a downward tail load that must be lifted by the wing in addition to the weight of the airplane. Also, a horizontal stabilizer of adequate size is necessary to provide the required forces.

If the c.g. is moved aft, close to the center of lift, two benefits result:

- A smaller and lighter tail surface is possible due to lower balancing and control forces.
- Less wing area is required due to reduced or zero download on the tail.

The net result is reduced drag which translates into less power required and therefore less fuel used. A fuel saving of approximately 3 percent is realized; however, such an inherently unstable force balance requires a stability augmentation (active control system) similar to that used for wingload alleviation. Using accelerometers and pitch rate sensing transducers, the active control system senses aircraft motion to provide signals to the pitch control surface (elevator or stabilator). All this is independent of pilot control inputs. Figure 11-7 shows the "active" tail relative to a conventional tail.

AERODYNAMIC IMPROVEMENTS

The aerodynamic improvements investigated during the NASA-led quest for the fuel-efficient airliner involved three basic technological areas (also discussed in other chapters):

- High-aspect-ratio wing, including the effects of winglets
- Supercritical wing
- Laminar flow control

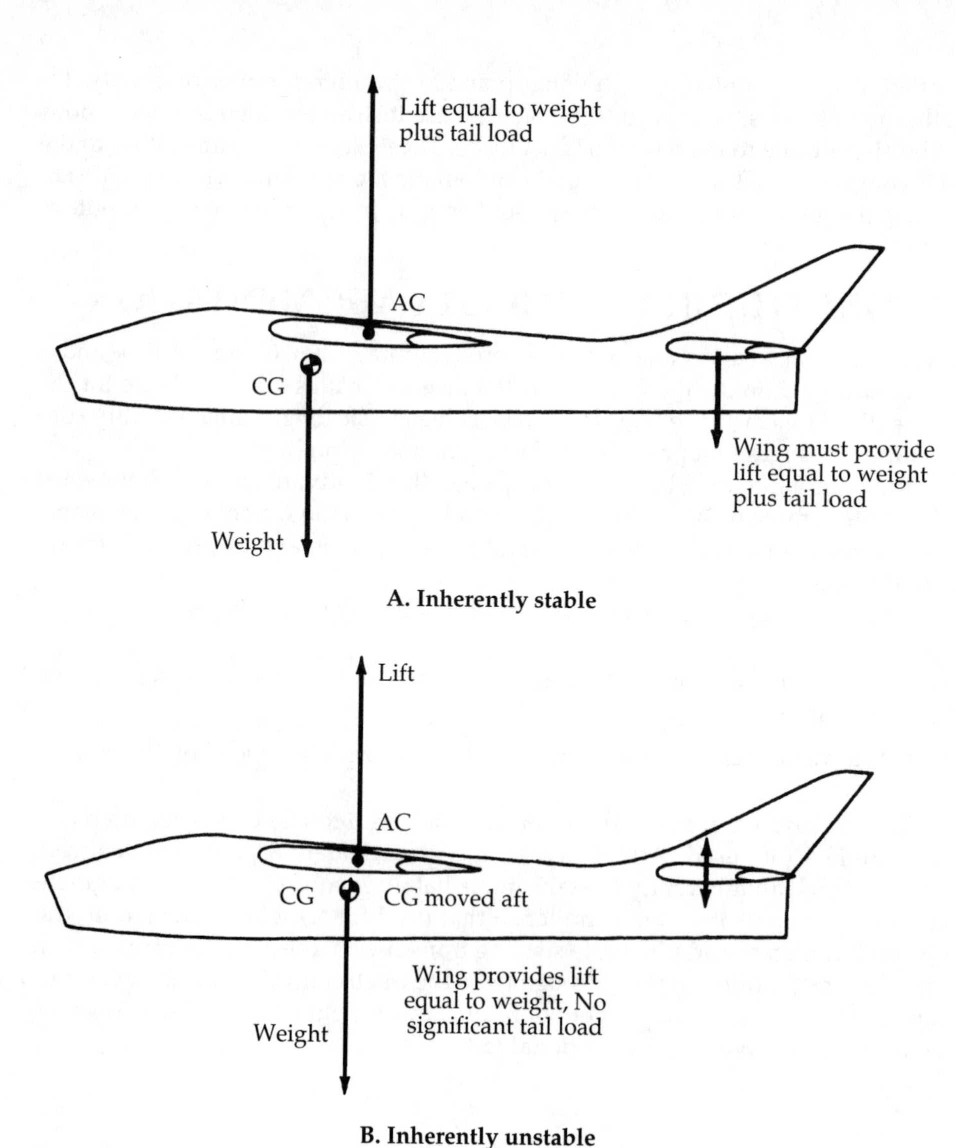

Fig. 11-6. For the inherently stable airplane, the center of gravity is ahead of the aerodynamic center with a download on the horizontal stabilizer to provide a force balance. Moving the center of gravity aft, closer to the aerodynamic center, reduces the tailload, which permits a smaller horizontal stabilizer and less wing area. This configuration is inherently unstable and requires active controls.

High-aspect-ratio wings/winglets

Recall that a high-aspect-ratio wing has less induced drag, which requires less power than a low-aspect-ratio wing; however, as aspect ratio is increased, the

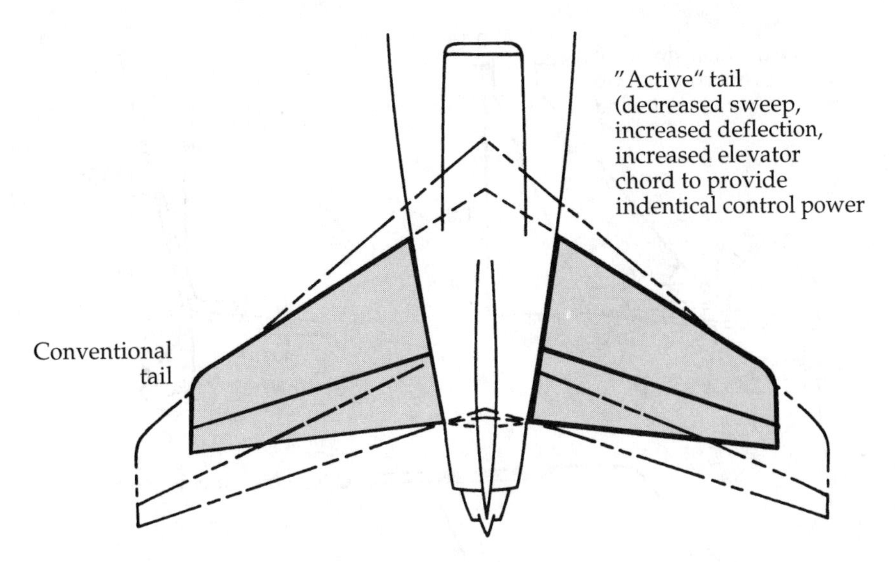

Fig. 11-7. An example of an active horizontal stabilizer compared to a conventional tail.

wing root bending moment is also increased, resulting in heavy structural weight, especially when using the thin airfoil section dictated by high-speed flight.

A winglet reduces induced drag (Fig. 11-8). The advantage of a winglet over a wingtip extension having the same drag reduction is a significantly smaller penalty on wing bending loads and a smaller weight penalty.

Winglets are small, nearly vertical aerodynamic surfaces that are designed to be mounted on the tips of aircraft wings. Unlike flat endplates, winglets are designed with the same careful attention to airfoil shape and local flow conditions as the wing itself. The primary component of the winglet configuration is a large winglet mounted rearward above the wingtip; the "upper surface" of this airfoil is the inboard surface. Some configurations require an additional small winglet, mounted forward, below the wingtip; the "upper surface" of the airfoil for this lower winglet is the outboard surface.

The winglets operate in the circulation field around the wingtip. Because of the pressure differential between the wing surfaces at the tip, the airflow tends to move outboard along the lower wing surface, around the tip, and inbound along the wing upper surface. This wingtip vortex produces crossflows at each winglet; thus, the winglets produce large side forces even at low aircraft angles of attack. Since the side force vectors are approximately perpendicular to the local flow, the side forces produced by the winglets have forward (thrust) components that reduce the wing-induced drag.

The side force produced by the winglets and the thrust that is produced is dependent upon the strength of the circulation around the wingtip. Because the circulation strength is a function of the lift loads near the wingtip, winglets are more effective on aircraft with higher loads near the tip.

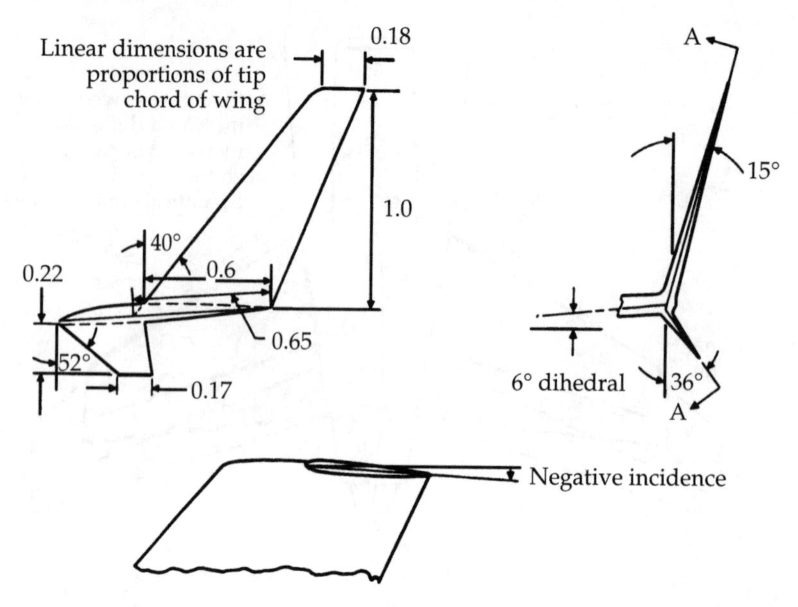

Fig. 11-8. Typical geometry of a winglet. The winglet provides an increase in the effective aspect ratio without increasing wing bending moments. The lower induced drag increases fuel efficiency.

The fuel flow reduction of high-aspect-ratio wings with or without winglets ranges from 2–10 percent depending on tip loading and the degree of wing redesign.

Supercritical wing

Recall that the supercritical airfoil delays the formation and strength of the shocks to a point closer to the trailing edge. Additionally, the shock-induced separation is greatly decreased. The critical Mach number is delayed even up to M = 0.99, which represents a major increase in aircraft performance.

A fuel saving of 3 percent is possible, but if the supercritical airfoil is combined with a high-aspect-ratio wing, fuel saving of up to 7 percent is possible.

Boundary layer control

Although no practical application of a complete laminar-flow wing exists, research indicates a sizable drag reduction that can reduce fuel consumption as much as 20 percent. As tested by Douglas Aircraft Company, the laminar flow system is a unique wing panel that has an outer skin perforated with more than a million tiny holes. The system uses suction to maintain smooth laminar flow over the wing. Airflow closest to the surface of a conventional aircraft wing becomes turbulent as it passes over the wing, greatly increasing aerodynamic drag. The reduced drag of the laminar flow wing translates into reduced fuel consumption especially on long-range flights.

PROPULSION SYSTEM IMPROVEMENTS

The NASA-directed, energy-efficient engine (E3) program had three objectives to reduce fuel consumption of engines powering subsonic airline aircraft:

- Improve existing jet engines
- Develop new fuel efficient jet engines
- Develop advanced turboprops

Existing jet engines

The engine components improvement project was begun first. It involved improving existing engine components by using improved aerodynamics and materials, applying clearance-control techniques, and increasing the bypass ratio for a projected fuel saving of 5 percent.

Along with the major producers of airline engines, Pratt & Whitney and General Electric, newer versions of existing engines such as the P&W JT8 and JT9, were improved with as much as 8 percent better fuel efficiency. The JT8 turbofan powers the Boeing 727 and 737 and McDonnell Douglas DC-9 medium and short range jetliners. Many Boeing 747s, McDonnell Douglas DC-10s, and Airbus Industries A300 transports are powered by P&W JT9 fanjet engines.

General Electric CF6 engines that power many DC-10s, Boeing 747s, and Airbus Industries' A300 and A310 jetliners were also improved.

Improvements involved in these engines consisted of a number of minor changes such as advanced flowpath contouring to reduce pressure losses, better clearance control, and redesigned fuel control. These component improvements resulted in a 5-percent fuel saving.

The component improvement project was followed by the energy-efficient engine project that incorporated the best fuel-saving technologies in the new engine designs for a projected fuel saving of 15–20 percent.

Results of both projects exceeded the goals, and many of the design features have been included in models of JT8D, JT9D, CF, and CFM-56 engines being produced by Pratt & Whitney and General Electric.

All-new-technology engines

These all-new engines obtained a fuel saving of 18 percent compared to existing engines with increased compressor pressure ratio, increased bypass ratio, higher turbine temperatures (due to improved turbine blade materials), and all-electronic engine controls. Examples of all-new-technology engines are the Pratt & Whitney PW2000 powering the Boeing 757 and the PW4000 powering late-model Boeing 747s and the McDonnell Douglas MD-11. The new-technology

General Electric CF6-80A powers versions of the Boeing 767, Airbus Industries A310, and others.

Advanced turboprop development

During the early stages of the NASA's energy efficient engine program, several concepts were identified and pursued, but the advanced turboprop promised the highest potential fuel saving for high-speed subsonic aircraft. It was, however, the most challenging concept technically and was initially resisted almost entirely by United States engine and airframe manufacturers, the airlines, and the military. In spite of the challenges, NASA decided to pursue the program because the potential payoff was too large to ignore. The Advanced Turboprop Office was formed at the NASA Lewis Research Center in Cleveland, Ohio, to manage and integrate the program.

Engineers at NASA, as well as the aircraft industry in general, were aware of the propeller's high efficiency; as well as its limitations. With the tremendous speed advantage of jet propulsion, airlines disregarded the fact that propellers, such as those on Lockheed's Electra, were the most efficient method of propulsion at speeds to Mach 0.6.

With fuel prices at 10–13 cents a gallon, the larger amount of fuel that turbojets and turbofans required seemed inconsequential in comparison with the quieter cabins and the faster speed, higher altitude, and farther distance that they promised. Propeller research ended in 1958 with the aircraft industry and NASA looking toward a future of fast subsonic and supersonic transports.

In conjunction with Hamilton Standard, the most experienced propeller designers and manufacturers, NASA engineers concluded that a highly loaded, multiblade, swept, variable-pitch propeller, which they called the *propfan*, could be combined with the latest in turbine-engine technology. The resulting advanced turboprop would offer a potential fuel saving of 50 percent (Fig. 11-9) over an equivalent-technology turbofan engine operating at competitive speeds and altitudes because of the turboprop's much higher installed efficiency.

The same principles involved in high-speed wing design (sweepback and thin airfoils) to reduce compressibility losses were used to develop a multibladed propfan design. An added benefit of sweeping the blades was the potential decrease in the noise levels that would result from the high blade tip speed. Sweeping the blade tip delays to a higher relative helical tip Mach number the sharp rise in drag and noise that occurs when the airflow over the blade approaches Mach 1 as a result of both airplane forward speed and propeller rotational speed.

The propfan was ready for flight tests only after an extensive wind tunnel and ground test program. Participants in the test program included propeller manufacturer Hamilton Standard; engine manufacturers Pratt & Whitney, Allison, and General Electric; and aircraft manufacturers Boeing, McDonnell Douglas, and Lockheed.

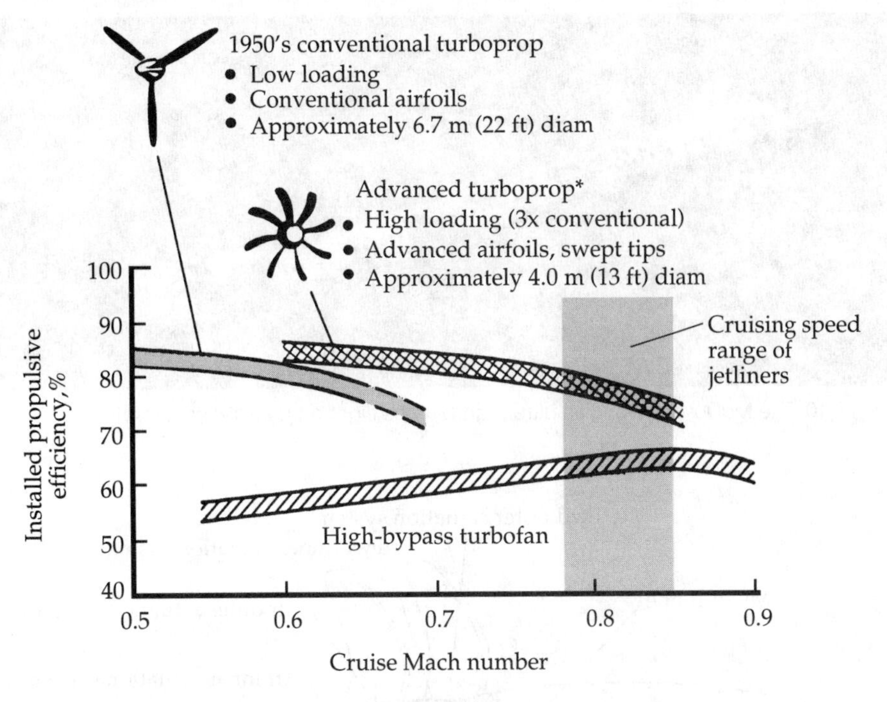

Fig. 11-9. A comparison of 1950s conventional turboprops and the high-bypass turbofan with the advanced turboprop. The advanced turboprop shows considerably higher efficiency than the high-bypass turbofan.

Although flight tests were conducted by NASA research aircraft and Boeing on a modified 727, the McDonnell Douglas program was carried out to the extent of offering a production propfan airplane to the airlines.

A General Electric UDF (unducted fan) engine was installed in place of one of the two conventional turbojets on a McDonnell Douglas MD80 testbed aircraft (Fig. 11-10). The GE UDF engine (Fig. 11-11) offered as much as a 70-percent improvement in fuel burn compared to older, low-bypass-ratio, turbofan engines, plus a 25–30 percent improvement compared to all-new-technology turbofans.

The counter-rotating turbine concept of the UDF allowed direct coupling of the large diameter, unducted, counter-rotating fans to the turbine, which eliminated the weight and complexity of a reduction gearbox.

After a 3-year development and test program, McDonnell Douglas proposed the MD-91 and MD-92 for delivery in 1992; the MD-91 would carry 114 passengers, and the MD-92 would carry 165 passengers. These were advanced technology aircraft (Fig. 11-12) that cruised at Mach 0.78, the same as equivalent jetliners. They would feature a 37-percent fuel saving compared to the similar turbofan MD-87.

Fig. 11-10. The McDonnell Douglas ultra-high-bypass flight demonstration aircraft.

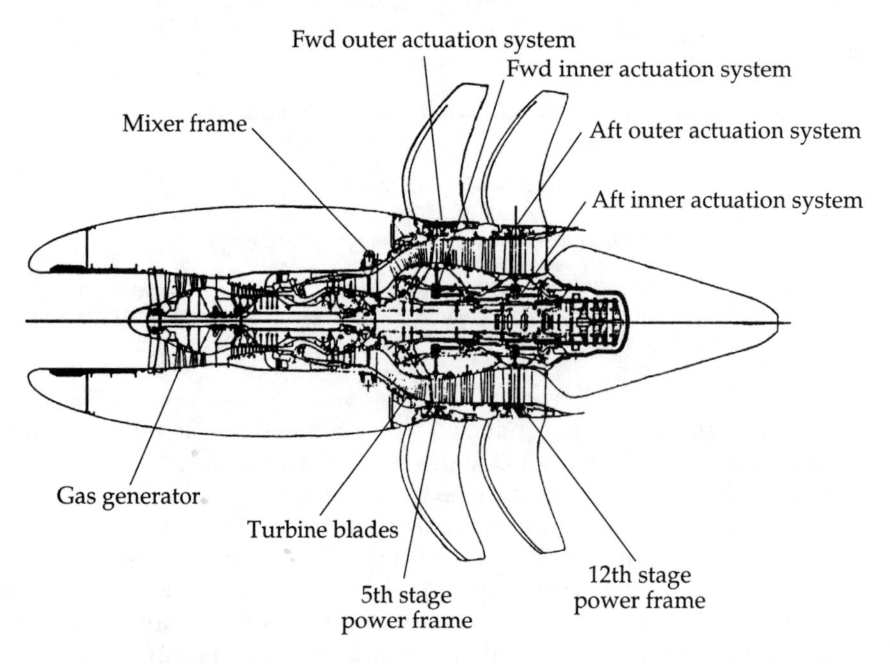

Fig. 11-11. The General Electric proof-of-concept demonstration unducted fan propulsion system as installed on the McDonnell Douglas flight demonstrator aircraft.

The MD-80 demonstrator aircraft conducted a series of demonstration flights for airline executives. McDonnell Douglas and GE began an extensive marketing drive for the proposed twin-engine, fuel efficient transports. This was the culmination of the extensive fuel-efficient airliner research and development program that originated in 1975.

Did the airlines beat a path to Douglas' door? Is Douglas producing this high-technology, fuel-efficient airliner by the hundreds?

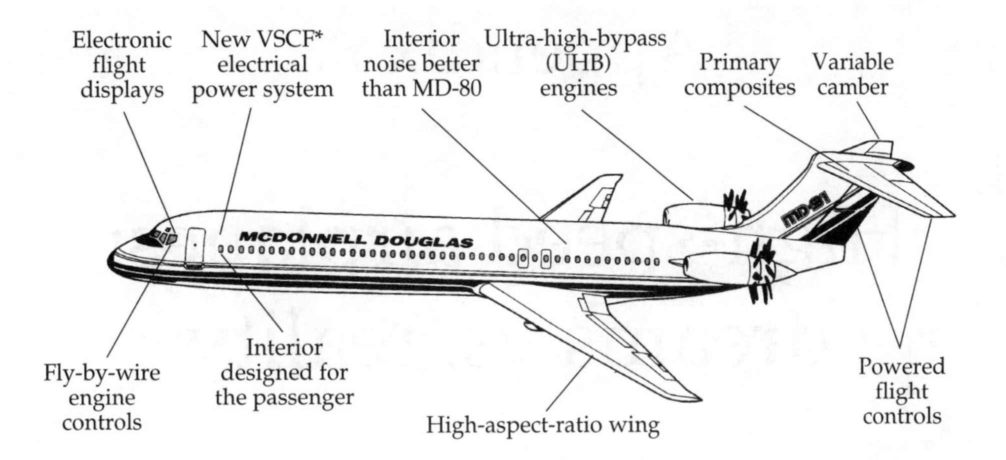

- 114 passengers (mixed class)
- Meets "exempt" noise standards at John Wayne Airport
- Maximum takeoff weight: 129,000 lb
- 37-percent better fuel burn than MD-87

*Variable speed, constant frequency

Fig. 11-12. The advanced technology MD-91 was one of two aircraft with unducted fan propulsion that were offered to the airlines in 1989. Although more fuel efficient than turbofan-powered aircraft, the airlines were not interested in a propeller-driven aircraft.

No airline ordered the MD-91 or MD-92 from Douglas. Boeing discontinued development of the advanced turboprop. Apparently the airlines were enamored with the all-jet airplane. Propellers were forsaken, apparently never to be revived, even for economic reasons.

Appendix A

High-speed airplanes: dreams vs. reality

SINCE WORLD WAR II, it was the military airplane that advanced the frontiers of flight ever forward in terms of supersonic speed. The civil jet transport reached Mach 0.8 and remained at this speed; increased efficiency was the goal of civil aviation.

During the Cold War period, the military services, mainly the U.S. Air Force, pursued many ambitious paths in order to obtain a "super-performance" airplane, some of which are discussed in this appendix. Also presented is an historical summary of the logical development of the jet airplane by example of successful, operational airplanes, thus indicating the contrast between dreams and reality.

DREAMS

Rocket-boost F-104

In an effort to increase performance of existing designs, a rocket booster engine was installed on a Lockheed F-104 turbojet powered fighter airplane. Public relations people called it "A missile with a man in it." The rocket boosted the F-104 into a ballistic trajectory at 130,000 feet. At the top of the trajectory, the aerodynamic controls were ineffective; therefore, control was achieved by hydrogen peroxide reaction nozzles in the nose, tail, and wingtips.

Rocket-boost F-101B

Design studies were conducted by McDonnell in 1957 at the request of the U.S. Air Force for installation of a rocket pod on the F-101B turbojet-powered interceptor. The rocket was supposed to boost its shorttime speed to faster than Mach 2.0 and its ceiling above 70,000 feet. The pod was almost as long as the fuselage and consisted of a rocket fuel tank with an integral rocket motor slung under the fuselage where a drop tank was normally located. Very little modification to the basic airplane was required.

Republic XF-103

Republic Aviation received a Phase I development contract in September 1951 for a highly sophisticated airplane listed on the books as "Weapon System 204A." A full-scale mockup was constructed. The airplane was reconfigured and the Phase I contract was extended 18 months from March 1953 as the company dealt with further state-of-the-art studies of titanium fabrication, high-temperature hydraulics, escape capsules, and periscopic sights to replace the canopy. Republic finally obtained a contract in July 1954 for three prototype airplanes, now called XF-103 (Fig. A-1).

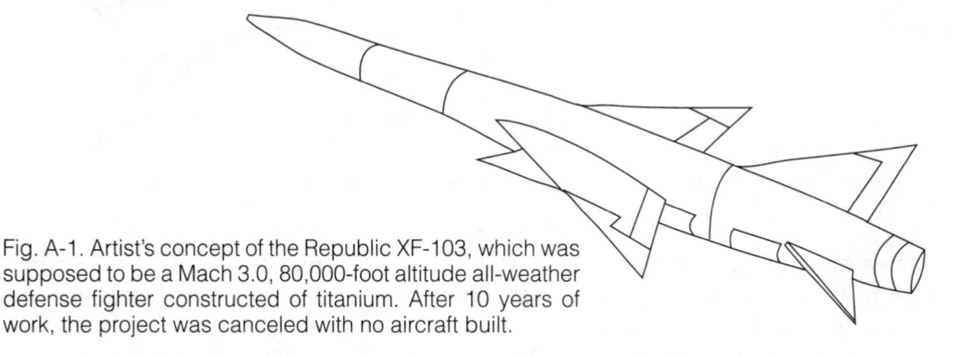

Fig. A-1. Artist's concept of the Republic XF-103, which was supposed to be a Mach 3.0, 80,000-foot altitude all-weather defense fighter constructed of titanium. After 10 years of work, the project was canceled with no aircraft built.

The XF-103 all-weather defense fighter was to fly Mach 3.0 at 80,000 feet altitude. It was to be constructed of titanium alloy and powered by the new Wright J-67 turbojet engine. This sophisticated airplane was to use the jet engine for low-speed flight and by a complex system of ducting and flaps, convert to a ramjet at high speeds. During takeoff and acceleration to Mach 2.0, the turbojet would be in afterburning. At ramjet-operating speed, a series of valves would direct inlet-air directly to the afterburner, bypassing the turbojet; thus, the afterburner became a ramjet.

The XF-103 was designed to intercept a hostile bomber approaching from the North Pole at 75,000 feet at Mach 2.0. The XF-103 would return to its base using the turbojet.

Bell XF-109

Military services started experimenting with vertical takeoff and landing (VTOL) aircraft as soon as the first jet airplanes were flying in the 1940s, requiring long runways for operations. Among the many VTOL projects was the Bell XF-109 (Fig. A-2), which was supposed to be a Mach-2.0 fighter capable of taking off vertically, transitioning to horizontal flight up to Mach 2.0, slowing to a hover, and then landing vertically.

To accomplish this amazing feat, the XF-109 was to be powered by eight General Electric J-85 turbojet engines with 2,600 pounds thrust each. Six of these engines were afterburning, which increased their thrust to 3,800 pounds. Two of the afterburning engines were conventionally installed in the fuselage to provide forward thrust. Two more nonafterburning engines were installed upright at the air-

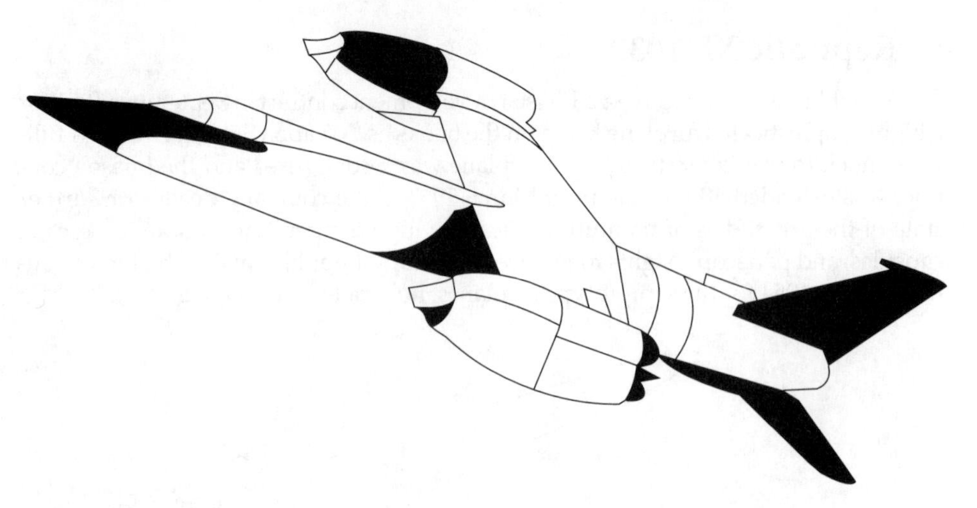

Fig. A-2. The Bell XF-109 vertical takeoff and landing interceptor. No flight vehicle materialized.

craft's center of gravity to provide vertical thrust. The remaining four afterburning engines were installed in pairs in two rotating wingtip pods to provide either vertical or forward thrust as required.

Takeoff was to be accomplished using the two vertical fuselage-mounted-non-afterburning engines, plus the four wingtip-mounted-afterburning engines rotated to provide vertical thrust. Transition to forward flight required rotation of the wingtip engines to a horizontal position, simultaneously starting the tail engines and shutting down the fuselage-mounted vertical engines. During the vertical and transitional flight phase, control would be provided by reaction jets using compressor bleed air. As the XF-109 gained speed, the normal aerodynamic controls would take over.

North American XB-70 Valkyrie

On December 23, 1957, in competition with Boeing, North American Aviation won a contract for development of a supersonic cruise bomber with Mach 3.0 dash capability at 70,000 feet. Power was supplied by six General Electric J-93 afterburning engines of 30,000 pounds thrust each (Fig. A-3).

The XB-70's skin was stainless-steel honeycomb with a stainless-steel and titanium internal structure to withstand the 550°F aerodynamic heating at Mach 3.0. It had a 542,000-pound gross weight and a large delta wing. The XB-70 had an awesome appearance.

North American F-108 Rapier

The North American XF-108 letter contract dated June 6, 1957, called for an all-weather two-man, twin-engine, long-range interceptor, with a combat speed of at least Mach 3.0 and an operating ceiling of 70,000 feet. The airplane would carry two

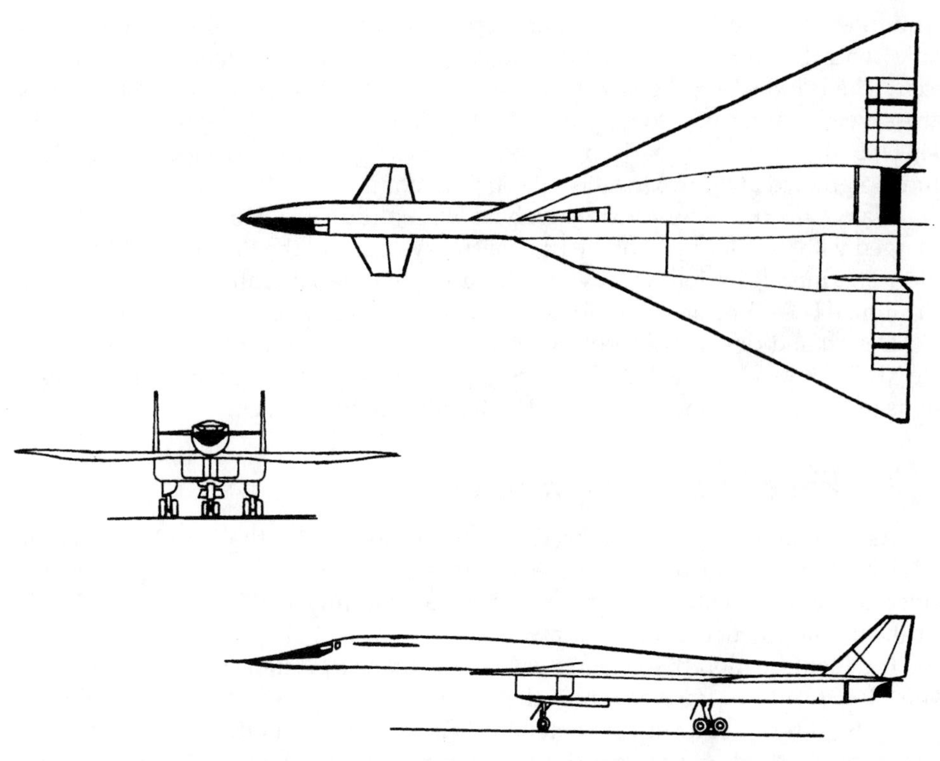

Fig. A-3. The stainless-steel North American XB-70, Mach 3.0 bomber. Two prototypes were constructed, but the XB-70 never became operational.

or more air-to-air missiles with nuclear or conventional warheads. The General Electric J-93 turbojet was to be the propulsion system. A stainless steel structure was proposed to withstand temperatures of more than 550°F at Mach 3.0 (Fig. A-4).

The U.S. Air Force expected a lot from this complex new fighter aircraft. It wanted an early 1963 operational date, 1,000-nautical-mile range at Mach 3.0 with

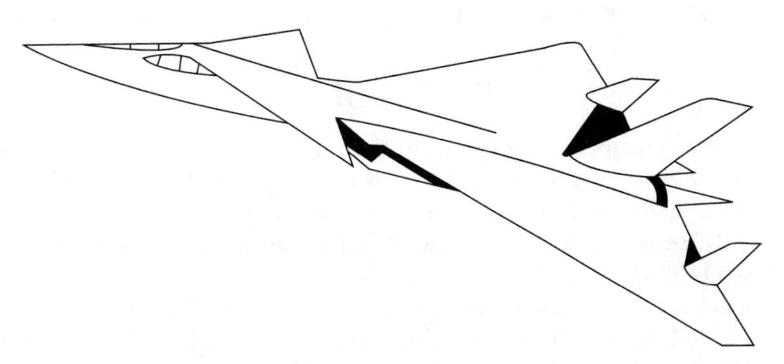

Fig. A-4. The North American XF-108 Rapier Mach 3.0 interceptor was to be fabricated with stainless steel. No flight vehicles were built.

5 minutes of combat at Mach 3.0, or a cruise speed of Mach 3.0 for 350 nautical miles with 10 minutes of combat at maximum power and supersonic return to base. The F-108 would be larger than a normal fighter airplane: 102,000 pounds gross weight, much of it coming from 7,100 gallons of internal fuel (approximately 44,000 pounds). For comparison, 1950s and 1960s fighters such as the F-101, F-105, and F-4 grossed at approximately 50,000 pounds.

Many subcontractors were involved with the F-108. Hughes Aircraft was charged with development of the fire-control system and the Phoenix AIM-54 missiles. This missile system was eventually used on other aircraft, notably the Navy Grumman F-14 Tomcat, carrier-based, swing-wing fighters.

Design and development of the F-108 proceeded simultaneously with that of the XB-70. Both were Mach 3.0, delta-wing, stainless-steel aircraft powered by versions of the same General Electric J-93 afterburning turbojet engines.

Lockheed YF-12 Interceptor

As a variation of the Lockheed design number A-11, the YF-12 interceptor originated in November 1959. The A-11 had a long and narrow fuselage, twin engines, and a fixed delta wing. Its first flight came in July 1962, only 32 months after the development contract was awarded.

President Lyndon B. Johnson revealed the plane's existence in February 1964. The YF-12A interceptor version of the almost all-titanium A-11 was unveiled at Edwards AFB in California in September. The Air Force in October set forth performance standards surpassing those first imposed on the North American F-108 Rapier: a combat radius up to 1,200 nautical miles, faster than Mach 3.0 speed, and high maneuverability at high altitude. Two YF-12A prototypes established nine world speed and altitude records in May 1965.

The YF-12 (Fig. A-5) was powered by two Pratt & Whitney J-58 engines that produced 32,500 pounds thrust each in afterburning. A complex sensing and flap system converted the afterburners into ramjets for speeds faster than Mach 2.0, similar to the F-103 system. A special fuel, JP-7, was developed for the YF-12. More than 13,000 gallons of fuel at 80,000 pounds were carried in five tanks. The YF-12 was a huge airplane for a fighter: gross weight 136,000 pounds. Its contemporaries, such as the F-101, F-105, and F-4 fighters, weighed approximately 50,000 pounds.

SYNOPSIS

All of the above described contracts were eventually canceled. No operational airplanes were produced. Only one rocket-boosted F-104 was built. The YF-12A program resulted in three flying prototypes, one of which was converted to a two-place SR-71 trainer designated SR-71C. Two flying prototypes of the XB-70 bomber were built and tested.

The rocket boosted F-104 was a one-only experiment. The F-101B rocket pod was only carried out through the preliminary design stage in 1958. A full-scale mockup of the Republic XF-103 was reviewed by the Air Force in March 1953. The XF-103 contract was canceled in September 1957 after 10 years work; development

Fig. A-5. The Lockheed SR-71, a development of the YF-12, Mach-3.0 interceptor, is constructed with titanium. Three YF-12 prototypes were built, but the F-12 never went into production. The SR-71 development, however, became operational.

was too slow to justify further expense. The contract for the Wright J-67 turbojet engine was also canceled due to development problems.

A mockup of the Bell XF-109 VTOL fighter was completed in 1961, but the contract was canceled before a prototype was built. The XF-108 mockup was inspected by the Air Force in January 1959. A funding pinch wiped out the whole project in September 1959. Development of the Lockheed F-12 program was discontinued in November 1967.

All of these super fighter airplanes were replaced by an airborne warning and control system (AWACS) coupled with a modernized Convair F-106 interceptor for a more cost-effective defense system. The price of super performance was just too high.

The YF-12 program, however, evolved into the SR-71 reconnaissance plane, which is the only operational Mach 3.0 aircraft to fly. The SR-71 story, as outlined in appendix B, illustrates the tremendous problems involved in the development of Mach 3.0 airplane.

The XB-70 evolved into two flying prototypes. The XB-70's first flight was on September 21, 1964, at Palmdale, California. It landed at Edwards AFB, its home base for the remainder of the test program. The second XB-70 flew on July 17, 1965, with a few changes from the number one aircraft. This second XB-70 was lost as the result of a collision with an F-104 chase plane. The first XB-70 was flown for 2 years gathering supersonic data for the United States' Supersonic Transport Program. The remaining XB-70 flew to Wright-Patterson AFB in Dayton, Ohio, on February 4, 1969, for display at the Air Force Museum.

REALITY

In contrast to the unsuccessful quest for super-performance, steady progress was made as exemplified by a series of successful, operational airplanes. The advent of the jet engine in the mid-1940s was the first step in solving the problem of high-speed flight.

Germany's Messerschmitt Me262 and Great Britain's Gloster Meteor (Fig. A-6) proved the concept of the jet airplane. These airplanes and the United States' Lockheed P-80 were all based on modified low-speed aerodynamics.

Wartime German research into high subsonic, transonic, and supersonic speeds was the second step in attaining high-speed flight. It is significant that German development of the axial-flow turbojet engine and research into high-speed aerodynamics were carried out simultaneously. The British Whittle engine featured the less-advanced centrifugal flow principle.

The North American F-86 Sabre was the first true high-speed airplane since it combined the axial-flow turbojet with captured German high-speed aerodynamic research data. The prototype F-86 first flew in October 1947. From that time on, high-speed aerodynamics and jet engine development advanced rapidly.

The fighter airplane was always at the forefront when it came to speed. Advances in aerodynamics and propulsion then filtered down to other types of aircraft such as bombers and commercial aircraft.

Fighter development from the P-80, which first flew in January 1944, to the F-22's first flight in 1990 graphically illustrates the advances in high-speed aerodynamics and propulsion. As jet engine performance increased, aircraft flight speed increased from the P-80's 500-knot speed to the F-22's Mach 3.0.

Wing design changed from the P-80's more conventional straight wing with aspect ratio of 6.0 to the F-22's highly swept wing with an aspect ratio of 2.25. The variable sweepback wing of the F-111 is in between.

The horizontal stabilizer design also illustrates the transition from subsonic to transonic and eventually supersonic flight. The subsonic P-80 and F-86 used conventional horizontal stabilizer and elevator for pitch control. Due to shock stall and the increased control forces during transonic flight, the F-100 employed the first all-movable horizontal stabilizer with irreversible power-operated control system. This pitch control concept continued through all supersonic aircraft to the F-22.

From 1944 through the 1990s, engine thrust increased dramatically: 4,000 pounds of thrust from the P-80's J33 engine to 35,000 pounds of thrust from each of the F-22's two P&W F119 engines. All transonic and supersonic airplanes from the F-100 through the F-15 achieved supersonic flight using the increased thrust of an afterburner. All these aircraft cruised subsonically without fuel-thirsty afterburning. Only the F-22 can "supercruise" supersonically without afterburning.

Thrust-to-weight ratio changed dramatically from the P-80's 1.0-to-4.2 (weight is 4.2 times thrust) to the F-22's thrust-to-weight ratio of 1.0-to-0.9 (thrust capability is more than the aircraft weight, which yields exceptional climb performance).

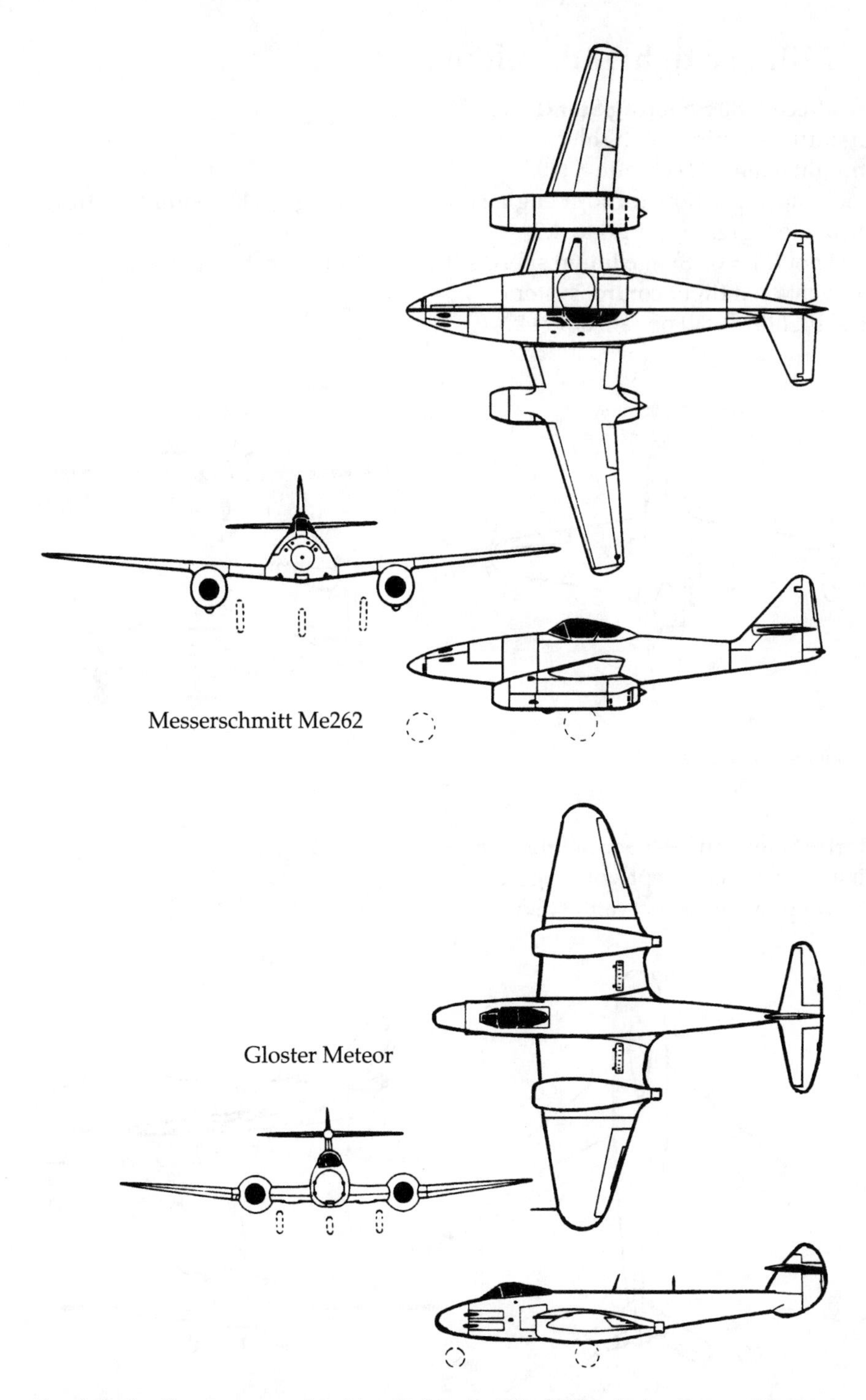

Fig. A-6. The first two operational jet airplanes of World War II, the German Messerschmitt Me262 (top) and the British Gloster Meteor (bottom). The Me262 was powered by two Junkers Jumo 004B, 1,980-pound-thrust axial-flow turbojet engines for a maximum speed of 540 mph at 20,000 feet. The Gloster Meteor was powered by two Rolls-Royce Derwent, 1,995-pound-thrust centrifugal-flow turbojet engines at a maximum speed of 493 mph at 30,000 feet.

Military fighter development

Lockheed F-80 Shooting Star (Fig. A-7)
First true American jet fighter
Straight wing, aspect ratio = 6.0
One centrifugal-flow turbojet engine (no afterburning), 4,000 pounds of thrust
Thrust-to-weight ratio 1.0-to-4.2
500 knot (M = 0.78) maximum speed subsonic at 7,000 feet altitude
Conventional flight control system
First flight (XP-80) on Jan. 8, 1944

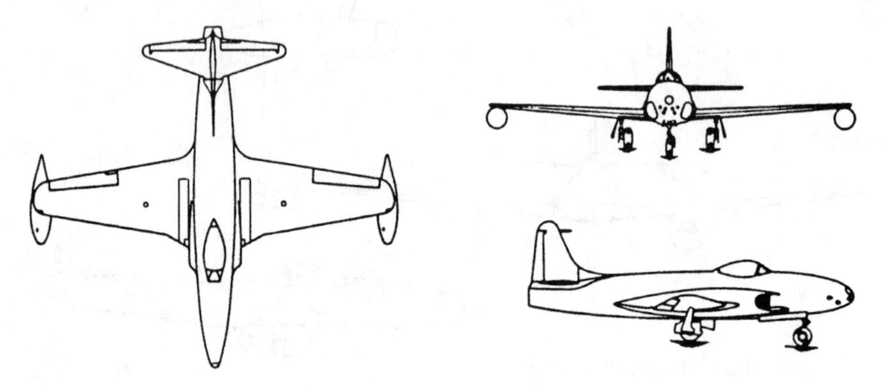

Fig. A-7. Lockheed F-80.

North American F-86 Sabre (Fig. A-8)
First operational swept-wing fighter
35° swept wing, aspect ratio = 6.5

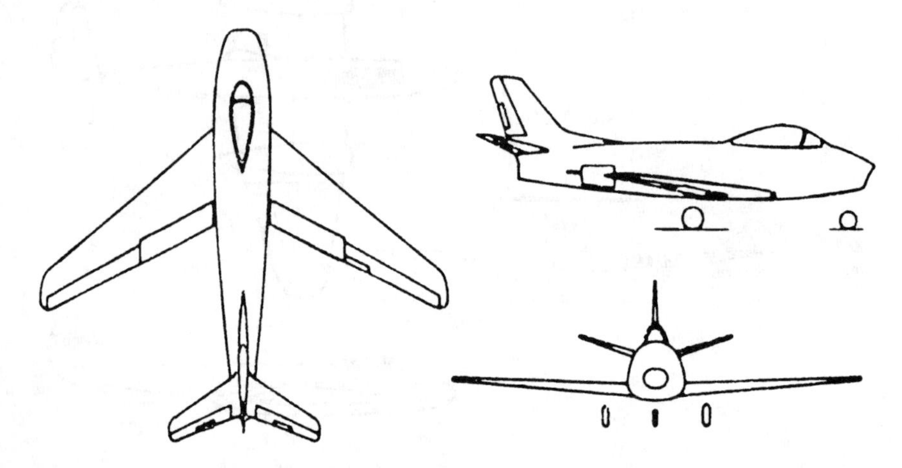

Fig. A-8. North American F-86 Sabre.

One axial flow turbojet engine (no afterburning), 5,900 pounds of thrust
Thrust-to-weight ratio 1.0-to-3.5
600-knot (M = 1.04) maximum speed at 35,000 feet
Mach 0.83 cruising speed
Conventional flight control system
First flight (prototype) on Oct. 1, 1947

North American F-100 Super Sabre (Fig. A-9)
World's first operational supersonic airplane
45° swept wing, aspect ratio = 4.0
710 knot (M = 1.24) maximum speed at 35,000 feet with afterburner
Subsonic cruise
One 10,000-pound-thrust axial-flow turbojet engine, 15,000 pounds of thrust in afterburning
Thrust-to-weight ratio, 1.0-to-2.5
Irreversible power-boost flight control system with all-movable horizontal stabilator
First flight (prototype) on May 25, 1953

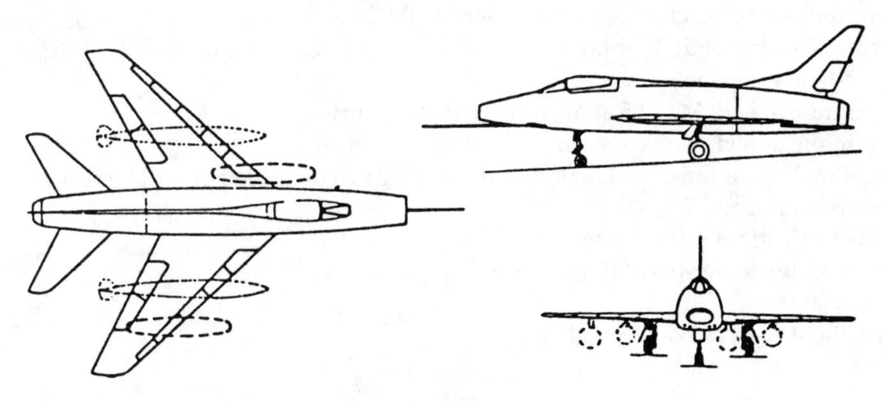

Fig. A-9. North American F-100 Super Sabre.

McDonnell F-4 Phantom II (Fig. A-10)
50° leading-edge sweepback, aspect ratio = 3.0
Two turbojets, 11,800 pounds of thrust each, 17,800 pounds of thrust each in afterburning
Thrust-to-weight ratio, 1.0-to-1.14
Maximum speed, M = 2.24
Cruise speed 500 knots subsonic
Irreversible power-boost flight control system with all-movable horizontal stabilator
First flight (Navy F4H-1) on May 27, 1958
First flight (Air Force F-4C) on May 27, 1963

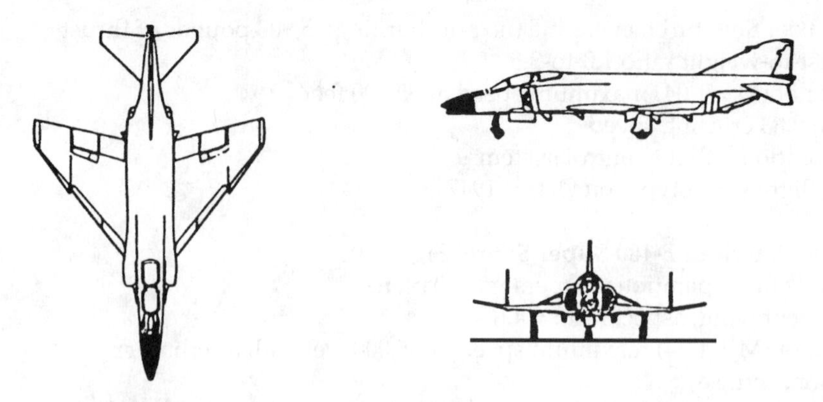

Fig. A-10. McDonnell F-4 Phantom II.

General Dynamics F-111 (Fig. A-11)

The variable-sweep wing could be positioned in flight at various angles between the full forward and aft position, which enabled all F-111 tactical fighters to operate from relatively short runways, fly at supersonic speeds at low altitudes, and reach Mach 2.5 above 60,000 feet.

Swing wing, extended for takeoff, landing, and slow speeds was highly swept (72°) for supersonic speed

Maximum speed, M = 2.5 at altitude with afterburner

Maximum speed, M = 1.2 at sea level with afterburner

Two low-bypass fanjets, 11,000 pounds of thrust each, 19,000 pounds of thrust each in afterburning

Thrust-to-weight ratio, 1.0-to-2.25

Irreversible, power-boost flight control system with all-movable horizontal stabilator

First flight (F-111A) on Dec. 21, 1964

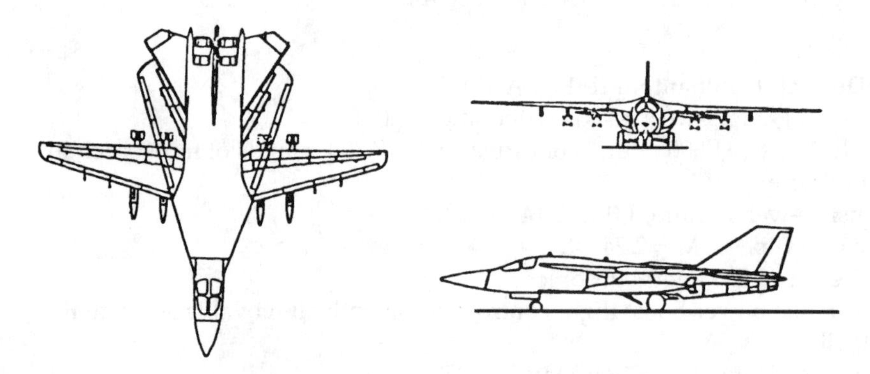

Fig. A-11. General Dynamics F-11.

McDonnell Douglas F-15 Eagle (Fig. A-12)
Sweepback 45° at leading edge, aspect ratio = 3.0
Maximum speed, M = 2.5 with afterburning
Subsonic cruise
Two low-bypass turbofan engines, 25,000 pounds of thrust each in afterburning
Thrust-to-weight ratio, 1.0-to-1.25
Irreversible, power-boost flight control system with all-movable horizontal stabilator
First flight (prototype) on July 27, 1972

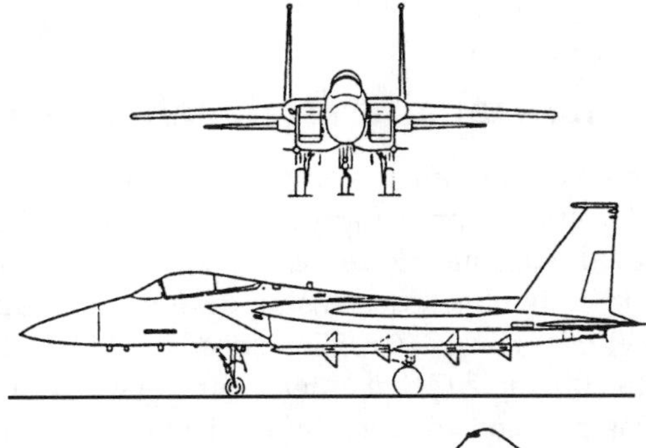

Fig. A-12. McDonnell Douglas F-15 Eagle.

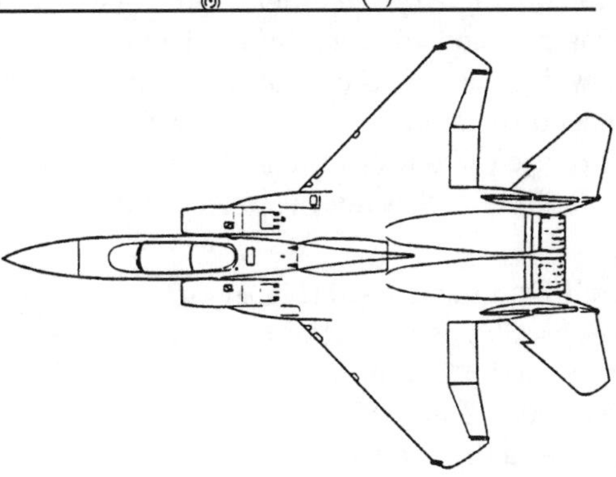

Lockheed (Boeing) F-22 (Fig. A-13)
Leading edge sweep = 48°, aspect ratio = 2.2
Maximum speed, M = 3+ dash with afterburners
Supersonic cruise M = 1.5 without afterburning
Two low-bypass engines, 35,000 pounds of thrust each in afterburning
Thrust-to-weight ratio, 1.0-to-0.9
Irreversible, power-boost flight control system with all-movable horizontal stabilator
First flight (prototype YF-22) on Sept. 29, 1990

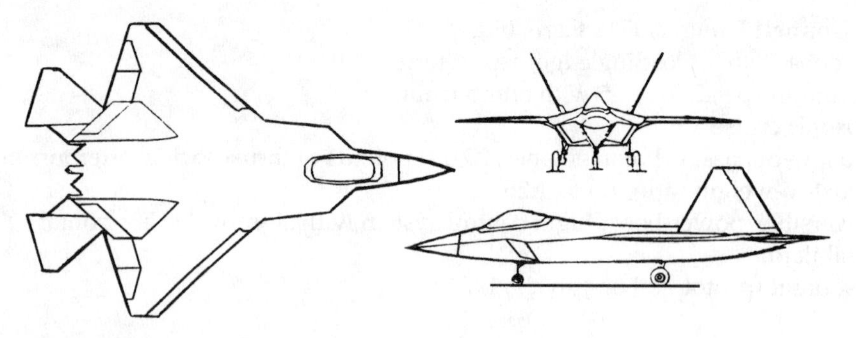

Fig. A-13. Lockheed/Boeing F-22.

Commercial aircraft development

Commercial aircraft development proceeded more conservatively than the military. Except for brief forays into supersonic transports, commercial aircraft speeds are all subsonic. Research for achieving supersonic flight was directed at delaying it, or in effect, decreasing drag so as to approach Mach one as close as possible with less thrust. Cruising speeds of M = 0.72 to M = 0.87 have changed little since the Boeing 707 first entered airline service in 1958. The high-bypass fanjet engine was a major advance in jet airliner development, providing high thrust with low weight, low fuel consumption and low noise. Economics rather than increased speed dictated development of the commercial jet airliner. Compare performance of the 1970s McDonnell Douglas DC-10 with that of the same company's MD-11. The difference is not in speed but in efficiency.

McDonnell Douglas DC-10 (Fig. A-14)
Wing sweep = 35°, aspect ratio = 7.0
Maximum speed = 593 mph
Best cruise speed, M = 0.82
Power-operated flight controls
Three high-bypass fanjet engines (no afterburning), 50,000 pounds of thrust each
Range 4,000 to 6,000 miles
Passenger capacity 250 to 380
First flight in 1970

Advanced-technology commercial aircraft

McDonnell Douglas MD-11 trijet airliner (Fig. A-15)
Wing sweep = 35°, aspect ratio = 7.9
Supercritical airfoil with winglets
Maximum speed, M = 0.87, 588 mph at 31,000 feet

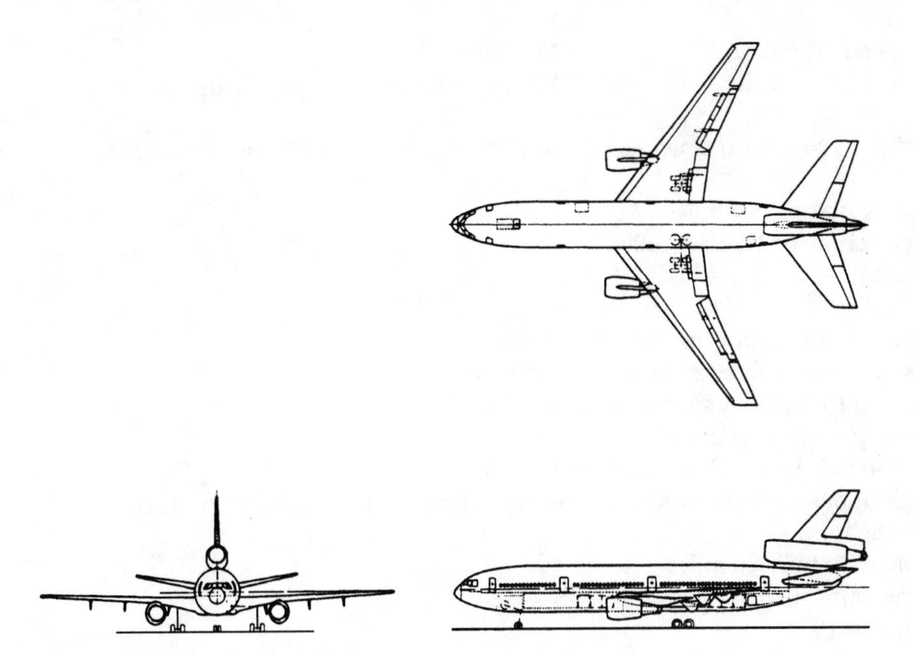

Fig. A-14. McDonnell Douglas DC-10.

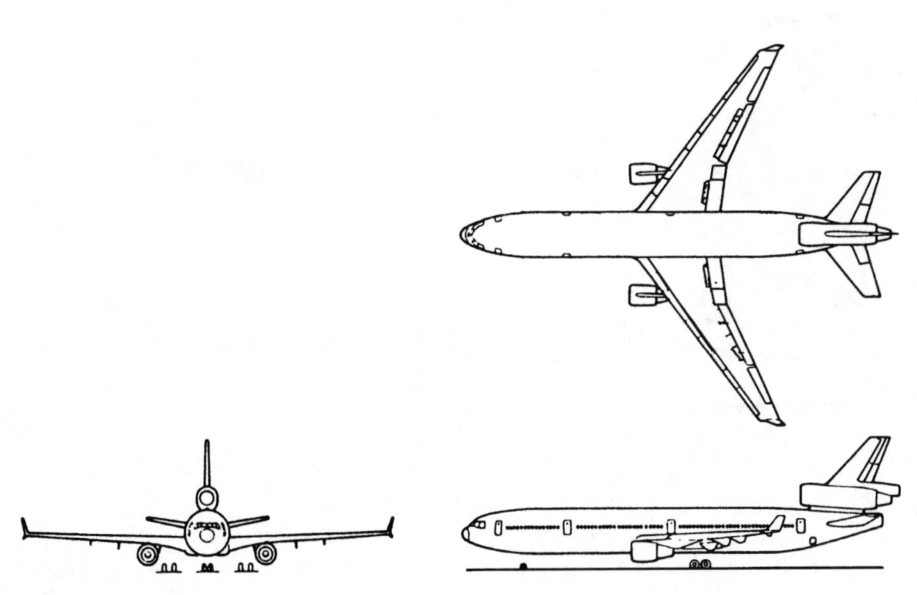

Fig. A-15. McDonnell Douglas MD-11.

Cruise speed, M = 0.82, 550 mph at 31,000 feet
Active controls, smaller horizontal stabilizer with elevators, power-operated flight controls
Three high-bypass ratio fanjet engines (no afterburning), 60,000 pounds of thrust each
Long range up to 9,000 miles
Passenger capacity 217 up to 400
First flight in 1989

Boeing 777 twin-engine airliner (Fig. A-16)
First airplane engineered entirely on computer
Boeing-modified NASA supercritical airfoil
Wing sweep = 35°, aspect ratio = 7.5
Cruising speed, M = 0.82, 550 mph at altitude
Two high-bypass ratio turbofan engines (no afterburning), 61,000 pounds of thrust each
Long range up to 9,000 miles
Passenger capacity 300 up to 440
First flight in 1994

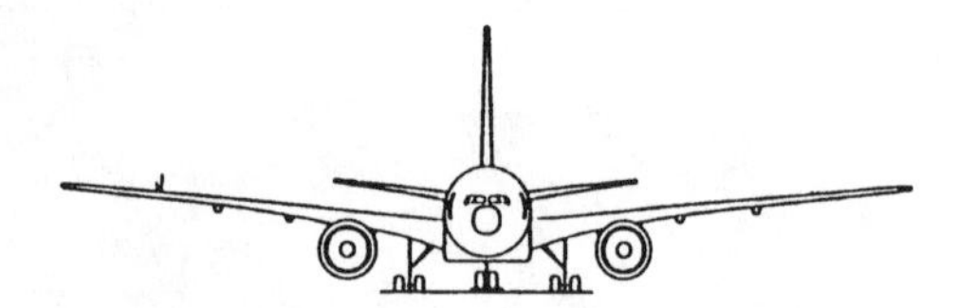

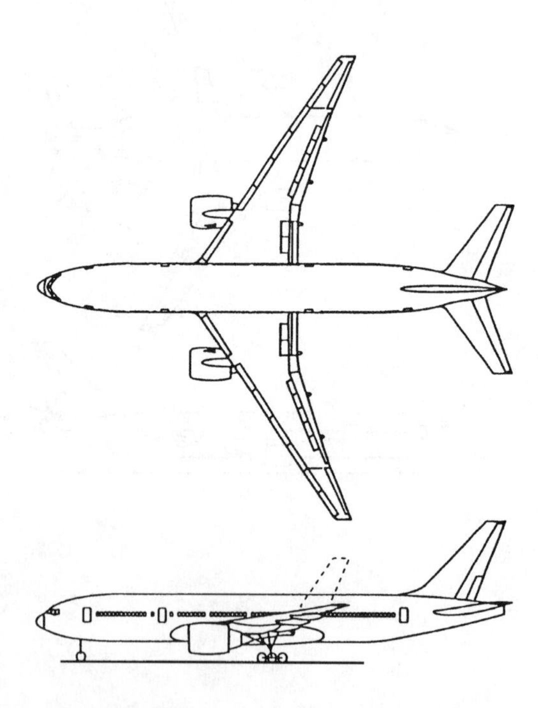

Fig. A-16. Boeing 777.

Appendix B

Development of the Lockheed SR-71 Blackbird

(THE SR-71 BLACKBIRD WAS A PRODUCT of Lockheed's "Skunk Works" headed by renowned aircraft designer Clarence L. (Kelly) Johnson and is only one of the many outstanding airplanes produced by this famed organization (Figs. B-1 and B-2). Among these airplanes are the YF-12 interceptor and the famous U-2 reconnaissance airplane. The following is a copyrighted article from Lockheed Horizons magazine (Winter 1981/1982) by Kelly Johnson and is reprinted with permission of the Lockheed Corporation. A minor amount of editing was required due to unavailability of some of the illustrations.) (Johnson's text starts on pg. 221.)

Fig. B-1. The Lockheed SR-71 Blackbird in flight.

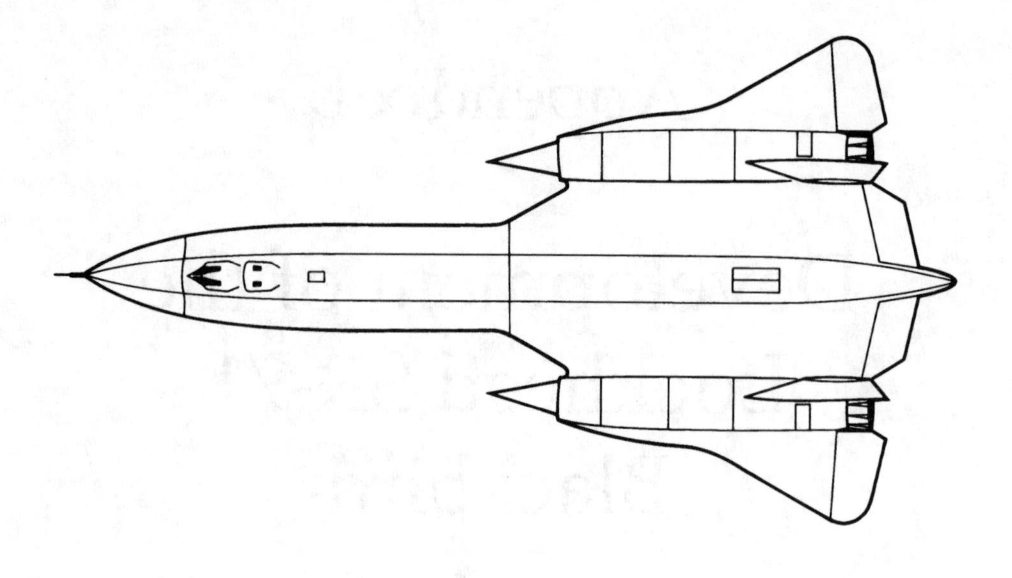

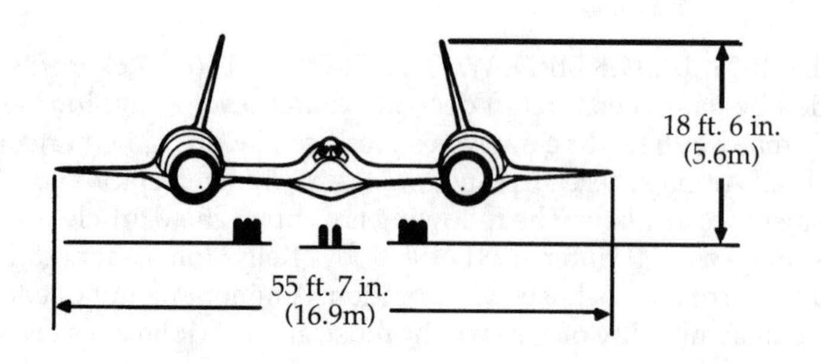

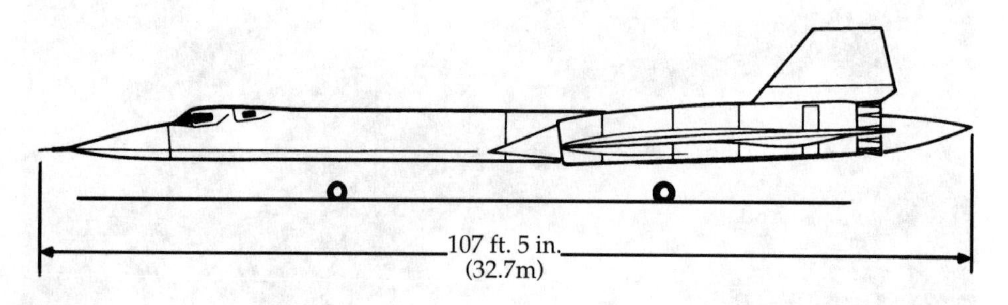

Fig. B-2. Three-view of the SR-71.

This paper has been prepared by the writer to record the development history of the Lockheed SR-71 reconnaissance airplane. In my capacity as manager of Lockheed's Advanced Development Division (more commonly known as the "Skunk Works") I supervised the design, testing, and construction of the aircraft referred to until my partial retirement five years ago. Because of the very tight security on all phases of the program, there are very few people who were ever aware of all aspects of the so-called "Blackbird" program.

Fortunately, I kept as complete a log on the subject as one individual could on a program that involved thousands of people, over three hundred subcontractors and partners, plus a very select group of Air Force and Central Intelligence Agency people. There are still many classified aspects of the design and operation of the Blackbirds but by my avoiding these, I have been informed that I can still publish many interesting things about the program.

In order to tell the SR-71 story, I must draw heavily on the data derived on two prior Skunk Works programs—the first Mach 3-plus reconnaissance type, known by our design number as the A-12, and the YF-12A interceptor, which President Lyndon Johnson announced publicly 1 March 1964. He announced the SR-71 on 24 July of the same year.

BACKGROUND FOR DEVELOPMENT

The Lockheed U-2 subsonic, high-altitude reconnaissance plane first flew in 1955. It went operational a year later and continued to make overflights of the Soviet Union until 1 May 1960. In this five-year period, it became obvious to those of us who were involved in the U-2 program that Russian developments in the radar and missile fields would shortly make the U-Bird too vulnerable to continue overflights of Soviet territory, as indeed happened when Francis Gary Powers was shot down on May Day of 1960.

Starting in 1956, we made many studies and tests to improve the survivability of the U-2 by attempting to fly higher and faster as well as reducing its radar cross-section and providing both infrared and radar jamming gear. Very little gains were forthcoming except in cruise altitude so we took up studies of other designs. We studied the use of new fuels such as boron slurries and liquid hydrogen. The latter was carried into the early manufacturing phase because it was possible to produce an aircraft with cruising altitudes well over 100,000 feet at a Mach number of 2.5. This design was scrapped, however, because of the terrible logistic problems of providing fuel in the field.

Continuing concern for having a balanced reconnaissance force made it apparent that we still would need a manned reconnaissance aircraft that could be dispatched on worldwide missions when required. From vulnerability studies, we derived certain design requirements for this craft. These were a cruising speed well over Mach 3, cruising altitude over 80,000 feet, and a very low radar cross-section over a wide band of frequencies. Electronic counter-measures and advanced communications gear were mandatory. The craft should have at least two engines for safety reasons.

GETTING A GRASP ON THE PROBLEM

Our analysis of these requirements rapidly showed the very formidable problems which had to be solved to get an acceptable design.

The first of these was the effect of operating at ram-air temperatures of over 800°F (Fig. B-3). This immediately ruled out aluminum as a basic structural material, leaving only various alloys of titanium and stainless steel to build the aircraft. It meant the development of high-temperature plastics for radomes and other structures, as well as a new hydraulic fluid, greases, electric wiring and plugs, and a whole host of other equipment. The fuel to be used by the engine had to be stable under temperatures as low as minus 90°F in subsonic cruising flight during aerial refueling, and to over 350°F at high cruising speeds when it would be fed into the engine fuel system. There it would first be used as hydraulic fluid at 600°F to control the afterburner exit flap before being fed into the burner cans of the powerplant and the afterburner itself.

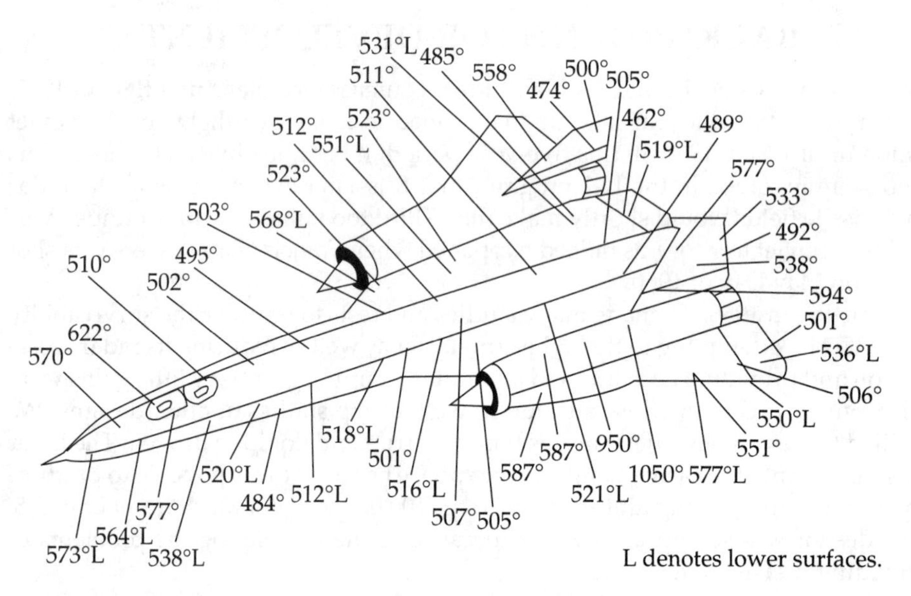

Fig. B-3. Surface temperatures at design cruising speed and altitude.

Cooling the cockpit and crew turned out to be seven times as difficult as on the X-15 research airplane which flew as much as twice as fast as the SR-71 but only for a few minutes per flight. The wheels and tires of the landing gear had to be protected from the heat by burying them in the fuselage fuel tanks for radiation cooling to save the rubber and other systems attached thereto.

Special attention had to be given to the crew escape system to allow safe ejection from the aircraft over a speed and altitude range of zero miles per hour at sea level to Mach numbers up to 4.0 at over 100,000 feet. New pilots' pressure suits, gloves, dual oxygen systems, high-temperature ejection seat catapults, and parachutes would have to be developed and tested.

The problems of taking pictures through windows subjected to a hot turbulent airflow on the fuselage also had to be solved.

HOW THE BLACKBIRD PROGRAM GOT STARTED

In the time period of 21 April 1958 through 1 September 1959, I made a series of proposals for Mach 3-plus reconnaissance aircraft to Mr. Richard Bissell of the CIA and to the U.S. Air Force. These airplanes were designated in the Skunk Works by design numbers of A-1 through A-12.

We were evaluated against some very interesting designs by the General Dynamics Corporation and a Navy in-house design. This latter concept was proposed as a ramjet-powered rubber inflatable machine, initially carried to altitude by a balloon and then rocket boosted to a speed where the ramjets could produce thrust. Our studies on this aircraft rapidly proved it to be totally unfeasible. The carrying balloon had to be a mile in diameter to lift the unit, which had a proposed wing area of ⅐ of an acre!

Convair's proposals were much more serious, starting out with a ramjet-powered Mach 4 aircraft to be carried aloft by a B-58 and launched at supersonic speeds. Unfortunately, the B-58 couldn't go supersonic with the bird in place, and even if it could, the survivability of the piloted vehicle would be very questionable due to the probability of ramjet blow-out in maneuvers. At the time of this proposal the total flight operating time for the Marquardt ramjet was not over 7 hours, and this time was obtained mainly on a ramjet test vehicle for the Boeing Bomarc missile. Known as the X-7, this test vehicle was built and operated by the Lockheed Skunk Works!

The final Convair proposal, known as the Kingfisher, was eliminated by Air Force and Department of Defense technical experts, who were given the job of evaluating all designs.

On 29 August 1959 our A-12 design was declared the winner and Mr. Bissell gave us a limited go-ahead for a four-month period to conduct tests on certain models and to build a full-scale mock-up. On 30 January 1960 we were given a full go-ahead on the design, manufacturing, and testing of 12 aircraft. The first one flew 26 April 1962.

The next version of the aircraft, an Air Defense long-range fighter, was discussed with General Hal Estes in Washington, D.C. on 16 and 17 March 1960. He and Air Force Secretary for Research and Development, Dr. Courtlandt Perkins, were very pleased with our proposal so they passed me on for further discussions with General Marvin Demler at Wright Field. He directed us to use the Hughes ASG 18 radar and the GAR-9 missiles which were in the early development stages for the North American F-108 interceptor. This we did, and when the F-108 was eventually canceled, Lockheed worked with Hughes in the development and flight testing of that armament system. The first YF-12A flew 7 August 1963.

In early January 1961 I made the first proposal for a strategic reconnaissance bomber to Dr. Joseph Charyk, Secretary of the Air Force; Colonel Leo Geary, our Pentagon project officer on the YF-12; and Mr. Lew Meyer, a high financial officer in the Air Force. We were encouraged to continue our company-funded studies on the aircraft. As we progressed in the development, we encountered very strong opposition in certain Air Force quarters on the part of those trying to save the North American B-70 program, which was in considerable trouble. Life became very interesting in that we were competing the SR-71 with an airplane five times its weight and size. On 4 June 1962 the Air Force evaluation team reviewed our design and the mock-up—and we were given good grades.

Our discussions continued with the Department of Defense and also, in this period, with General Curtis LeMay and his Strategic Air Command officers. It was on 27 and 28 December 1962 that we were finally put on contract to build the first group of six SR-71 aircraft.

One of our major problems during the next few years was in adapting our Skunk Works operating methods to provide SAC with proper support, training, spare parts, and data required for their special operational needs. I have always believed that our Strategic Air Command is the most sophisticated and demanding customer for aircraft in the world. The fact that we have been able to support them so well for many years is one of the most satisfying aspects of my career.

Without the total support of such people as General Leo Geary in the Pentagon and a long series of extremely competent and helpful commanding officers at Beale Air Force Base, we could never have jointly put the Blackbirds into service successfully.

BASIC DESIGN FEATURES

Having chosen the required performance in speed, altitude, and range, it was immediately evident that a thin delta-wing planform was required with a very moderate wing loading to allow flight at very high altitude. A long, slender fuselage was necessary to contain most of the fuel as well as the landing gear and payloads. To reduce the wing trim drag, the fuselage was fitted with lateral surfaces called chines, which actually converted the forward fuselage into a fixed canard which developed lift.

The hardest design problem on the airplane was making the engine air inlet and ejector work properly. The inlet cone moves almost three feet to keep the shock wave where we want it. A hydraulic actuator, computer controlled, has to provide operating forces of up to 31,000 pounds under certain flow conditions in the nacelles. To account for the effect of the fuselage chine air flow, the inlets are pointed down and in toward the fuselage.

The use of dual vertical tails canted inward on the engine nacelles took advantage of the chine vortex in such a way that the directional stability improves as the angle of attack of the aircraft increases.

AERODYNAMIC TESTING

All the usual low-speed and high-speed wind tunnel tests were run on the various configurations of the A-12 and YF-12A, and continued on the SR-71. Substantial efforts went into optimizing chine design and conical camber of the wing leading edge. No useful lift increase effect was found from the use of wing flaps of any type so we depend entirely on our low wing-loading and powerful ground effect to get satisfactory takeoff and landing characteristics.

Correlation of wind tunnel data on fuselage trim effects was found to be of marginal value because of two factors: structural deflection due to fuselage weight distribution; and the effect of fuel quantity and temperature. The latter was caused by fuel on the bottom of the tanks, keeping that section of the fuselage cool, while the top of the fuselage became increasingly hotter as fuel was burned, tending to push the chines downward due to differential expansion of the top and bottom of the fuselage. A full-scale fuel system test rig was used to test fuel feed capability for various flight attitudes.

By far the most tunnel time was spent optimizing the nacelle inlets, bleed designs, and the ejector. A quarter-scale model was built on which over 250,000 pressure readings were taken. We knew nacelle air leakage would cause high drag so an actual full-size nacelle was fitted with end plugs and air leakage carefully measured. Proper sealing paid off well in flight testing.

With the engines located halfway out on the wing span, we were very concerned with the very high yawing moment that would develop should an inlet stall. We therefore installed accelerometers in the fuselage that immediately sensed the yaw rate and commanded the rudder booster to apply 9 degrees of correction within a time period of 0.15 seconds. This device worked so well that our test pilots very often couldn't tell whether the right or left engine blew out. They knew they had a blowout, of course, by the bad buffeting that occurred with a "popped shock." Subsequently, an automatic restart device was developed which keeps this engine-out time to a very short period.

POWERPLANT DEVELOPMENT

Mr. Bill Brown of Pratt & Whitney presented a fine paper on this subject 13 May 1981 to the American Institute of Aeronautics and Astronautics in Long Beach, California. Mr. Brown's paper is reproduced herewith.

(J58/SR-71 Propulsion Integration or The Great Adventure Into the Technical Unknown. By William H. Brown, retired engineering manager, Government Product Division of the Pratt & Whitney Aircraft Group United Technologies Corporation.)

Successful integration of the J58 engine with the SR-71 aircraft was achieved by inherently compatible engine cycle, size, and characteristics, (plus) intensive and extensive design/development effort.

Propulsion integration involved aerodynamic compatibility, installation and structural technology advances, development of a unique mechanical power takeoff drive, and fuel system tailoring. All four areas plowed new ground and uncovered unknowns that were identified, addressed, and resolved. Interacting airframe systems, such as the variable mixed compression inlet, exhaust nozzle, and fuel system were ground tested with the J58 engine prior to and coincident with flight testing. Numerous iterative redesign-retest-resolution cycles were required to accommodate the extreme operating conditions. Successful propulsion operation was primarily the result of:

- Compatible conceptual designs.
- Diligent application of engineering fundamentals.
- Freedom to change the engine and/or aircraft with a minimum of contractual paperwork.
- A maximum of trust and team effort with engineer-to-engineer interchange.

The centerline of the basic J58 engine was laid down in late 1956. It was to be an afterburning turbojet rated at 26,000-lb maximum takeoff thrust and was to power a Navy attack aircraft which would have a dash capability of up to Mach 3 for several seconds. By the time Pratt & Whitney Aircraft, along with Lockheed and others, began to study the SR-71 "Blackbird" requirements several years later, we had completed approximately 700 hours of full-scale engine testing on the J58. In the "Blackbird" joint studies, the attitude of open cooperation between Lockheed and Pratt & Whitney Aircraft personnel seemed to produce better results than if a more "arms-length" attitude were adopted.

This open cooperation resulted in a more complete study which identified the enormous advances in the state-of-the-art and the significant amount of knowledge which had to be acquired to achieve a successful engine/airframe integration. The completeness of this study was probably instrumental in Lockheed and Pratt & Whitney Aircraft winning the competition. The Government stated that the need for the "Blackbird" was so great that the program had to be conducted despite the risks and the technological challenge. Furthermore, the Government expected the risks to be reduced by fallout from the X-15 and B-70 programs. Unfortunately, there was no meaningful fallout.

Figure B-4 indicates some of the increased requirements of the "Blackbird" engine compared to the requirements for the previous J75 engine. As it turned out, even these requirements didn't hold throughout the "Blackbird's" actual mission.

	J57 and J75	JT11D-20
Mach number	2.0 for 15 min (J75 only)	3.2 (continuous)
Altitude	55,000 ft.	100,000 ft.
Compressor inlet temperature	250°F (J75 only)	800°F
Turbine inlet temperature	1750°F (takeoff) 1550°F (cruise)	2000°F
Maximum fuel inlet temperature	110–130°F	300°F
Maximum oil inlet temperature	250° F	550°F
Thrust/weight ratio	4.0	5.2
Military operation	30-min time limit	Continuous
Afterburner operation	Intermittent	Continuous

Fig. B-4. Comparison of J-58 development objectives with the then-current production engines.

For example, the engine inlet air temperature exceeded 800°F under certain conditions. The fuel inlet temperature increased to 350°F at times and the fuel temperature ranged from 600°F to 700°F at the main and afterburner fuel nozzles. Lubricant temperatures rose to 700°F and even to 1000°F in some localized parts of the engine. Because of these extremely hostile environmental conditions, the only design parameters that could be retained from the Navy J58-P2 engine were the basic size and the compressor and turbine aerodynamics. Even these were modified at a later date.

The extreme environment presented a severe cooling problem. It was vital to cool the pilot and aircraft electronics; but this left little or no heat sink in the fuel available to cool the rest of the aircraft or the engine. Because of this, the only electronics on the engine was a fuel-cooled solenoid which was added later and a trim motor buried inside the engine fuel control. To keep cooling requirements to a minimum, we even had to provide a chemical ignition system using tetraethyl borane (T.E.B.) for starting both the main engine and the afterburner. A new fuel and a chemical lubricant had to be developed to meet the temperature requirements. Pratt & Whitney Aircraft together with the Ashland, Shell, and Monsanto Companies took on the task of developing these fluids.

Early in the development, we found that a straight turbojet cycle did not provide a good match for the inlet nor the required net thrust at high Mach number operating conditions. To overcome these problems, we invented the bleed bypass cycle with which we could match the inlet airflow requirements, as shown in Fig. B-5. Another advantage of this cycle was that above Mach 2, the corrected airflow could be held constant at a given Mach number regardless of the throttle position. The bleed bypass cycle also provided more than 20-percent additional thrust during high Mach number operation (Fig. B-6).

Fabrication and materials technology presented one of the greatest challenges. We had to learn how to form sheet metal from materials which previously had been used only for forging turbine blades. Once we had achieved this, we had to learn how to weld it successfully. Disks, shafts, and other components also had to be fabricated from high-strength, temperature-resistant turbine-blade-like materials to withstand temperatures and stresses encountered. I do not know of a single part, down to the last cotter key, that could be made from the same materials as used on previous engines. Even the lubrication pump was a major development.

The newly developed special fuel was not only hot, but it had no lubricity. A small amount of fluorocarbon finally had to be added to allow the airframe and engine pumps and servos to work. Fuel was used as the engine hydraulic fluid to actuate the bleeds, afterburner nozzle, etc. Because there was nothing to cool the fuel, it just made one pass through the hydraulic system and then was burned. If the foregoing were not enough, developmental testing problems also had to be overcome. There were no test facilities which had the capabilities to provide steady-state temperature and pressure conditions required for testing at maximum operating conditions nor could they provide for performing transients.

(A partial solution was a test stand with) the exhaust of a J75 engine run through and around the J58 to simulate transients of the temperature environment. In addition, there was essentially no instrumentation rugged enough to obtain accurate real-time measurements. As Pratt & Whitney Aircraft developed

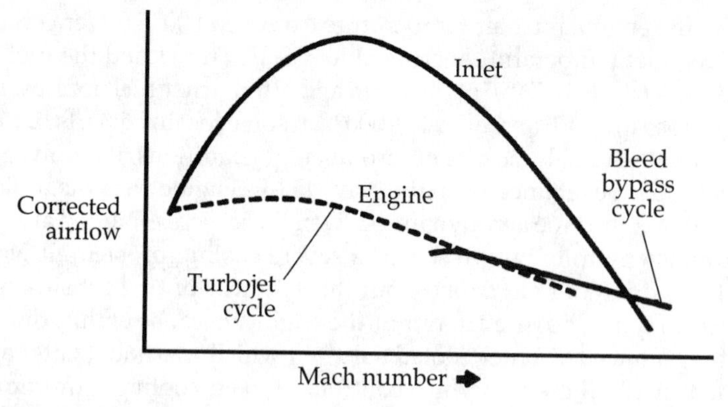

Fig. B-5. Inlet and engine airflow match.

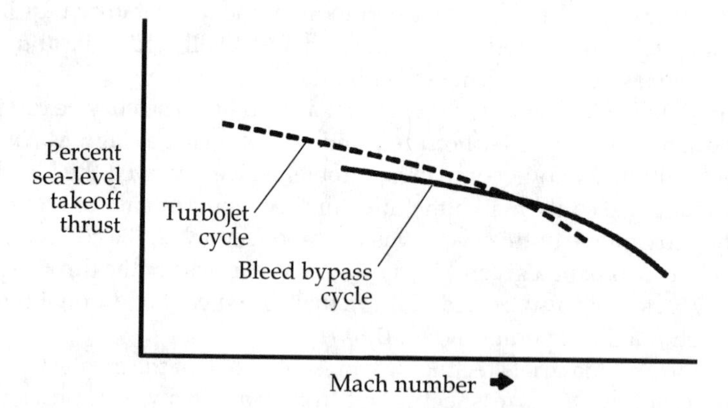

Fig. B-6. Net thrust comparison of bleed-bypass cycle vs. turbojet cycle.

more rugged instrumentation and better calibration facilities, improved data were gradually obtained. Lockheed, of course, was kept up-to-date as we obtained better data. A good part of the time Lockheed and Pratt & Whitney Aircraft jointly ran fuel system rigs, inlet distortion rigs, etc., as well as some engine calibration tests and wind tunnel testing of the ejector.

(It's important to recall technology of the era.) Although Pratt & Whitney Aircraft had a very large computer system for its day (the IBM 710), it was no more sophisticated than (some hand-held calculators). Consequently, the J58 engine, in effect, was a slide-rule design. Despite all of the testing and faired curves, we knew we had to solve many of our mutual integration problems through flight test.

Approximately three months before Pratt & Whitney finished the Pre Flight Rating Test which was 3 years and 4 months after go-ahead (the Model Qualification Test was completed 14 months later), the first "Blackbird" took to the air.

It was powered by two afterburning J75 turbojet engines to wring out the aircraft subsonically. As soon as Lockheed felt comfortable with the aircraft, a J58 was installed in one side. After several months of subsonic flight tests, J58 engines were installed in both sides, and we started flight testing for real. Naturally there were problems. Here are a few notable ones and the solutions.

The first problem happened very early—the engine wouldn't start! The small inlet wind tunnel model did not show the inlet being so depressed at the starting J58 airflows. In fact, instead of air flowing out of the compressor 4th-stage through the bleed ducts into the afterburner, it flowed the other way! As a temporary fix, Lockheed removed an inlet access panel for ground starts. They later added two suck-in doors (Fig. B-7) and Pratt & Whitney Aircraft added an engine bleed to the nacelle. These two changes eliminated the ground starting problem.

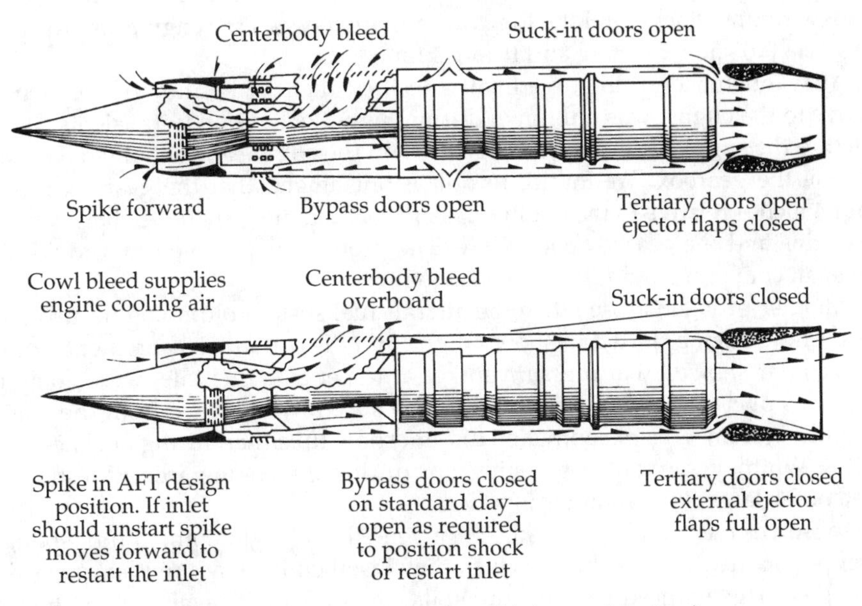

Fig. B-7. Airflow patterns: static aircraft (top) and high speed (bottom).

Originally, the blow-in door ejector or convergent-divergent nozzle was built as part of the engine. It was subsequently decided jointly by Lockheed and Pratt & Whitney Aircraft that it would save weight if it was built as part of the airframe structure. This was deemed appropriate, particularly as the main wing spar structure had to go around the throat of the ejector. Pratt & Whitney Aircraft, however, would still be responsible for nozzle performance in conjunction with the engine primary nozzle. In addition, we would perform all of the wind tunnel testing. In exchange, Pratt & Whitney Aircraft would build the remote gearbox because Lockheed's gearbox vendor had no experience with gear materials or bearings and seals that would withstand the temperatures required. As a matter of fact neither did we, but we were already committed to learn.

A problem partially related to the ejector was that the airplane burned too much fuel going transonic. To help solve the problem, thrust measurements were taken in flight, movies of ejector operation in flight were made, local Mach numbers were measured, etc. Two fundamental mistakes were uncovered. The back end of the nacelle (the ejector) went supersonic long before the airplane did, and the fairing of the aircraft transonic wind tunnel drag data was not accurate. While we were puzzling out the solution, some pilot decided to go transonic at a lower altitude and higher Keas. This for all intents and purposes solved the problem.

From this we learned not to run nacelle wind tunnel tests unless the model contains at least a simulation of the adjacent aircraft surfaces. We also learned to take enough data points so that transonic drag wind tunnel data does not have to be faired. As flight testing increased to the higher Mach numbers, new problems arose. One, which today may be considered simple with our modern computer techniques, concerned the remote gearbox. The gearbox mounts started to exhibit heavy wear and cracks, and the long drive shaft between the engine and the gearbox started to show twisting and heavy spline wear.

After much slide-ruling, we finally decided that the location of the gearbox relative to the engine was unknown during high Mach number transients. We resorted to the simple test of putting styluses on the engine and mounted a scratch plate on the gearbox. We found, to our astonishment, that the gearbox moved about 4 inches relative to the engine. This was much more than the shaft between the engine and the gearbox could take. The problem was solved by providing a new shaft containing a double universal joint.

Another problem arose when the aircraft fuel system plumbing immediately ahead of the engine started to show fatigue and distortion. Measurements with a fast recorder showed that pressure spikes at the engine fuel inlet were going off scale. This overpressuring was found to be caused by feedback from the engine hydraulic system. This phenomenon did not show up either during Lockheed's or Pratt & Whitney Aircraft's rig testing nor during the engine ground testing because of the large fluid volumes involved.

To solve the problem Lockheed invented a "high-temperature sponge" (promptly named "the football") which they installed in an accumulator ahead of the engine. This reduced the pressure spikes to a tolerable level. A mounting-related problem occurred under certain conditions of down load on the wing. At these conditions, the outer half of the nacelle would rotate into the engine and crush the engine plumbing and anything else in the way. Originally, the engine was mounted on a stiff rail structure at the top of the nacelle with a stabilizing link from the top of the engine rear mount ring to the aircraft structure as shown in Fig. B-8.

To solve the crushing problem Pratt & Whitney Aircraft designed the rear mount ring so that a tangential link could be installed between the engine and the outboard side of the nacelle. This maintained a finite distance between the nacelle and engine under all conditions. As mentioned previously, there was a minimum of electronics in the engine control system because electronics would not survive the environment and the fuel was already too hot to provide cooling. Consequently, control adjustments normally made automatically had to be made manually.

For example, the pilot operated a vernier trimmer to make fine adjustments in the EGT (Exhaust Gas Temperature) as conditions varied from standard (one such device was used successfully in the U-2). The pilot was provided with a curve of EGT versus engine inlet temperature to make the required manual adjustments. However, unexpectedly sharp atmospheric changes were encountered. These, in

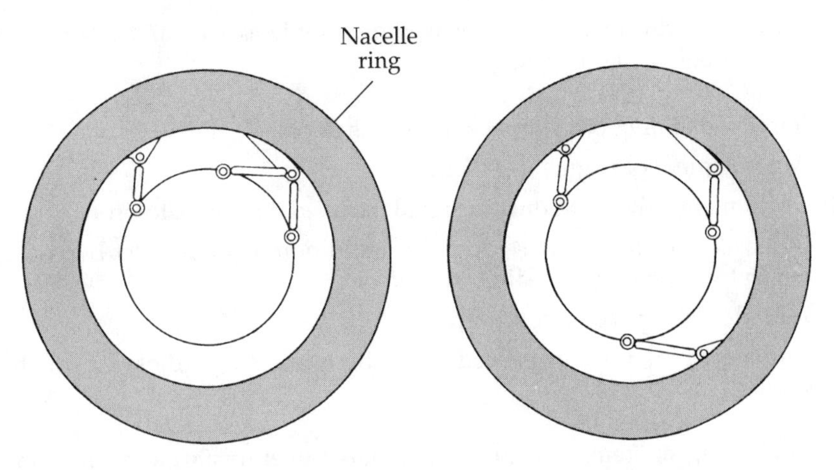

Fig. B-8. Original engine mount at left; modified engine mount at right.

combination with the speed of the aircraft, resulted in changes too fast for the pilot to handle. By the time he read the engine inlet temperature and adjusted the EGT, the inlet temperature had changed. This caused some inlet unstarts (highly reduced inlet airflow) and other undesirable results.

To correct this unacceptable state of affairs, Pratt & Whitney Aircraft proposed to revise the aircraft EGT gage by feeding in an engine inlet temperature signal and adding some additional gadgetry to trim automatically. The digital EGT readout was retained as was an override manual trim in case of failure. This modification has worked well ever since.

The most sensational and most confusing problem at the high Mach number condition was inlet unstarts. These occurred without warning and were seemingly inconsistent. To add to the confusion, the pilots consistently reported the unstart occurring on the wrong side of the airplane. This anomaly was solved rather quickly when Lockheed found that the Stability Augmentation System (SAS) slightly overcompensated for the sudden one-sided drag. This led the pilot to believe that the wrong side had unstarted, and consequently, his corrective action usually resulted in worsening the problem.

Oddly enough, the engine did not blow out. It just sat there and overheated because the inlet airflow was so reduced that the engine minimum fuel flow was approximately twice that required. Worst of all, the inlet would not restart until the pilot came down to a much lower altitude and Mach number. A great many tests and investigations were conducted including the possibility of engine surge being the initiator. This was not the case.

Three major causes were finally isolated:

1. Manual trimming of engine.
2. High, inconsistent nacelle leakage at the approximately 40:1 pressure ratio.
3. Alpha signal (angle of attack from noseboom) to inlet control subject to G-loading.

The following improvements were incorporated by Lockheed and Pratt & Whitney Aircraft essentially as a package:

1. Improved sealing of the inlet and bypass doors.
2. Auto-trimmer of engine installed.
3. Derichment valve with unstart signal installed on engine to protect turbine.
4. Increased area inlet bypass doors and addition of an aft inlet bypass door which bypassed inlet air direct to ejector.
5. Added a "G" bias on inlet control.
6. Automated inlet restart procedure on both inlets regardless of which unstarted.

The foregoing six items essentially eliminated inlet unstart as a problem. An additional benefit was also realized by the ability to use the aft inlet bypass door in normal flight instead of dumping all inlet bypass air overboard. As this air became heated as it passed over the engine to the ejector instead of going overboard, drag was substantially reduced. Also better sealing of the nacelle reduced drag further.

As you have probably noticed, I have had difficulty in differentiating between "we" Pratt & Whitney Aircraft and "we" Lockheed. But that is the kind of program it was. In any complicated program of this magnitude we all do something dumb and we both did our share. Here is one from each of us: "We" (Pratt & Whitney), became so obsessed with the problems of hot fuel and hot environment that we neglected the fact that sometimes the fuel was cold when the environment was hot and vice versa. When this occurred, the engine fuel control did not track well.

To correct this, we had to insulate the main engine control body from the environment and make all the servos, etc., respond only to fuel temperature. Eventually, we had to make a major redesign of the control. Lockheed and Pratt & Whitney Aircraft spent many hours coordinating the inlet and engine arrangement so that doors, bleeds, air conditioner drive turbine discharge, etc., would not affect any of the engine control sensors in the engine inlet. In fact, the air conditioner turbine discharge was located 45 degrees on one side of the vertical centerline and the engine temperature bulb was located 45 degrees on the opposite side.

To save design time, Lockheed built one inlet as a mirror image of the other. It is now easy to conclude where the 1200°F air conditioner turbine discharge turned out to be!! For a while the fact that one engine always ran faster than the other was a big mystery!

That this complex, difficult program was successful is attributable, in large part, to the management philosophy adopted by the Government people in charge. Their approach was that both the engine and airframe contractors must be free to take the actions which in their judgment were required to solve the problems. The Government management of the program was handled by no more than a dozen highly qualified and capable individuals who were oriented toward understanding the problems and approaches to solutions, rather than toward substituting their judgment for that of the contractors.

Requirements for Government approval as a prerequisite to action were minimal and were limited to those changes involving significant cost or operational impact. As a result, reactions to problems were exceptionally quick. In this manner, the time from formal release of engineering paperwork to the conversion to hardware was drastically shortened. This not only accelerated the progress of the program but saved many dollars by incorporating the changes while the number of units were still relatively small.

On this program, the Government fully recognized that many of the problems involving either the engine or airframe manufacturer, or both, could be solved most effectively by a joint engineering effort and the contracts were written to allow this activity without penalties. As a result, an extremely close working relationship between the engineering groups was developed and flourished until the SR-71 became fully operational.

This method of operation led to prompt solutions of many problems which, under a more cumbersome management system, could have severely impeded the program by introducing very costly delays or forcing inappropriate compromises because of contractual interpretations.

In summary, the method of managing this program by the Government resulted in shorter development time, faster reaction to field problems, reduced retrofit costs, and earlier availability of production systems incorporating corrections for problems uncovered by operations in the field. The result was an operating system incorporating a magnum step in the state-of-the-art at an earlier time and at less cost to the Government than would otherwise have been possible.

I have little to add to Mr. Brown's fine paper except to record an interesting approach to the problem of ground starting the J-58. We learned that it often required over 600 horsepower to get the engine up to starting RPM. To obtain this power, we took two Buick racing car engines and developed a gear box to connect them both to the J-58 starter drive. We operated for several years with this setup, until more sophisticated air starting systems were developed and installed in the hangars.

STRUCTURAL PROBLEMS

The decision to use various alloys of titanium for the basic structure of the Blackbirds was based on the following considerations:

1. Only titanium and steel had the ability to withstand the operating temperatures encountered.
2. Aged B-120 titanium weighs one half as much as stainless steel per cubic inch but its ultimate strength is almost up to stainless.
3. Conventional construction could be used with fewer parts involved than with steel.
4. High strength composites were not available in the early 1960s.

We did develop a good plastic which has been remarkably serviceable but it was not used for primary structure. Having made the basic material choice, we decided to build two test units to see if we could reduce our research to practice. The first unit was to study thermal effects on our large titanium wing panels. We heated up this element with the computed heat flux that we would encounter in flight. The sample warped into a totally unacceptable shape.

To solve this problem we put chordwise corrugations in the outer skins and reran the tests very satisfactorily. At the design heating rate, the corrugations merely deepened by a few thousandths of an inch and on cooling returned to the basic shape. I was accused of trying to make a 1932 Ford Trimotor go Mach 3 but the concept worked fine.

The second test unit was the forward fuselage and cockpit, which had over 6,000 parts in it of high curvature, thin gages, and the canopy with its complexity. This element was tested in an oven where we could determine thermal effects and develop cockpit cooling systems. We encountered major problems in manufacturing this test unit because the first batch of heat-treated titanium parts was extremely brittle. In fact, you could push a piece of structure off your desk and it would shatter on the floor. It was thought that we were encountering hydrogen embrittlement in our heat-treat processes.

Working with our supplier, Titanium Metals Corporation, we could not prove that the problem was in fact hydrogen. It was finally resolved by throwing out our whole acid pickling setup and replacing it with an identical reproduction of what TMCA had at their mills. We developed a complex quality control program. For every batch of ten parts or more we processed three test coupons which were subjected to the identical heat treatment of the parts in the batch.

One coupon was tensile tested to failure to derive the stress-strain data. A quarter-of-an-inch cut was made in the edge of the second coupon by a sharp scissor-like cutter and it was then bent around a mandrel at the cut. If the coupon could not be bent 180 at a radius of X times the sheet thickness without breaking, it was considered to be too brittle. (The value of X is a function of the alloy used and the stress/strain value of the piece.) The third coupon was held in reserve if any reprocessing was required.

For an outfit that hates paperwork, we really deluged ourselves with it. Having made over 13 million titanium parts to date we can trace the history of all but the first few parts back to the mill pour and for about the last 10 million of them even the direction of the grain in the sheet from which the part was cut has been recorded.

On large forgings, such as landing gears, we trepanned out 12 sample coupons for test before machining each part. We found out the hard way that most commercial cutting fluids accelerated stress corrosion on hot titanium so we developed our own. Titanium is totally incompatible with chlorine, fluorine, cadmium, and similar elements.

For instance, we were baffled when we found out that wing panels which we spot welded in the summer, failed early in life, but those made in the winter lasted indefinitely. We finally traced this problem to the Burbank water system which had heavily chlorinated water in the summer to prevent algae growth but not in the winter. Changing to distilled water to wash the parts solved this problem.

Our experience with cadmium came about by mechanics using cadmium-plated wrenches working on the engine installation primarily. Enough cadmium was left in contact with bolt heads which had been tightened so that when the bolts became hot (over 600°F) the bolt heads just dropped off! We had to clean out hundreds of tool boxes to remove cadmium-plated tools. Drilling and machining high strength titanium alloys, such as B-120, required a complete research program to determine best tool cutter designs, cutting fluids, and speeds and feeds for best metal removal rates.

We had particular trouble with wing extrusions, which were used by the thousands of feet. Initially, the cost of machining a foot out of the rolled mill part was $19.00 which was reduced to $11.00 after much research. At one time we were approaching the ability at our vendor's plants to roll parts to net dimensions, but the final achievement of this required a $30,000,000 new facility which was not built.

Wyman Gordon was given $1,000,000 for a research program to learn how to forge the main nacelle rings, on a 50,000-ton press which was successful. Combining their advances with our research on numerical controls of machining and special tools and fluids, we were able to save $19,000,000 on the production program.

To prevent parts from going under-gage while in the acid bath, we set up a new series of metal gages two thousandths of an inch thicker than the standard gages and solved this problem. When we built the first Blackbird, a high-speed drill could drill 17 holes before it was ruined. By the end of the program we had developed drills that could drill 100 holes and then be resharpened successfully.

Our overall research on titanium usage was summarized in reports which we furnished not only to the Air Force but also to our vendors who machined over half of our machined parts for the program. To use titanium efficiently required an on-going training program for thousands of people—both ours in manufacturing and in the Air Force in service.

Throughout this and other programs, it has been crystal clear to me that our country needs a 250,000-ton metal forming press—five times as large as our biggest one available (at the time). When we have to machine away 90 percent of our rough forgings both in titanium (SR-71 nacelle rings and landing gears) and aluminum (C-5 fuselage side rings) it seems that we are nationally very stupid! My best and continuing efforts to solve this problem have been defeated for many years. Incidentally, the USSR has been much smarter in this field in that they have more and larger forging presses than we do.

FLUID SYSTEMS

Very difficult problems were encountered with the use of fuel tank sealants and hydraulic oil. We worked for years developing both of these, drawing as much on other industrial and chemical companies as they were willing to devote to a very limited market. We were finally able to produce a sealant which does a reasonable job over a temperature range of minus 90°F to over 600°F.

Our experience with hydraulic oil started out on a comical situation. I saw ads in technical journals for a "material to be used to operate up to 900°F in service." I contacted the producer who agreed to send me some for testing. Imagine my surprise when the material arrived in a large canvas bag. It was a white powder at room temperature that you certainly wouldn't put in a hydraulic system. If you did, one would have to thaw out all the lines and other elements with a blow torch! We did finally get a petroleum-based oil developed at Penn State University to which we had to add several other chemicals to maintain its lubricity at high temperatures. It originally cost $130 per gallon so absolutely no leaks could be tolerated.

Rubber O-rings could not be used at high temperatures so a complete line of steel rings was provided which have worked very well. Titanium pistons working in titanium cylinders tended to gall and seize until chemical coatings were invented which solved the problem.

THE FLIGHT TEST PHASE

The first flight of the A-12 took place 26 April 1962 or thirty months after we were given a limited go-ahead on 1 September 1959. We had to fly with Pratt & Whitney J75 engines until the J58 engine became available in January 1963. Then our problems really began!

The first one was concerned with foreign object damage (FOD) to the engines—a particular problem with the powerful J58 and the tortuous flow path through the complicated nacelle structure. Small nuts, bolts, and metal scraps not removed from the nacelles during construction could be sucked into the engines on starting with devastating results. Damage to the first-stage compressor blades from an inspector's flashlight used to search for such foreign objects: engine damage—$250,000!

Besides objects of the above type, the engine would suck in rocks, asphalt pieces, etc., from the taxi-ways and runways. An intensive campaign to control FOD at all stages of construction and operation—involving a shake test of the forward nacelle at the factory, the use of screens, and runway sweeping with double inspections prior to any engine running—brought FOD under reasonable control.

The hardest problem encountered in flight was the development of the nacelle air inlet control. It was necessary to throw out the initial pneumatic design after millions of dollars had been spent on it and go to a design using electronic controls instead. This was very hard to do because several elements of the system were exposed to ram-air temperatures over 800°F and terrific vibration during an inlet duct stall. This problem and one dealing with aircraft acceleration between Mach numbers of 0.95 to 2.0 are too complex to deal with in this paper.

Initially, air temperature variations along a given true altitude would cause the Blackbird to wander up and down over several thousand feet in its flight path. Improved autopilots and engine controls have eliminated this problem.

There are no other airplanes flying at our cruising altitude except an occasional U-2 but we were very scared by encountering weather balloons sent up by the FAA. If we were to hit the instrumentation package while cruising at over 3,000 feet per second, the impact could be deadly!

Flight planning had to be done very carefully because of sonic boom problems. We received complaints from many sources. One such stated that his mules on a pack-train wanted to jump off the cliff trail when they were "boomed." Another complained that fishing stopped in lakes in Yellowstone Park if a boom occurred because the fish went down to the bottom for hours. I had my own complaint when one of my military friends boomed my ranch and broke a $450 plate glass window. I got no sympathy on this, however.

OPERATIONAL COMMENTS

The SR-71 first flew 23 December 1964. It was in service with the Strategic Air Command a year later. In-flight refueling from KC-135s turned out to be very routine. Over eighteen thousand such refuelings have been made to date by all versions of the Blackbirds and they have exceeded Mach 3 over 11,000 times.

The SR-71 has flown from New York to London in 1 hour 55 minutes then returned nonstop to Beale Air Force Base, including a London/Los Angeles time of 3 hours 48 minutes.

It has also flown over 15,000 miles with refueling to demonstrate its truly global range. It is by far the world's fastest, highest flying airplane in service. I expect it to be so for a long time to come.

The Air Force retired its fleet of SR-71s from service in January 1990. In March of 1990, the Smithsonian Institution received an SR-71 for permanent display at its facility at Dulles International Airport, Virginia, near Washington, D.C.

Appendix C

Evolution of Lockheed's SST design

ALTHOUGH LOCKHEED LOST THE SST competition to Boeing and no airplane was built, Lockheed's efforts to launch an SST during the United States' design competition illustrate the supersonic transport's high degree of sophistication and the vast number of technologies that were dealt with at that time. Lockheed's efforts also illustrate in a general sense the techniques and procedures used by designers to arrive at a high-speed airplane design.

(The following was taken from the spring 1965 issue of *Lockheed Horizons* magazine, Roy Blay, editor, and written by its editors. The copyrighted article is reproduced with permission of Lockheed Corporation. This is an edited version of the original article, and selected technical illustrations have been omitted. This article was written in the 1960s prior to the FAA's final evaluation of the project, but the technical information is still relevant. Refer to chapter 9 for additional details about the SST design competition.)

For some years, work on the SST was either company-funded or consisted of preliminary research studies for government agencies, including the Supersonic Commercial Air Transport (SCAT) feasibility studies for NASA. In August 1963, the U.S. government officially launched the SST program with the FAA's issuance of a request for proposal (RFP).

The response to the RFP was designated Phase I of what was, in effect, an FAA design competition. Three airframe manufacturers and three engine manufacturers submitted proposals. After deliberation by the FAA, by the President's Advisory Committee, and by the president himself, Phase IIA contracts were awarded to Boeing, Lockheed, General Electric, and Pratt & Whitney. The feasibility of the SST was firmly established in Phase IIA, and early in 1965, Phase IIB was initiated, with the same airframe and engine manufacturers participating. It was anticipated that Phase III would begin some time in the latter part of 1965.

DESIGN DEVELOPMENT FROM 1956 TO 1963

Lockheed began conducting research and development on supersonic transport configurations in 1956. From that time, until submittal of the Phase I bid proposal to the FAA in January 1964, more than 100 different basic designs were investigated, some of which are shown in Fig. C-1. Arrow wings, multiple wings, fixed wings, variable-sweep wings, and many others were thoroughly investigated to determine their ability to meet the basic requirements of a long-range commercial transport with high-altitude cruise speeds of about 2,000 mph, and with landing and takeoff capabilities equivalent to those of current large subsonic jets.

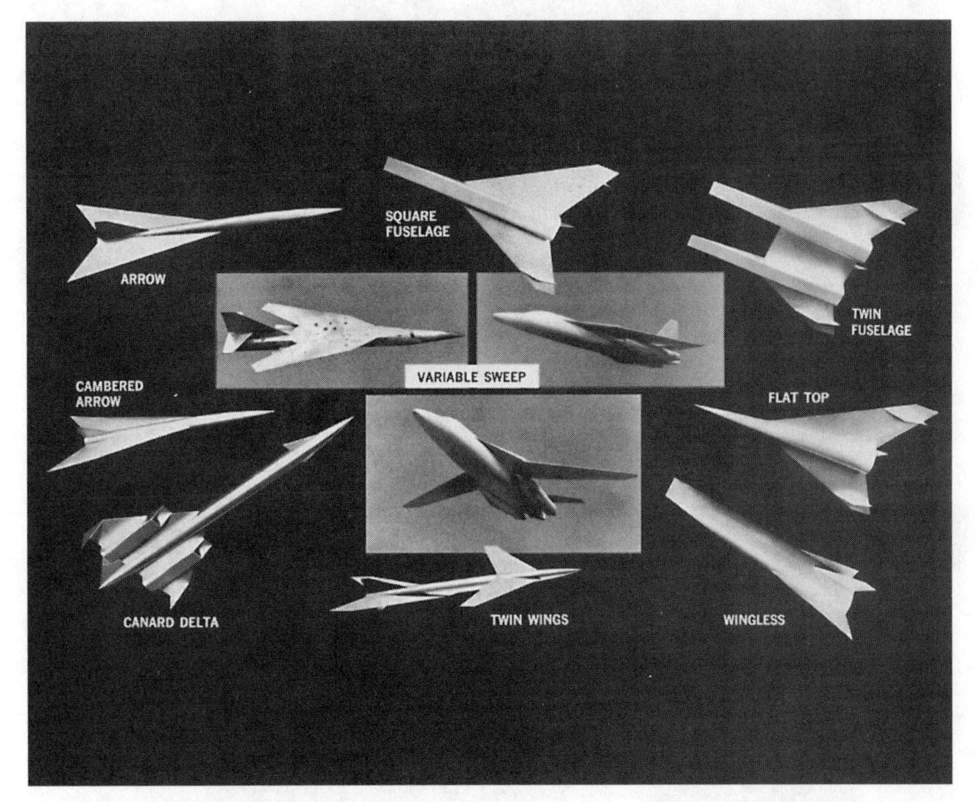

Fig. C-1. Configuration research.

The pattern of configuration development from 1958 to the Phase IIA configuration in 1964 is shown in Fig. C-2. Control over the movement of the aerodynamic center has always been the primary concern; the center of lift normally tends to move a large distance aft between subsonic and supersonic speeds. From the outset it was realized that this control could be achieved by using a variable-sweep wing in which the aerodynamic characteristics can be varied during flight. But, although the variable-sweep wing configuration is attractive from the standpoint of

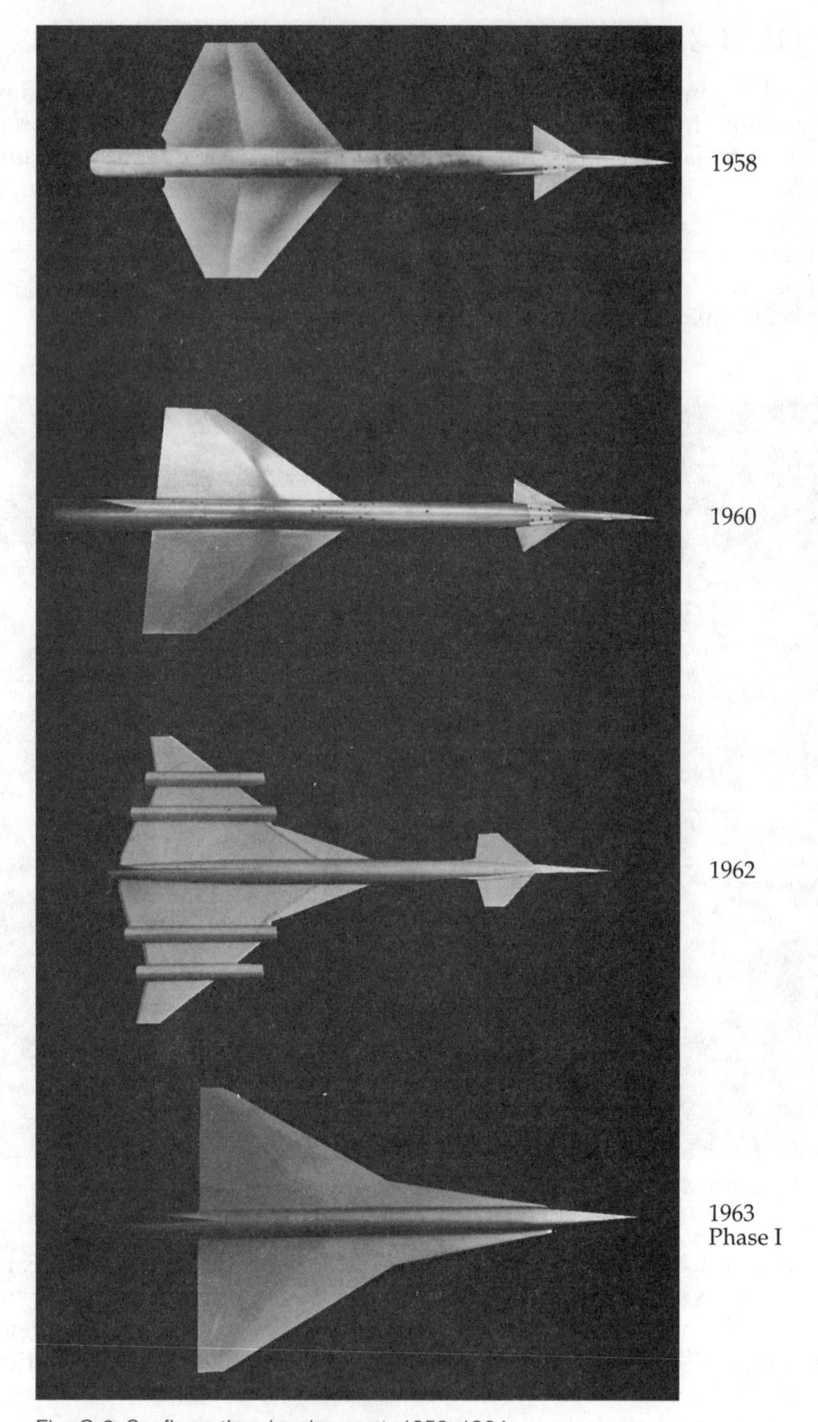

Fig. C-2. Configuration development, 1958–1964.

Fig. C-2. Continued.

controlling the shift of the aerodynamic center, it is complex and gives rise to serious problems in low-speed handling, stability, control, and weight.

[Note: Boeing won the FAA SST competition with a swing-wing design; however, after further development by Boeing, the swing-wing concept was unworkable and they reverted to a fixed-wing design, vindicating Lockheed's case.]

Alternatively, a fixed-wing configuration could have been adopted in which fuel is transferred aft during the transonic speed regime, moving the aircraft center of gravity to compensate for the movement of the center of lift. If there were no other choice, either of these methods was considered to be acceptable, but it was also considered that such complications and marginally satisfactory configurations should be avoided if at all possible. Development after 1958 concentrated on evolving a fixed-wing configuration in which the control of aerodynamic center shift during speed changes would be achieved aerodynamically as an inherent part of the basic design.

In 1958, during work with the trapezoidal wing and canard configuration (forward horizontal stabilizer), the shift was substantial and posed stability and control problems. By 1960, it was clear that a delta-shaped wing with a canard surface was a considerable improvement. In 1962, configurations in which the wing inboard leading edges of canard-delta designs were extended forward in a highly swept bat-wing showed even more promise.

This work led to the adoption of the tailless double-delta configuration in which the bat-wing was extended even farther forward to eliminate the canard surface entirely. As shown in Fig. C-2, this basic double-delta design was refined during the period between Phases I and IIA.

The results of seven years of research demonstrated that the large fixed double-delta wing could meet or exceed all the requirements of a supersonic transport. Several other configurations were tested, including variable-sweep designs, and also proved to have potential for a successful SST, but the double-delta was the simplest configuration meeting the requirements. Moreover, in contrast to the serious low-speed problems associated with the variable-sweep wing, the double-delta's low-speed characteristics were outstanding.

It also had many important advantages over other designs in that a light and strong structure could be more easily attained, no horizontal stabilizer was necessary, and the requirement for slots, slats, trailing edge flaps, and artificial stability devices was eliminated. These factors and others constituted substantial improvements in weight, economics, maintenance, and inherent safety.

It is pertinent to this discussion to note that the development of the basic double-delta planform paralleled the similar basic configuration development of the YF-12A, which initially carried the company designation A-11. Little information has been released regarding this military aircraft produced by the Lockheed-California Company, but it has been described officially as successful in "flying at a sustained speed of more than 2,000 mph and at altitudes in excess of 70,000 feet." It is, in a sense, a flying scale model of the double-delta SST.

PHASE I AIRPLANE CONFIGURATION

Lockheed Aircraft Corporation's January 1964 SST proposal described a fixed double-delta wing airplane with a cruising speed of Mach 3.0 and a passenger capacity of approximately 200. The general arrangement of the Phase I airplane is shown in Fig. C-3.

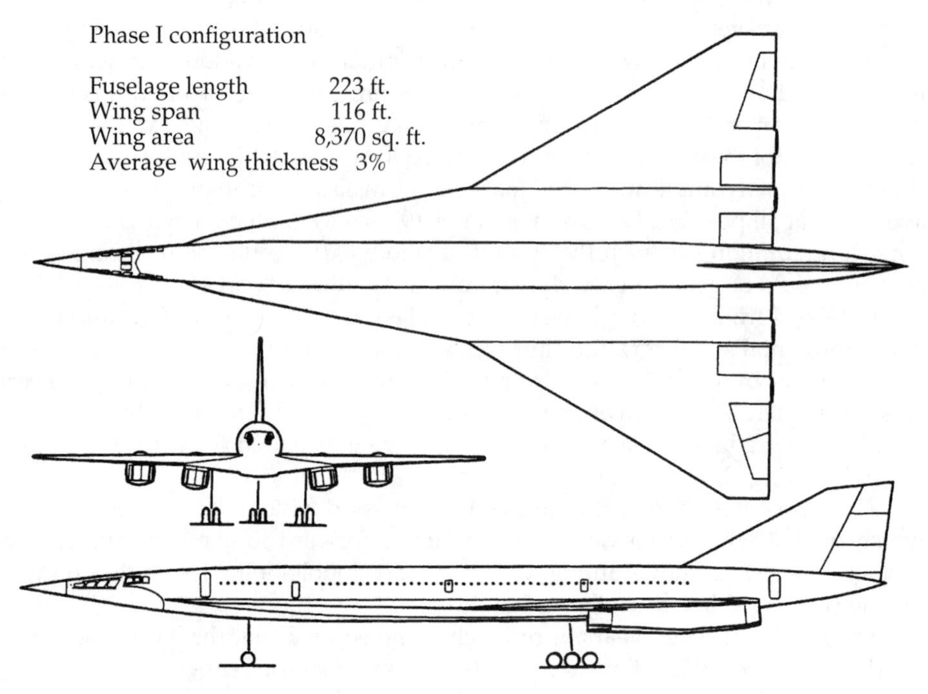

Fig. C-3. General arrangement of Phase I configuration.

The large, fixed wing of the Phase I design had an area of 8,370 square feet disposed between the large aft delta surface and the smaller forward delta surface. The sweepback angle of the leading edge of the forward delta was 80°; the main delta wing leading edge was swept 60°. The wing trailing edge incorporated no sweepback and the wing span was 116 feet. Thin supersonic airfoils averaged approximately 3.0 percent in thickness ratio. The wing surface was twisted slightly, but incorporated essentially no camber. Wind tunnel testing of this configuration completed at the time of the Phase I proposal indicated that the airplane would achieve a maximum cruising lift-drag ratio of 7.25 at a speed of Mach 3.0.

The engine selected for Phase I airplane was the Pratt & Whitney JT11F-4 duct burning turbofan. This engine was recommended at that time because with certain improvements it appeared to offer a sufficient level of performance with minimum development risk. It also had a favorable schedule position because it was a modification of the military J58 engine.

Discussions with United States airlines during the preparation of the Phase I proposal led Lockheed to believe that a passenger capacity of 200 or more would be necessary to realize acceptable operating economics; therefore, the Phase I airplane incorporated a 132-inch-diameter fuselage capable of seating more than 200 passengers in a 5-abreast, 34-inch seat pitch arrangement.

The results of the government evaluation of the Phase I proposal generally validated the basic design, but also concluded that improvement in sonic boom characteristics and payload-range capability were necessary to attain desired goals of operational economy. It also appeared that a more advanced engine than the fan adaptation of the J58 engine was required to meet all requirements.

DESIGN DEVELOPMENT DURING 1964

Intensive efforts were initiated under the FAA Phase IIA program on June 1, 1964, to achieve an airplane of improved characteristics and increased economic potential. To accomplish this, the FAA set a requirement for an international airplane with a range of 4,000 statute miles and a minimum payload of 30,000 pounds. Emphasis was placed on sonic boom characteristics, aerodynamic lift-to-drag ratio, definition and selection of optimum powerplant characteristics, airplane weight development, and identification of optimum fuselage sizes for international and domestic operation.

Sonic boom

Lockheed focused attention at the outset of the Phase IIA program upon an aspect of climb and acceleration sonic boom that had not been previously considered, and which it was felt might give rise to sonic boom intensities not slightly, but seriously in excess of the FAA's established maximum of 2.0 pounds per square foot.

The techniques of sonic boom estimation used for the Phase I program predicted that at 40,000 feet, boom overpressures would not reach the ground at aircraft speeds between Mach 1.0 and 1.15. Acceleration to Mach 1.2 before attaining an altitude of 40,000 feet was considered to be an adequate margin to avoid excessive overpressures on the ground.

The Phase I climb profile passed through an altitude of 42,000 feet at Mach 1.2 under standard atmospheric conditions (Fig. C-4); however, since the Phase I air-

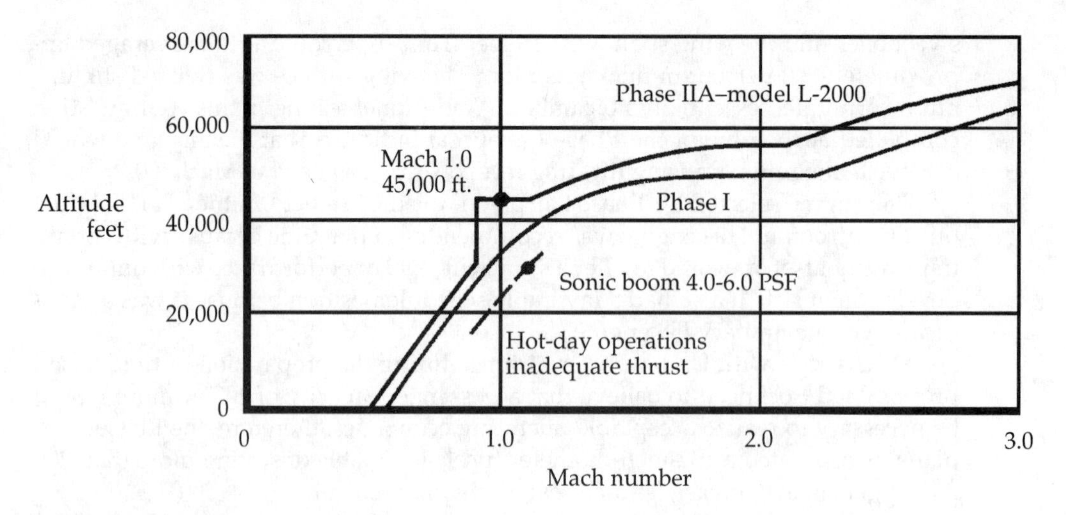

Fig. C-4. Climb profile for minimum sonic boom.

plane incorporated only a moderate amount of excess thrust for acceleration in the transonic region, there was a possibility that, during hot-day operations, boom overpressures of 4–6 pounds per square foot might be produced at altitudes of about 30,000 feet and at speeds only slightly exceeding Mach 1.0, considering magnification effects.

Therefore, at the outset of the Phase IIA program, it was decided to seek airplane and propulsion system design changes to make it possible to remain at subsonic speeds to an altitude of 45,000 feet. At this altitude, adequate thrust margins would enable acceleration through Mach 1.0 for a wide variety of atmospheric temperature conditions.

Further, the altitude of the entire transonic climb and acceleration flight path was raised above that used for boom reduction in Phase I. The climb profile adopted for the Phase IIA design (Model L-2000) is also shown in Fig. C-4.

To meet the new sonic boom design goal, the wing area of the L-2000 was increased over that of the Phase I airplane, together with about a 25 percent increase in engine thrust. In addition, careful contouring of the wing and fuselage was accomplished to reduce boom source strength. These changes made routing operations possible in the transonic area at altitudes 10,000 to 20,000 feet higher than were possible with the Phase I airplane and thus minimized the possibility of excessive sonic boom.

Aerodynamic improvements

During Phase IIA, effort was concentrated on identifying and developing a wing shape for the L-2000, which offered improved aerodynamic performance over that of the Phase I wing without significant weight penalty. A coordinated airplane design and wind tunnel program was initiated to study all aspects of wing optimization within the environment of practical airplane requirements.

Figure C-5 shows a few of the 44 wing models designed, fabricated, and tested during the first five months of the program. Specific parametric design and wind tunnel investigations were carried out on leading edge sweep, planform, camber and twist, dihedral, and airfoil cross section.

Fig. C-5. Phase IIA wind tunnel development.

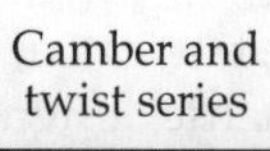

Fig. C-5. Continued.

Throughout the search for optimum drag characteristics in cruise flight, other aspects—such as stability and control characteristics, propulsion system installation, and low-speed handling qualities—were investigated in additional low- and high-speed wind tunnel tests.

Typical of the optimization studies were those involving the effect of the wing trailing edge sweep angle upon payload, the results are in Fig. C-6. As the sweep is increased (trailing edge notched out), considerable payload penalties occur.

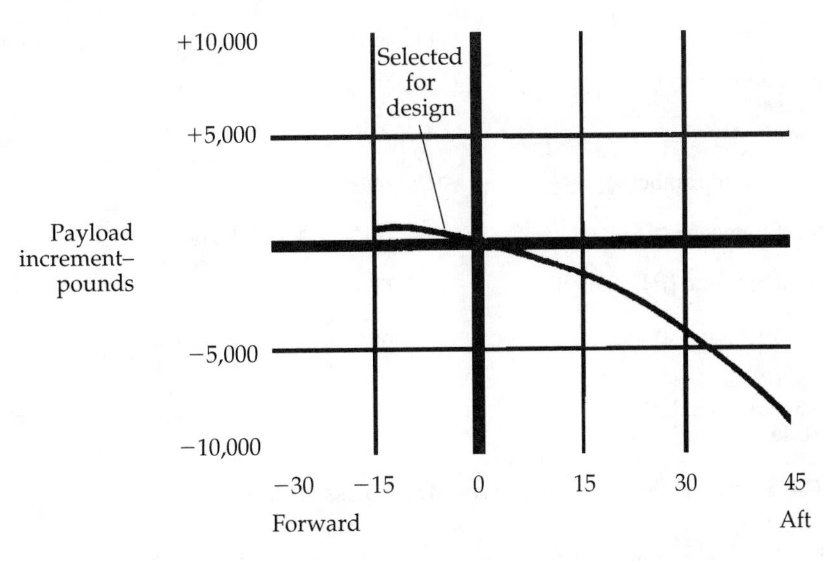

Fig. C-6. Effect of trailing edge sweep on payload at 4,000-statute-mile range and 450,000 pounds gross weight.

For example, a trailing edge aft sweep of 45°, resulting in an arrow-type wing, had 8,000 pounds less payload than the Phase I wing, despite the fact that it had substantially improved aerodynamic characteristics; thus, although the maximum lift-drag ratio for cruising flight was found from the wind tunnel data to increase to as much as 8.5 at Mach 3.0 with a 45° trailing edge sweep, the increase in empty weight to build this wing and to provide satisfactory takeoff, landing, and aeroelastic characteristics more than offset the aerodynamic gain.

Early results from this parametric investigation suggested that payload could be increased over the Phase I values by slight forward sweep of the trailing edge of the wing; hence, activities during the latter portion of Phase IIA were concentrated in this area. A 10° forward sweep of the trailing edge of the wing was selected for design of the L-2000 supersonic transport (Fig. C-6).

Results of the extensive wind tunnel tests demonstrated that the wing selected for the L-2000 during the Phase IIA program substantially improved the aerodynamic efficiency in cruising flight. The full-scale airplane maximum lift-drag ratio of 7.25 established for the Phase I airplane compared to a value of 7.94 for the Model L-2000.

Figure C-7 compares the Phase I and Phase IIA maximum lift-drag ratios at Mach 3.0 and breaks down the source of these improvements as they were developed during the Phase IIA work. Changes to wing thickness ratio, area, planform, twist, and camber resulted in gains totalling 0.68 in the lift-drag ratio. Refinements in other areas, such as the fuselage and nacelle contours, resulted in a further gain of 0.20.

	Phase 1	L-2000	D L/D
Wing thickness ratio	3.0	2.3	.30
Wing area	8370	9026	.12
Wing planform	L.E. 60° T.E. 0°	L.E. 62° T.E. −10°	.10
Wing twist and camber	$1^1/_2$° twist	Modified T & C	.16
Other refinements	–	Fuselage nacelles	.20
Mach 3 maximum L/D (smooth)	7.25	8.13	.88
Roughness & emissivity	none	Rational increment	−.19
Mach 3 maximum L/D full-scale airplane	7.25	7.94	+.69

Fig. C-7. Summary of Phase IIA maximum L/D progress at Mach 3.0.

The total gain over the Phase I maximum lift-drag ratio of 7.25 was about 0.88, or 12 percent; however, rational allowances for roughness and the effect on the boundary layer of the thermal emissivity coating, which is utilized to reduce structural temperatures, reduced this increment by 0.19, resulting in the L-2000 maximum lift-drag ratio of 7.94 for Mach 3.0 flight at 75,000 feet. A net improvement of almost 10 percent was achieved in aerodynamic cruise efficiency.

While this high-speed improvement was attained by configuration changes, low-speed aerodynamic characteristics and handling qualities were also substantially improved over those of the Phase I design. For example, low-speed stability was increased and pitch-up was completely eliminated, even at extremely high angles of attack. The improvement in aerodynamic efficiency contributed to improve payload-range and economic characteristics.

PHASE IIA AIRPLANE DESCRIPTION

The basic configuration

The basic configuration of the L-2000 supersonic transport that emerged from the Phase IIA program is shown in Fig. C-8 and can also be seen in Fig. C-9. In Fig. C-8, a gray overlay of the Phase I airplane is superimposed to display clearly the areas in which principal changes were made.

Average wing thickness ratio was reduced from 3.0 percent to approximately 2.3 percent. Wing area was increased from 8,370 square feet to 9,026 square feet. Leading and trailing edge sweep angles were changed; the size of the large, basic delta wing was increased, and the size of the forward delta was decreased. The spanwise juncture between the forward delta and the aft delta was moved inboard as a result of the increase from 80° to 85° in the leading edge sweep of the forward delta. The wing incorporated improved twist and camber, and a special Lockheed-developed supersonic airfoil. The nose gear was shortened, and hence the ground attitude was also changed.

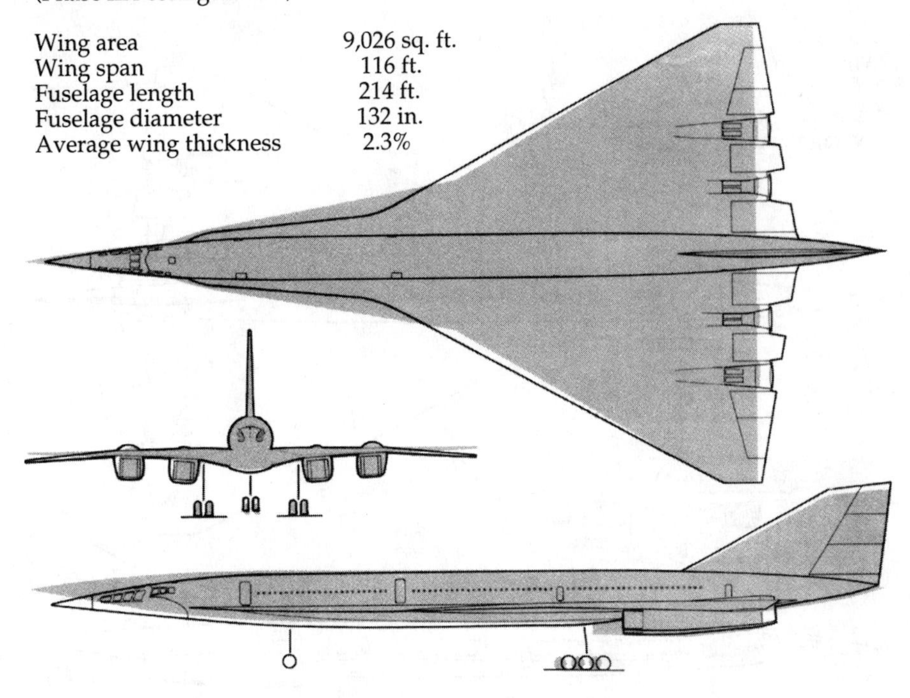

Fig. C-8. Comparison of Phase I and Phase IIA configurations.

The original five-abreast, 132-inch-diameter fuselage was retained in the Model L-2000 because it represented the most promising choice considering the requirements of drag, economics, and seating capacity. The general arrangement of four engines disposed in independent nacelles across the wing was also retained, but the design was made capable of maximum performance with either a Pratt & Whitney turbofan or a General Electric turbojet engine.

Comparison of the wing geometry of the Phase I airplane and the L-2000, as shown in Fig. C-8, raises the question of whether the improved configuration retained the same control over aerodynamic center shift from subsonic to supersonic speeds as did the Phase I airplane. This unique feature of the double-delta fixed-wing design was, in fact, retained. Deflection of the elevons/elevators for trim throughout the speed range from subsonic flight to Mach 3.0 cruise showed almost the same pattern as for the Phase I airplane. Twist and camber were incorporated so that the elevons/elevators were in the faired position for both subsonic and supersonic cruise. Only moderate deflection is required for transient flight in the transonic and low-supersonic speed ranges.

Low-speed characteristics of the L-2000 configuration were improved substantially; wind tunnel tests to an angle of attack of 36° showed increased longitudinal stability for landing and takeoff and complete elimination of pitch-up. The powerful wing leading edge vortex provided directional stability at angles of at-

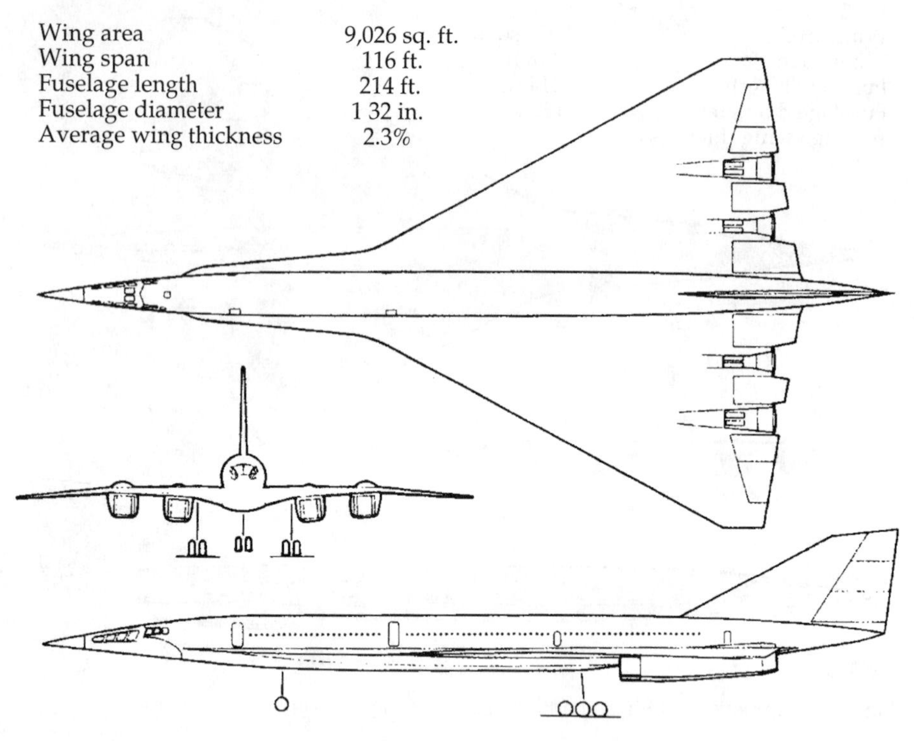

Fig. C-9. L-2000-1 General arrangement.

tack of 20° and sideslip angles as high as 30°. Elevator and aileron effectiveness was retained undiminished to extremely high attitudes in pitch and yaw.

Three general arrangements of the L-2000 were designed (designated L-2000-1, -2, and -3), each of which could utilize either Pratt & Whitney or General Electric engines. The three general arrangements have fuselages of differing length and capacity, but use the same wing, powerplants, systems, landing gear, flight station, and vertical tail. Fuselage capacity varies from 170 to 250 passengers in these versions, but design takeoff gross weight remains the same for all.

Figure C-10 shows the relationship among the three configurations by identifying that part of each fuselage that is unique to that version; thus, the aft body of the -1 configuration is unique, while the aft body of the -2 and -3 are identical.

The -2 includes a fuselage barrel extension ahead of the wing, which increases fuselage length from the 214 feet of the -1 to 225 feet 7 inches. This change in length, combined with the space made available by the aft body shape and a rearrangement of the interior, allows maximum coach passenger capacity to increase from 170 to 221.

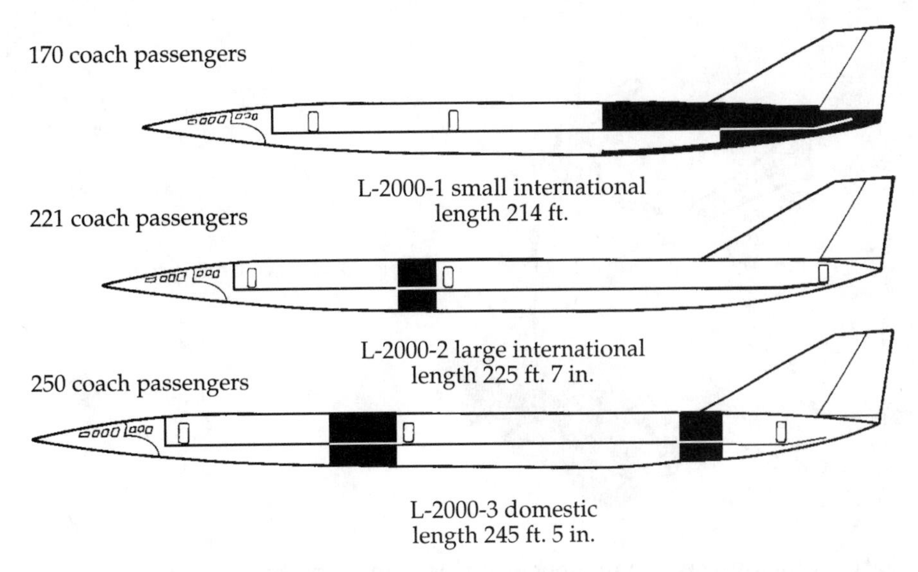

Fig. C-10. Fuselage comparison of models L-2000-1, -2, and -3.

The L-2000-3 is derived from the -2 by the addition of three passenger row barrel sections both ahead of and behind the wing. These additions increase fuselage length to 245 feet 5 inches while further increasing maximum coach passenger capacity to 250.

Because ground clearance angle for takeoff and landing is determined by the outboard engines, the lengthening of the fuselage to the rear in no way compromises operation of the airplane. The extensions of the forward fuselage result in increased length of the forward delta surface, but have a negligible effect on longitudinal stability. The increased vertical tail length on the -3 airplane compensates for the longer nose and results in equal directional stability. All configurations are in balance and have the same aft center-of-gravity limit.

The basic general arrangement

The basic general arrangement described below is applicable to all Phase IIA versions of the L-2000. Wing area is 9,026 square feet and the wing span is 116 feet. The large wing size and long root chord length provide large structural depth even though wing airfoil sections are as thin as 2 percent in the chordwise direction. This structural depth makes the large wing and light wing loading possible within a normal wing weight allowance. The light wing loading makes takeoff and landing possible at the same speeds as current subsonic jets without use of wing flaps.

The only movable surfaces incorporated are the flight controls, consisting of elevators, ailerons, and elevons along the wing, and a three-piece rudder on the vertical tail. The main landing gear (Fig. C-11) is a six-wheel bogie utilizing exactly the same wheel and tire size as the DC-8 jet transport, but with lower tire pressures. This landing gear arrangement spreads the increased gross weight of the

Fig. C-11. Main landing gear.

L-2000 over the runway in a manner ensuring compatibility with existing airports throughout the world.

Superior visibility for takeoff, landing, and subsonic flight, together with direct forward vision, minimum drag, and minimum flight station noise in supersonic flight are provided by the "Weather-Vision" movable fuselage nose. The flight station has a basic arrangement for three cockpit crew members, allowing safe operation even in the event of incapacitation of one crew member.

The location of entrance doors in the front and midfuselage positions permits rapid turnaround with nose-in loading at the terminal (Fig. C-12). Cabin windows afford a view of the dark blue sky, the curvature of the Earth, and, in the northern latitudes, the aurora borealis. Cooling air circulates within the side walls and maintains

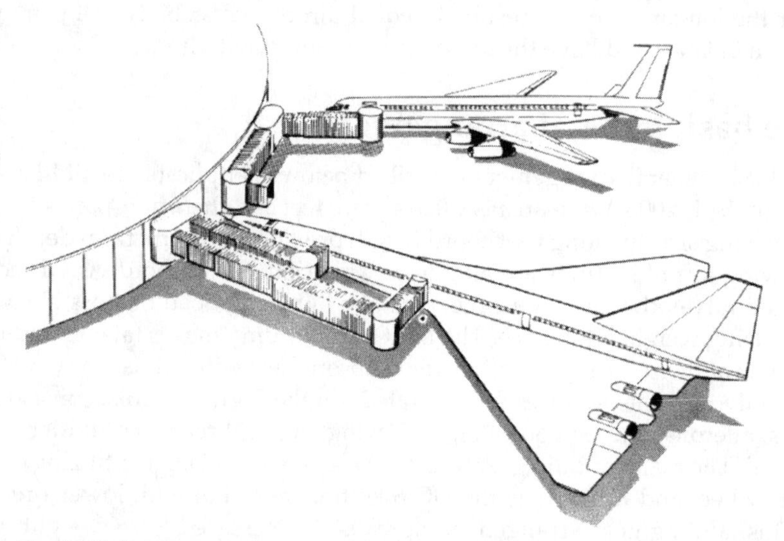

Fig. C-12. Passenger loading.

the cabin at normal comfort levels despite the ovenlike temperatures of the outside skin. The fuel in the wing center and stub sections serves as a heat sink to absorb the heat dissipated by the cooling system.

Location of the engines in four independent nacelles spaced well out along the wing span is in the pattern of previous commercial transport aircraft practice and provides complete individual powerplant integrity. Particular attention has been placed upon location of the engines to ensure freedom from ingestion and foreign object damage originating at either the nose wheels or main landing gear wheels. Placement of the engines outboard on the wing and with the jet exhaust plane well to the rear removes serious noise sources as far as possible from the passenger cabin, improving the internal environment and reducing sonic fatigue.

The propulsion system incorporates a new Lockheed-developed mixed-compression inlet located beneath and well behind the wing leading edge. This location shields the inlets from excessive changes in angle of inflow during various flight conditions and maneuvers. The inlet configuration offers low drag through minimum external cowl angles and high-pressure recovery. An important feature of the inlet under development is a self-starting characteristic, which is of great importance in the event of transient disturbances generated either internally by the engine or externally by the atmosphere. This inlet concept eliminates the need for impractical, fast-acting, variable-geometry devices within the inlet to reestablish the flow. The exhaust nozzles incorporate thrust reversal systems for use on the ground during landing and in the air for rapid descent at speeds below Mach 1.2.

Interior arrangements

An all-coach configuration with five-abreast seating and 34-inch seat spacing was provided for the three configurations shown in Fig. C-10. More space is provided for each passenger than in current economy-class jets. A five-abreast seating arrangement was adopted rather than the present industry economy standard of six abreast to suit the aerodynamic requirements. A long slender fuselage was needed to avoid increasing the supersonic drag to an intolerable extent.

The 225-foot fuselage of the L-2000-2 provides for an all-coach passenger capacity of 221. Two entrance doors are located on the left side of the forward fuselage and a third one aft of the wing. There are five lavatories: two forward and three in the aft fuselage. Coat space, emergency equipment, and seats for six cabin attendants are provided. A cargo volume of 950 cubic feet is available beneath the floor in the forward fuselage, with 420 cubic feet available above the floor in the aft fuselage. Total cargo capacity of the L-2000-2 is 1,370 cubic feet. With seat pitch varying from 32 to 40 inches, numerous alternate arrangements are possible for mixed-class service, resulting in passenger capacities both below and above the nominal 221 figure for the all-coach interior.

Aircraft structure

The entire structure of the Model L-2000 is conventional except perhaps for the basic materials of construction, which are drawn from the family of titanium alloys, Ti 8-1-1 predominating. The wing covers are stiffened by chordwise corrugations, and the wing skins have preformed chordwise depressions that result in controlled contour smoothness during periods of flight when thermal gradients occur between the surface and substructure. This type of titanium wing construction has already been proven in flight.

Very little honeycomb material is used throughout the airplane; even the leading and trailing edge areas of the wing and fin use integrally stiffened structure. In the nacelle inlet duct area, where the temperatures are somewhat higher than on the rest of the structure, Ti 6-4 material is used rather than Ti 8-1-1 because of the greater stress corrosion resistance of this alloy at elevated temperatures.

The fuselage shell is of conventional design and utilizes rings and longitudinal stiffeners. Ring spacing is 17 inches with a 6-inch-diameter window provided in each bay.

The efficient alignment of major structural members is most noticeable in the wing-to-fuselage juncture, in which each of the 59 wing beams mates directly with a fuselage ring. The large number of wing beams results in substantial structural redundancy; and the wing-to-fuselage juncture progressively transmits loads between wing and fuselage while using the wing structure to complete the fuselage shell. Advantages are apparent in weight, wing, and fuselage stiffness, fail-safe multiple-load paths, minimum concentration of loads, producibility, and crashworthiness.

Vertical tail structure, without the complication of a horizontal tail, is easily coordinated with the aft fuselage frames, and the load paths are relatively short and direct because fin loads are progressively transferred from the fuselage shell into the multiple wing beams.

The wing-fuselage arrangement resulting from application of the double-delta, fixed-wing concept provides outstanding capability for fuel containment. All the fuel is contained in integral wing tanks in which 56 percent of the total fuel is inherently protected from the thermal environment of Mach 3.0 flight by the cool cabin floor above and the lower fuselage fairing shield below.

Tests conducted during the Phase IIA program show that suitable nonmetallic materials for use in a Mach 3.0 commercial airplane are or will be readily available to meet the needs of the supersonic transport program. These materials include cabin sealants, fuel tank sealants, elastomeric seals, oils, lubricants, glass, hydraulic oil, and hydraulic seals.

The flight profile

The flight profile used in the Phase IIA program for evaluation of the various versions of the Model L-2000 is shown in Fig. C-13. In this flight profile, 5 minutes air maneuver time is provided at low altitude after takeoff for departure from the terminal area and alignment with airways. Subsonic climb is carried out to an altitude of 45,000 feet at which point acceleration through Mach 1.0 is accomplished. The airplane then climbs and accelerates along a path in which performance can be limited by the requirement to not exceed sonic boom overpressures of 2.0 pounds per square foot.

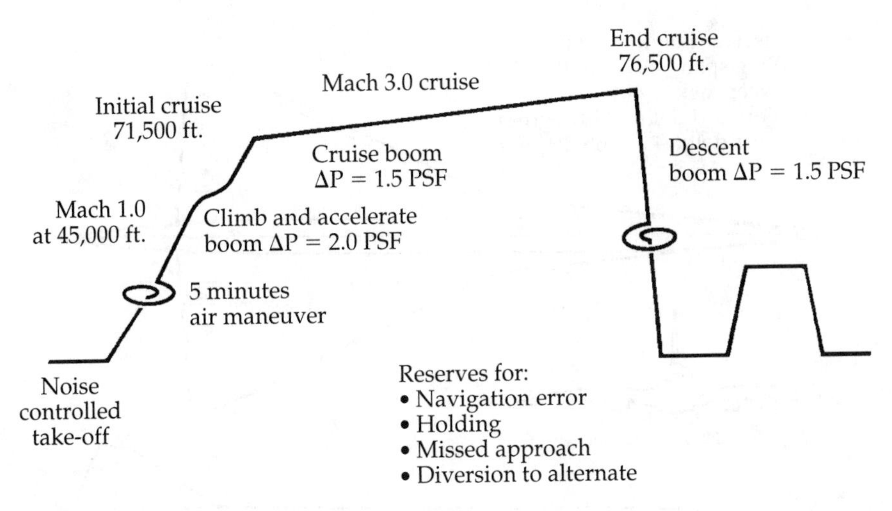

Fig. C-13. Phase IIA flight profile.

Initial Mach 3.0 cruise is at 71,500 feet where the sonic boom overpressure is about 1.5 pounds per square foot, conforming to sonic boom cruise requirements. At the end of the Mach 3.0 cruise, altitude had increased to 76,500 feet so that boom overpressure is reduced to approximately 1.3 pounds per square foot. Descent from cruise altitude is made in a manner not to exceed a sonic boom overpressure of 1.5 pounds per square foot. Reserve fuel is provided for navigation error en-route, for holding at low altitude after arriving in the terminal area, for a missed approach, and for diversion to an alternate airfield 300 miles away.

The payload-range characteristics

By the end of the Phase IIA program in November 1964, each airplane was capable of international range. The full space limit payload of the L-2000-1 was 39,850 pounds, the L-2000-2 was 49,000 pounds, and the L-2000-3 was 58,000 pounds.

Conclusion

At the end of 1964, improvements to the basic design of both the airframe and engines had progressed to the point where extremely attractive international and domestic versions of the L-2000 could be described using either of the two engines available for consideration. In Phase IIA, the feasibility of the SST was firmly established, as well as confidence that desirable performance and economic characteristics could be attained.

Early in 1965, Phase IIB was initiated, and under this program, further intensive design work, research, and development will explore all possible means of improving the Phase IIA configuration to increase performance of the L-2000. Subsequently, an intensive design program refined this basic design concept until it evolved as the Lockheed L-2000-7A and L-2000-7B as shown in Fig. C-14.

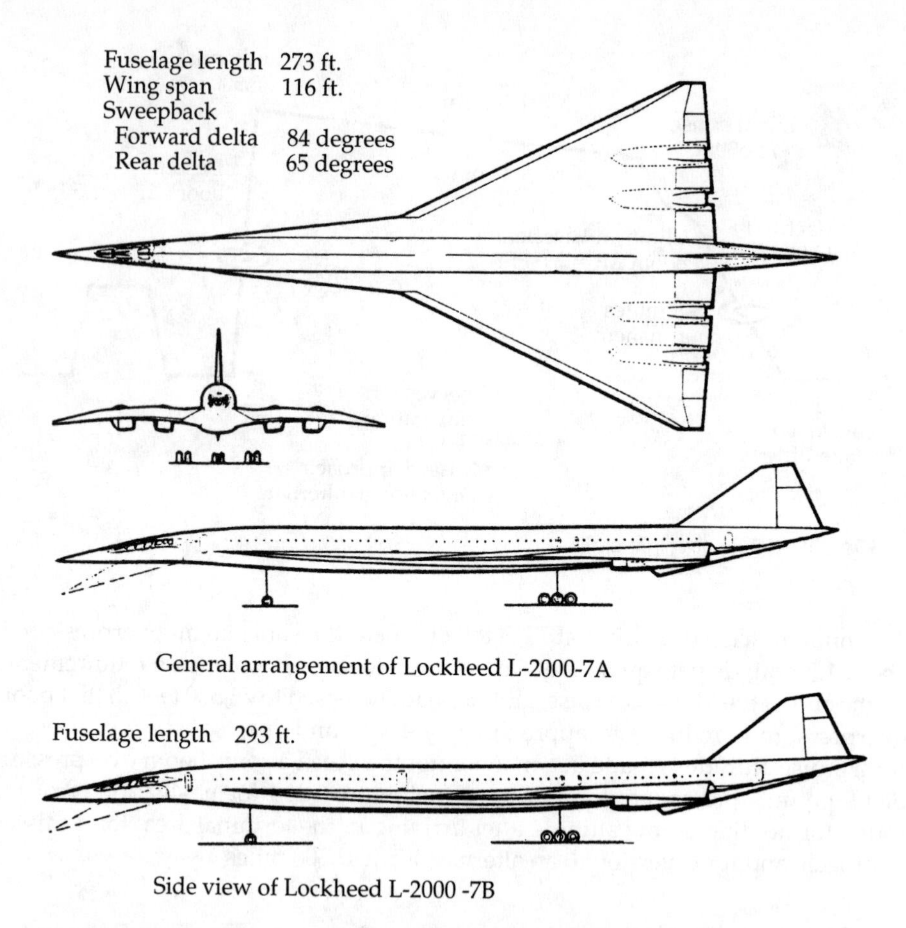

Fig. C-14. After an intensive design development program, the Lockheed SST evolved as the L-2000-7A and L-2000-7B. The -7B was essentially the same basic airplane as the -7A with a lengthened fuselage for additional payload and passengers.

Glossary

(Excerpted from *Aviation/Space Dictionary* by Larry Reithmaier, published by TAB Books, a division of McGraw-Hill, Inc.)

acoustic velocity (symbol $\propto$) Speed of sound.

active controls Automatic movement of airplane control surfaces in response to various sensors for load alleviation, stability augmentation or flutter load control.

aerodynamic center of a wing section A point located on the wing chord approximately one-quarter of the chord length back of the leading edge about which the moment coefficient is practically constant for all angles of attack.

aerodynamic coefficient Any nondimensional coefficient relating to aerodynamic forces or moments, such as a coefficient of drag, a coefficient of lift, and the like.

aerodynamic force The force exerted by a moving gaseous fluid upon a body completely immersed in it.

aerodynamic heating The heating of a body produced by passage of air or other gases over the body; caused by friction and by compression processes and significant chiefly at high speeds.

aerodynamics The science that deals with the motion of air and other gaseous fluids, and of the forces acting on bodies when the bodies move through such fluids, or when such fluids move against or around the bodies.

aerodynamic twist Variation of the zero-lift line along the span of a wing or other airfoil. See washin; washout.

aerodynamic vehicle A device, such as an airplane, glider, etc. that is capable of flight only within a sensible atmosphere and relying on aerodynamic forces to maintain flight. Used when the context calls for discrimination from space vehicle.

aeroelasticity The study of the response of structurally elastic bodies to aerodynamic loads.

aileron reversal Aileron deflection can cause twisting of a wing producing opposite lateral motion (roll) of the airplane if the wing is not structurally rigid enough. Many high-speed jet airplanes incorporate outboard ailerons for lateral control at low speeds. At high speeds, outboard ailerons are locked out, and lateral control is provided by inboard ailerons.

airfoil A structure, piece, or body, originally likened to a foil or leaf in being wide and thin, designed to obtain a useful reaction of itself in its motion through the air.

airfoil section See wing section.

airspeed The speed of an aircraft relative to its surrounding air mass.

angle of attack The angle between the chord of a wing or the reference line in a body and the direction of the undisturbed flow or relative wind in the absence of sideslip.

area rule A prescribed method of design for obtaining minimum zero-lift drag for a given aerodynamic configuration, such as a wing-body configuration, at a given speed. For a transonic body, the area rule is applied by subtracting from, or adding to, its cross-sectional area distribution normal to the airstream at various stations so as to make its cross-sectional area distribution approach that of an ideal body of minimum drag. For a supersonic body, the sectional areas are frontal projections of area intercepted by planes inclined at the Mach angle.

artificial feel A control feel simulated by mechanisms incorporated in the control system of an aircraft where the forces acting on the control surfaces are not transmitted to the cockpit controls, as in the case of an irreversible control system or a power-boosted system.

aspect ratio Of an airfoil wing, the ratio of span to mean chord of an airfoil, i.e., the ratio of the square of the maximum span to the total area of an airfoil. In a simple rectangular airfoil, it is the ratio of the span to the chord.

boundary layer The layer of fluid in the immediate vicinity of a bounding surface. (A.) In fluid mechanics, it is the layer affected by viscosity of the fluid, referring ambiguously to the laminar boundary layer, turbulent boundary layer, planetary boundary layer, or surface boundary layer. (B.) In aerodynamics, the boundary-layer thickness is measured from the surface to an arbitrarily chosen point, e.g. where the velocity is 99 percent of the stream velocity. Thus, in aerodynamics, boundary layer by selection of the reference point can include only the laminar boundary layer or the laminar boundary layer plus all, or a portion of, the turbulent boundary layer.

boundary-layer control (of a wing) Control by artificial means of the development of the boundary layer with the object of affecting transition or separation. An example is withdrawing air from the boundary layer through the surface (suction) or injecting air or another gas into the boundary layer (blowing). Use of boundary-layer control provides a greater range of usable lift coefficients.

buffeting The beating of an aerodynamic structure or surfaces by unsteady flow, gusts, etc. Also, the irregular shaking or oscillation of a vehicle component owing to turbulent air or separated flow.

camber 1. Curvature of the median line of an airfoil section; more generally, the curvature of a surface. 2. The ratio of the maximum height of the median line above the chord to the chord length.

canard Pertaining to an aerodynamic vehicle in which horizontal surfaces used for trim and control are forward of the main lifting surface; the horizontal trim and control surfaces in such an arrangement.

Celsius temperature scale (C) Same as centigrade temperature scale. The Ninth General Conference on Weights and Measures (1948) replaced the designation degree centigrade with degree Celsius.

center of gravity The point within an aerospace vehicle through which, for balance purposes, the total force due to gravity is considered to act.

center of gravity limits The limits within which an aircraft's center of gravity must lie to ensure safe flight.

center of lift The mean of all the centers of pressure on an airfoil.

center of pressure The point on some reference line (e.g., the chord of an airfoil) about which the pitching moment is zero.

centigrade temperature scale (C) A temperature scale with the ice point at 0° and the boiling point of water at 100°. Now called Celsius temperature scale. Conversion to the Fahrenheit temperature scale is according to the formula: $°C=5/9(°F-32)$.

coefficient 1. A number indicating the amount of some change under certain specified conditions, often expressed as a ratio. 2. A constant in an algebraic equation.

composite materials Structural materials of metals, ceramics, or plastics with built-in strengthening agents that might be in the form of filaments, foils, powders, or flakes of a different compatible material.

compressibility The property of a substance, as air, by virtue of which its density increases with increase in pressure. In aerodynamics, this property of the air is manifested especially at high speeds (speeds approaching that of sound and higher speeds). Compressibility of the air about an aircraft might give rise to buffeting, aileron buzz, shifts in trim, and other phenomena not ordinarily encountered at low speeds, known generally as compressibility effects.

compressibility drag The increase in drag arising from the compressibility of the air that occurs at high speeds.

compressible flow In aerodynamics, flow at speeds sufficiently high that density changes in the fluid cannot be neglected.

control-configured vehicle (CCV) Control-configured vehicles utilize active controls technology that implies the use of automatic feedback control systems that sense aircraft control motion and provide signals to a control surface to reduce vehicle weight or to improve performance.

control feel The impression of the stability and control of an aircraft that a pilot receives through the cockpit controls, either from the aerodynamic forces acting on the control surfaces or from forces simulating these aerodynamic forces. See artificial feel; feel.

convergence A decrease in area or volume.

critical Mach number The free-stream Mach number at which a local Mach number of 1.0 is attained at any point on the body under consideration. For example, an airplane traveling at a Mach number of 0.8 with respect to the undisturbed flow might attain a Mach number of 1.0 in the flow about the wing; the critical Mach number would thus be 0.8.

delta wing A triangularly shaped wing of an aircraft.

density Weight or mass per unit volume expressed in pounds per cubic feet, grams per cubic centimeter, and the like.

density altitude Density altitude is pressure altitude corrected for nonstandard temperature. Density altitude must be computed for high altitude and/or high temperature conditions to determine takeoff run, for example.

Department of Defense (DOD) The department of the executive branch of the government including the office of the Secretary of Defense and the Army, Navy, and Air Force departments.

dihedral angle The acute angle between two intersecting planes or between lines representative of planes. An airplane wing design incorporating dihedral is used to improve lateral stability.

direct lift control system (DLC) A system utilizing spoilers to vary wing lift for flight path control such as precision glide-slope tracking. The result is translational movement without angular (pitch) motion.

divergence 1. The expansion or spreading out of a vector field; also a precise measure thereof. In mathematical discussion, divergence is considered to include convergence, i.e., negative divergence. 2. A static instability of a lifting surface or of a body on a vehicle wherein the aerodynamic loads tending to deform the surface or body are greater than the elastic restoring forces.

drag (symbol D) A retarding force acting upon a body in motion through a fluid, parallel to the direction of motion of the body. It is a component of the total fluid forces acting on the body. See aerodynamic force.

drag coefficient (symbol C_D) A coefficient representing the drag on a given airfoil or other body, or a coefficient representing a particular element of drag.

Dutch roll (aircraft) A lateral oscillation with a pronounced rolling component.

dynamic pressure (symbol q) The pressure of a fluid resulting from its motion, equal to one-half the fluid density times the fluid velocity squared ($\frac{1}{2}\rho V^2$). In incompressible flow, dynamic pressure is the difference between total pressure and static pressure. Also called kinetic pressure. Compare impact pressure.

dynamic stability The characteristics of a body, such as an aircraft, that causes it, when disturbed from an original state of steady flight or motion, to damp the oscillations set up by restoring moments and gradually return to its original state; specifically, the aerodynamic characteristics. See stability.

F Military mission designation for fighter aircraft.

FAA Abbreviation for Federal Aviation Administration.

Fahrenheit temperature scale (°F) A temperature scale with the ice point at 32° and the boiling point of water at 212°. Conversion with the Celsius (centigrade) temperature scale (abbreviated C) is by the formula: °F = 9/5 °C + 32.

Federal Aviation Administration (FAA) The Federal Aviation Administration is the arm of the Department of Transportation responsible for the promotion, regulation and safety of civil aviation, and for safe and efficient use of airspace that is shared by both civil and military aircraft.

feel The sensation or impression that a pilot has or receives as to his, or his craft's attitude, orientation, speed, direction of movement or acceleration, or proximity to nearby objects, or as most often used, as to the aircraft's stability and responsiveness to control. See control feel.

fence A stationary plate or vane projecting from the upper surface of an airfoil, substantially parallel to the airflow, used to prevent spanwise flow.

fighter (F) Aircraft designed to intercept and destroy other aircraft and/or missiles. (Includes multipurpose aircraft designed for ground support and interdiction.)

free stream 1. The stream of fluid outside the region affected by a body in the fluid. 2. Pertaining to the free stream, sense 1, as in free-stream dynamic pressure, free-stream flow, free-stream Mach number, free-stream static pressure, free-stream temperature, free-stream turbulence, and free-stream velocity.

free-stream Mach number The Mach number of the total airframe (entire aircraft) as contrasted with local Mach number of a section of the airframe.

hinge moment The moment about the hinge axis of a control or other hinged surface due to aerodynamic forces.

hypersonic 1. Pertaining to hypersonic flow. 2. Pertaining to speeds of Mach 5.0 or greater.

hypersonic flow In aerodynamics, flow of a fluid over a body at speeds much greater than the speed of sound and in which the shock waves start at a finite distance from the surface of the body. Compare supersonic flow.

impact pressure 1. That pressure of a moving fluid brought to rest that is in excess of the pressure the fluid has when it does not flow, i.e., total pressure less static pressure. Impact pressure is equal to dynamic pressure in incompressible flow, but in compressible flow, impact pressure includes the pressure change owing to the compressibility effect. 2. A measured quantity obtained by placing an open-ended tube, known as an impact tube or pitot tube, in a gas stream and noting the pressure in the tube. Since the pressure is exerted at a stagnation point, the impact pressure is sometimes referred to as the stagnation pressure or total pressure.

incompressible fluid A fluid in which the density remains constant for isothermal pressure changes, i.e., for which the coefficient of compressibility is zero. Expansion and contraction of an incompressible fluid under adiabatic heating or cooling is thus allowed for. See compressible flow.

induced drag That part of the drag induced by the lift, due to the tilting of the lift vector by the downwash to produce a component in the streamwise, or drag, direction.

ionization The process of adding one or more electrons to, or removing one or more electrons from, atoms or molecules, thereby creating ions. High temperatures, electrical discharges, or nuclear radiations can cause ionization.

ionosphere The region of the atmosphere, extending from roughly 40–250 miles altitude, in which there is appreciable ionization. The presence of charged particles in this region profoundly affects the propagation of electromagnetic radiations of long wavelengths (radio and radar waves).

irreversible control system A flight control in which the control surface can be moved freely by the pilot but cannot be moved by aerodynamic forces alone.

kilometer (km) A unit of distance in the metric system equal to 3,280.8 feet, or 1,093.6 yards, or 1,000 meters, or 0.62137 statute miles, or 0.53996 nautical miles.

knot (kt) A nautical mile per hour; 1.1508 statute miles per hour.

laminar boundary layer In fluid flow, layer next to a fixed boundary. The fluid velocity is zero at the boundary, but the molecular viscous stress is large because the velocity gradient normal to the wall is large.

laminar flow In fluid flow, a smooth flow in which no cross flow of fluid particles occurs between adjacent stream lines; hence, a flow conceived as made up of layers, commonly distinguished from turbulent flow.

L/D ratio The ratio of lift-to-drag.

lift (symbol L) That component of the total aerodynamic force acting on a body perpendicular to the undisturbed airflow relative to the body.

lift coefficient (symbol C_L) A coefficient representing the lift of a given airfoil or other body. The lift coefficient is obtained by dividing the lift by the free-stream dynamic pressure and by the representative area under consideration.

lift-to-drag ratio The ratio of lift-to-drag obtained by dividing the lift by the drag or dividing the lift coefficient by the drag coefficient. Also called L/D ratio.

longitudinal stability A measure of the tendency of the airplane to return to trim after a disturbance and the characteristic motion in doing so. See stability.

MAC or mac Abbreviation for mean aerodynamic chord.

Mach cone 1. The cone-shaped shock wave theoretically emanating from an infinitesimally small particle moving at supersonic speed through a fluid medium. It is the locus of the Mach lines. 2. The cone-shaped shock wave generated by a sharp-pointed body, as at the nose of a high-speed aircraft.

Machmeter An instrument that measures and indicates speed relative to the speed of sound, i.e., that indicates the Mach number. Also called Mach indicator.

Mach number (symbol M) (Pronounced mock, after Ernst Mach, 1838–1916, Austrian scientist.) 1. A number expressing the ratio of the speed of a body or of a point on a body with respect to the surrounding air or other fluid, or the speed of a flow, to the speed of sound in the medium. 2. The speed represented by this number. If the Mach number is less than 1.0, the flow is called subsonic, and local disturbances can propagate ahead of the flow. If the Mach number is greater than 1.0, the flow is called supersonic, and disturbances cannot propagate ahead of the flow with the result that shock waves form.

Mach number, critical See critical Mach number.

Mach tuck The destabilization effect in pitch occurring when the airplane exceeds its critical Mach number. As flight speed increases above the critical Mach number, supersonic flow, with its attendant shock waves, occurs over the wing, the center of lift starts to move rearward, causing a gradually increasing nose down pitching moment, the magnitude of which is dependent on airplane stability, as well as other factors.

mean aerodynamic chord (MAC) The chord of an imaginary rectangular airfoil that would have pitching moments throughout the flight range the same as those of an actual airfoil or combination of airfoils under consideration, calculated to make equations of aerodynamic forces applicable.

moment (symbol M) A tendency to cause rotation about a point or axis, as of a control surface about its hinge or of an airplane about its center of gravity; the measure of this tendency, equal to the product of the force and the perpendicular distance between the point of axis of rotation and the line of action of the force.

NACA Abbreviation for National Advisory Committee for Aeronautics, which was the predecessor of NASA.

National Aeronautics and Space Administration (NASA) An agency of the United States government directed to provide for research into the problems of flight within and outside the Earth's atmosphere.

ozone A nearly colorless, but faintly blue, gaseous form of oxygen with a characteristic odor like that of weak chlorine. It is found in trace quantities in the atmosphere, primarily above the tropopause.

ozone layer Ozonosphere.

ozonosphere The general stratum of the upper atmosphere in which there is an appreciable ozone concentration and in which ozone plays an important part in the radiation balance of the atmosphere. This region lies roughly between 10–50 kilometers (6–30 miles), with maximum ozone concentration at about 20–25 kilometers (12–16 miles). Also called ozone layer.

parasite drag Total drag minus induced drag. Parasite drag consists of form drag (due to shape), skin friction drag, interference drag, and other special definitions of drag not associated with the production of lift.

pitching moment A moment about a lateral axis of an aircraft, rocket, airfoil, etc. This moment is positive when it tends to increase the angle of attack or to nose the body upward.

pitch-up Unstable flight condition usually at high subsonic speeds, due to adverse airflow over horizontal control surfaces, resulting in uncontrollable nose up moment.

pressure altimeter An altimeter that utilizes the change of atmospheric pressure with height to measure altitude. It is commonly an aneroid altimeter. Also called barometric altimeter.

pressure altitude 1. Altitude in the Earth's atmosphere above the standard datum plane, standard sea level pressure, measured by a pressure altimeter. 2. The altitude in a standard atmosphere corresponding to atmospheric pressure encountered in a real atmosphere. 3. The simulated altitude created in an altitude chamber.

ram recovery The percentage of actual pressure obtained from a ram air inlet. If the total pressure is completely converted to static pressure, ram recovery is 100 percent. See impact pressure; stagnation pressure; dynamic pressure.

relative wind The direction of an airflow with respect to an airfoil.

relaxed static stability A reduction of static stability to increase maneuverability of fighter aircraft or reduce tail surface drag of transport aircraft. Usually involves a center of gravity farther aft than normal and an automatic flight control system that uses various sensors to provide artificial stability.

schlieren (German, streaks, striae) 1. Regions of different density in a fluid, especially as shown by special apparatus. 2. Pertaining to a method or apparatus for visualizing or photographing regions of varying density in a field of flow. See schlieren photography.

schlieren photography A method of photography for flow patterns that takes advantage of the fact that light passing through a density gradient in a gas is refracted as though it were passing through a prism.

shock wave A surface or sheet of discontinuity (i.e., of abrupt changes in conditions) set up in a supersonic field of flow, through which the fluid undergoes a finite decrease in velocity accompanied by a marked increase in pressure, density, temperature, and entropy, as occurs, e.g., in a supersonic flow about a body. Sometimes called a shock.

sonic In aerodynamics, of or pertaining to the speed of sound; that which moves at acoustic velocity as in sonic flow; designed to operate or perform at the speed of sound as in sonic leading edge.

sonic barrier A popular term for the large increase in drag that acts upon an aircraft approaching acoustic velocity; the point at which the speed of sound is attained and existing subsonic and supersonic flow theories are rather indefinite. Also called sound barrier.

sonic boom An explosion-like sound heard when a shock wave, generated by an aircraft flying at supersonic speed, reaches the ear. The principal shock waves are approximately conical in shape and originate at the front and rear of the aircraft or object. The shock wave cone angle depends upon aircraft speed and the speed of sound in the surrounding medium. To the observer, who senses the shock wave with her ear, the arrival of each pressure wave is manifested as a booming sound.

speed brakes Movable aerodynamic devices on aircraft that reduce airspeed during descent and landing. Also called dive brakes.

stabilator A movable horizontal surface of an airplane used to increase longitudinal stability as well as provide a pitching moment. A stabilator combines the actions of a stabilizer and elevator.

stability The property of a body, as an aircraft or rocket, to maintain its attitude or to resist displacement, and, if displaced, to develop forces and moments tending to restore the original condition.

stagnation point A point in a field of flow about a body where the fluid particles have zero velocity with respect to the body.

stagnation pressure 1. The pressure at a stagnation point. 2. In compressible flow, the pressure exhibited by a moving gas or liquid brought to zero velocity by an isentropic process. 3. Total pressure. 4. Impact pressure. Because of the lack of a standard meaning, stagnation pressure should be defined when it is used.

static pressure (symbol p) 1. The pressure with respect to a stationary surface tangent to the mass-flow velocity vector. 2. The pressure with respect to a surface at rest in relation to the surrounding fluid.

streamline A line whose tangent at any point in a fluid is parallel to the instantaneous velocity vector of the fluid at that point. In steady-state flow, the streamlines coincide with the trajectories of the fluid particles; otherwise, the streamline pattern changes with time.

streamline flow Laminar flow.

subsonic In aerodynamics, of or pertaining to, or dealing with speeds less than acoustic velocity, as in subsonic aerodynamics.

subsonic airplane An airplane incapable of sustained level flight speeds exceeding Mach 1.

subsonic flow Flow of a fluid, as air over an airfoil, at speeds less than acoustic velocity. Aerodynamic problems of subsonic flow are treated with the assumption that air acts as an incompressible fluid.

supercritical wing The supercritical wing employs a supercritical airfoil section with a relatively flat upper surface that develops a weak shock wave at transonic speeds in contrast to the very strong upper surface shock associated with the more conventional airfoil. A substantial increase in drag-rise Mach number results from use of a supercritical wing.

supersonic Of or pertaining to, or dealing with, speeds greater than the acoustic velocity.

supersonic diffuser A diffuser designed to reduce the velocity and increase the pressure of fluid moving at supersonic velocities.

supersonic flow In aerodynamics, flow of a fluid over a body at speeds greater than the acoustic velocity and in which the shock waves start at the surface of the body. Compare hypersonic flow.

supersonic nozzle A converging-diverging nozzle designed to accelerate a fluid to supersonic speed.

supersonic tunnel A wind tunnel capable of producing supersonic flow at the test section.

thrust 1. The pushing or pulling force developed by an aircraft powerplant or a rocket engine. 2. The force exerted in any direction by a fluid jet.

titanium A silver-gray, metallic element found combined with ilmenite and rutile. Alloys of titanium are used extensively in aircraft. It is light, strong, and resistant to stress-corrosion cracking. Titanium is approximately 60 percent heavier than aluminum and about 50 percent lighter than stainless steel.

total pressure 1. Stagnation pressure. 2. Impact pressure. 3. The pressure a moving fluid would have if it were brought to rest without losses.

transonic Pertaining to what occurs or is occurring within the range of speed in which flow patterns change from subsonic to supersonic or vice versa, about Mach 0.8 to 1.2, as in transonic flight, transonic flutter. Also that which operates in this regime, as in transonic aircraft and transonic wing. Characterized by transonic flow or transonic speed, as in transonic region, transonic zone.

transonic flow In aerodynamics, flow of a fluid over a body in the range just above and just below the acoustic velocity. Transonic flow presents a special problem in aerodynamics in that neither the equations describing subsonic flow nor the equations describing supersonic flow can be applied in the transonic range.

transonic trim change A nosedown pitching moment of a subsonic airplane as flight speed increases above the critical Mach number requiring increasing nose-up trim. See Mach tuck.

trim The condition of static balance about one of the major axes of an airplane.

trimming To set or adjust the flying controls and/or trimming devices so that the aircraft will maintain a desired attitude in steady flight. An airplane is normally trimmed by the pilot so that control forces are reduced to zero.

tropopause The boundary between the troposphere and stratosphere, usually characterized by an abrupt change of lapse rate. The change is in the direction of increased atmospheric stability from regions below to regions above the tropopause. Its height varies from 15–20 kilometers (9–12 miles) in the tropics to about 10 kilometers (6 miles) in polar regions.

troposphere That portion of the atmosphere from the Earth's surface to the stratosphere; that is, the lowest 10–20 kilometers (6–12 miles) of the atmosphere. The troposphere is characterized by decreasing temperature with height, appreciable vertical wind motion, appreciable water-vapor content, and weather.

turbulent flow Fluid motion in which random motions of parts of the fluid are superimposed upon a simple pattern of flow. All or nearly all fluid flow displays some degree of turbulence. The opposite is laminar flow.

washin A greater angle of incidence (and attack) in one wing, or part of a wing, to provide more lift; usually used to overcome torque effects. Also see washout; aerodynamic twist.

washout A lesser angle of incidence to decrease lift. Also see washin; aerodynamic twist.

wind tunnel A tubelike structure or passage, sometimes continuous, together with its adjuncts, in which a high-speed movement of air or other gas is produced, as by a fan, and within which objects such as engines or aircraft, airfoils, rockets (or models of these objects), etc., are placed to investigate the airflow about them and the aerodynamic forces acting upon them. Tunnels are designated by the means used to produce the gas flow, as hot-shot tunnel, arc tunnel, blow-down tunnel, or by the speed range, as supersonic tunnel and hypersonic tunnel.

wing fence See fence.

wing loading Gross weight of an airplane divided by gross wing area.

wing section The cross-sectional shape of a wing. Also called airfoil and airfoil section.

yaw The rotational or oscillatory movement of an aircraft, rocket, or the like about a vertical axis.

yaw damper Use of a yaw-rate gyro sensing system combined with the automatic flight control system (AFCS) or autopilot to sense directional instability and make corrections without pilot attention.

Bibliography

General Electric, Aircraft Engine Group: *Aircraft Gas Turbine Guide*

Hager, Roy D. and Vrabel, Deborah: *Advanced Turboprop Project*, NASA SP-000

Hurt, H.H. Jr.: *Aerodynamics for Naval Aviators*, NAVAIR 00-80T-80

Kermore, A.C.: "Mechanics of Flight," Pitman Publishing

Lockheed Corporation: Airpower, also various issues of *Lockheed Horizons*

Reithmaier, Larry: *The Aviation/Space Dictionary, 7th-Edition*, TAB Books, division of McGraw-Hill, Inc.

Smith, Hubert: *The Illustrated Guide to Aerodynamics, 2nd Edition*, TAB Books, division of McGraw-Hill, Inc.

Talay, Theodore A.: *Introduction to Aerodynamics of Flight*, NASA SP-367

United Technologies, Pratt & Whitney: *The Aircraft Gas Turbine Engine and its Operation*, P&W Part No. P&W 182408

Index

Illustration page numbers are in **boldface**.

T